The Genus *Syzygium*
Syzygium cumini and Other Underutilized Species

Traditional Herbal Medicines for Modern Times

Each volume in this series provides academia, health sciences, and the herbal medicines industry with in-depth coverage of the herbal remedies for infectious diseases, certain medical conditions, or the plant medicines of a particular country.

Series Editor: Dr. Roland Hardman

The Genus *Syzygium*
Syzygium cumini and Other Underutilized Species

Edited by
K. N. Nair

CRC Press
Taylor & Francis Group
Boca Raton London New York

CRC Press is an imprint of the
Taylor & Francis Group, an **informa** business

CRC Press
Taylor & Francis Group
6000 Broken Sound Parkway NW, Suite 300
Boca Raton, FL 33487-2742

First issued in paperback 2021

© 2017 by Taylor & Francis Group, LLC
CRC Press is an imprint of Taylor & Francis Group, an Informa business

No claim to original U.S. Government works

ISBN 13: 978-1-03-209697-1 (pbk)
ISBN 13: 978-1-4822-4972-9 (hbk)

To the loving memory of my teacher, guide, and mentor

Dr. Madhavan Parameswaran Nayar (1932–2016)

Contents

Series Preface

Global warming and global travel are contributing factors in the spread of infectious diseases such as malaria, tuberculosis, hepatitis B, and HIV. These are not well controlled by the present drug regimes. Antibiotics are also failing because of bacterial resistance. Formerly less well-known tropical diseases are reaching new shores.

A whole range of illnesses, such as cancer, occur worldwide. Advances in molecular biology, including methods of *in vitro* testing for a required medical activity, give new opportunities to draw judiciously on the use and research of traditional herbal remedies from around the world. The reexamining of the herbal medicines must be done in a multidisciplinary manner.

Since the start of the book series "Medicinal and Aromatic Plants—Industrial Profiles" in 1997, there have been 51 volumes published. The series continues.

The same series editor, Dr. Roland Hardman, is also covering a second series entitled "Traditional Herbal Medicines for Modern Times." Each volume of this series reports on the latest developments and discusses key topics relevant to interdisciplinary health sciences and research by ethnobiologists, taxonomists, conservationists, agronomists, chemists, pharmacologists, clinicians, and toxicologists. The series is relevant to all these scientists and will enable them to guide business, government agencies, and commerce in the complexities of these matters. The background to the subject is outlined next.

Over many centuries, the safety and limitations of herbal medicines have been established by their empirical use by the "healers" who also took a holistic approach. The healers are aware of the infrequent adverse effects and often know how to correct these when they occur. Consequently and ideally, the preclinical and clinical studies of an herbal medicine need to be carried out with the full cooperation of the traditional healer. The plant's composition of the medicine, the stage of development of the plant material, when it is to be collected from the wild or from its cultivation, its postharvest treatment, the preparation of the medicine, the dosage and frequency, and such other essential information are required. A consideration of the intellectual property rights and appropriate models of benefit sharing may also be necessary.

Wherever the medicine is being prepared, the first requirement is a well-documented reference collection of dried plant material. Such collections are encouraged by organizations including the World Health Organization and the United Nations Industrial Development Organization. The Royal Botanic Gardens at Kew (United Kingdom) is now increasing its collection of traditional Chinese dried plant material relevant to its purchase and use by those who sell or prescribe traditional Chinese medicine in the United Kingdom.

In any country, control of the quality of plant raw material, its efficacy, and its safety in use is essential. The work requires sophisticated laboratory equipment and highly trained personnel. This kind of control cannot be applied to the locally produced herbal medicines in the rural areas of many countries, on which millions of people depend. Local traditional knowledge of the healers has to suffice.

Conservation and protection of plant habitats are required, and breeding for biological diversity is important. Gene systems are being studied for medicinal exploitation.

There can never be too many seed conservation "banks" to conserve genetic diversity. Unfortunately, such banks are usually dominated by agricultural and horticultural crops, with little space for medicinal plants. Developments such as DNA markers enable the genetic variability of a species to be checked. This can be helpful in deciding whether specimens of close genetic similarity warrant storage.

From ancient times, a great deal of information concerning diagnosis and the use of traditional herbal medicines has been documented in the scripts of China, India, and elsewhere. Today, modern formulations of these medicines exist in the form of powders, granules, capsules, and tablets. They are prepared in various institutions, such as government hospitals in China and North and South Korea and by companies such as the Tsumura Company of Japan, with good quality control. Similarly, products are produced by many other companies in India, the United States, and elsewhere with a varying degree of quality control. In the United States, the Dietary Supplement and Health Education Act of 1994 recognized the class of physiotherapeutic agents derived from medicinal and aromatic plants. Furthermore, under public pressure, the U.S. Congress set up an Office of Alternative Medicine. In 1994, this office assisted in the filing of several Investigational New Drug (IND) applications required for clinical trials of some Chinese herbal preparations. The significance of these applications was that each Chinese preparation involved several plants and yet was handled with a single IND. A demonstration of the contribution to efficacy, of *each* ingredient of *each* plant, was not required. This was a major step forward toward more sensible regulations with regard to phytomedicines.

The subject of Western herbal medicines is now being taught again to medical students in Germany and Canada. Throughout Europe, the United States, Australia, and other countries, pharmacy and health-related schools are increasingly offering training in phytotherapy. Traditional Chinese medicine clinics are now common outside of China. An Ayurvedic hospital now exists in London, with a BSc Honors degree course in ayurvedic medicine being available: Professor Shrikala Warrier, Registrar/Dean, MAYUR, Ayurvedic University of Europe, 81 Wimpole Street, London, WIG 9RF, email sw@unifiedherbal.com. This is a joint venture with a university in Manipal, India.

The term *integrated medicine* is now being used, which selectively combines traditional herbal medicine with "modern medicine." In Germany, there is now a hospital in which traditional Chinese medicine is integrated with Western medicine. Such comedication has become common in China, Japan, India, and North America by those educated in both systems. Benefits claimed include improved efficacy, a reduction in toxicity and the period of medication, and a reduction in the cost of the treatment. New terms, such as *adjunct therapy, supportive therapy*, and *supplementary medicine*, now appear as a consequence of such comedication. Either medicine may be described as an adjunct to the other, depending on the communicator's view. Great caution is necessary when traditional herbal medicines are used by doctors not trained in their use, and likewise when modern medicines are used by traditional herbal doctors. Possible dangers from drug interactions need to be stressed.

Roland Hardman, BPharm, BSc (Chemistry), PhD (London), FRPharmS
Head of Pharmacognosy (retired), School of Pharmacy and Pharmacology
University of Bath, United Kingdom

Foreword

Traditional medicine (TM) is an important and affordable healthcare system in most countries across the world. Encompassing a diverse array of herbal health products, practices, and practitioners, TM is emerging as a new stream of "alternative" or "complementary" medicine to cope with the rising chronic noncommunicable diseases and lifestyle disorders. The increasing acceptance of TM has opened up new avenues for a globally competitive business regime, centered on numerous products such as herbal drugs, nutraceuticals, cosmeceuticals, and food supplements. TM is finding new frontiers of research and development (R&D) through the intervention of an "omics"-centered system biology approach involving genomics, transcriptomics, proteomics, functional proteomics, and metabolomics. The synergy among the herbal industries, R&D institutions, and the governments is a prerequisite for developing a global regulatory framework to ensure the quality, efficacy, and safety of TM. Such synergy is also vital for effective delivery of TM and achieving holistic human health and wellness.

Plants have been an integral component of the traditional medicare practices in many countries and cultures. The time-tested traditional uses and properties of many medicinal plants have led to the discovery of modern synthetic drugs. Bioprospecting for new and novel biomolecules of commercial or industrial importance is emerging as a major facet of R&D in many publicly and privately funded organizations in the developed and developing countries. Scientific validation of traditional herbal medicines and their standardization is an important component in medicinal plant research and natural and/or herbal products development. Therefore, R&D on medicinal plants today includes such diverse activities as systematic documentation of the phytochemicals and metabolomic diversity, pharmacology, pharmacodynamics, pharmacokinetics, toxicity, and biosafety of the target plants, leading to the development of standard herbal products, including drugs and other formulations. In India, China, South Korea, North Korea, and Japan, the traditional herbal medicine assumes a prime place in their healthcare and R&D systems.

The book entitled *The Genus Syzygium: Syzygium cumini and Other Underutilized Species* presents a comprehensive account of a well-known multipurpose plant, *Syzygium cumini*, and other important *Syzygium* species. *S. cumini* is an Indo-Malaysian tree and now cultivated in many tropical and subtropical countries for its multiple uses as an effective traditional medicine for diabetes management, as an edible fruit, and also as a source of timber and dye. *Syzygium* in general, and *S. cumini* in particular, is reported to contain a wide array of phytomolecules with a broad spectrum of biological activities that include antihypoglycemic, antioxidant, antiinflammatory, anticarcinogenic, and antihypertensive properties. There are increasing evidences on the potential application of *S. cumini* extracts in synthesis of gold and silver nanoparticles as bioactives, as biocontrol and bioremediation agents, and also as food additives and natural colorants. In many respects, *S. cumini* represents one of the most valuable multipurpose trees with great promises for pharmaceutical and horticultural trade.

The volume covers a broad range of topics including botany, traditional medicinal uses, phytochemical constituents, pharmacology, pharmacopeia standards, horticulture, genetic resource conservation, biocontrol, and bioremediation values. The book, in fact, is a monograph presenting the state of the art on the current wealth of knowledge on *S. cumini* and related species. I appreciate the sincere efforts of the editor and all chapter contributors for having brought out this important and timely publication on such a highly valued and promising plant resource as *Syzygium*.

I am glad that as many as 14 out of the 24 authors/coauthors, including the editor of this book, are currently working with the three premier biological laboratories (CSIR-NBRI, CSIR-CDRI, and CSIR-IHBT) of the Council of Scientific and Industrial Research (CSIR) in India. This reflects the importance that the CSIR attaches to promoting integrated research for generating new wealth of knowledge on Indian plant diversity and traditional herbal medicines for the benefit of science, industry, and society. I am confident that the book will serve as a useful reference to researchers and will provide fresh impetus to conduct more intensive research on *S. cumini* and other underutilized *Syzygium* species helping their better utilization.

S. K. Barik
Director, CSIR-National Botanical Research Institute
Lucknow, India

Preface

Syzygium, with about 1200 species, is the largest genus in the myrtle family (Myrtaceae). It is an Old World genus of evergreen trees or shrubs distributed from Africa to South and Southeast Asia, south China, Malaysia, Australia, and New Caledonia. Some species of *Syzygium* are economically important and are widely cultivated for their edible fruits and medicinal and aromatic properties. One of the most widely cultivated species is *Syzygium aromaticum*, the source of the globally traded clove and clove oil. (Clove is being dealt with in a volume in preparation for Dr. Hardman's book series "Medicinal and Aromatic Plants—Industrial Profiles.") The other popular species that are cultivated for their colorful edible fruits include *Syzygium aqueum* (water apple; native to India through Malaysia), *S. cumini* (Java plum; Sri Lanka, India, and Malaysia), *S. jambos* (rose apple; Malaysia and Southeast Asia), *S. malaccense* (mountain apple; Malaysia), and *S. samarangense* (wax apple; Malaysia and Southeast Asia). There are also several other little known species of *Syzygium* with great prospects for their cultivation for edible and medicinal uses.

Among the economically important *Syzygium* species, *S. cumini* (jambolan or jambul) is an Indo-Malaysian tree. It is widely cultivated in many tropical and subtropical countries for its multifunctional use as food, timber, landscape tree, dye, and medicine. The seed of *S. cumini* is considered a very useful medicine in folklore, as well as in traditional systems of medicine, specifically for diabetes. Different parts of the plant, including seeds, bark, leaves, and fruits, have been extensively studied for their active compounds and medicinal values. *S. cumini* works at several levels to yield hypoglycemic effects. It inhibits the carbohydrate-hydrolyzing enzyme, increases insulin secretion, and decreases glucose absorption in the small intestine. The plant is rich in phenolic compounds, including flavonoids, lignans, phenolic acids, and tannins. Scientific studies have revealed that the extract of different parts of *S. cumini* showed significant biological actions, such as antibacterial, antifungal, antiviral, antigenotoxic, antiallergic, anti-inflammatory, antiulcerogenic, antidiarrheal, chemopreventive, cardioprotective, radioprotective, hepatoprotective, anticancer, antioxidant, free radical scavenging, and antidiabetic effects. Phytochemicals present in the leaf and seed extract also have the potential for synthesis of nanoparticles. *S. cumini* can also functionally help improve soil and plant health, and hence is a potential candidate for bioremediation and biocontrol.

Despite its great economic prospects and an increasing global interest in pharmaceutical trade, no comprehensive or consolidated account is available to date on various aspects of *S. cumini*. Studies on the quantification, characterization, and standardization of *S. cumini* extracts from different parts will be pivotal for setting standards in future research. A review on the phytochemical constituents and their pharmacological actions with respect to different human ailments is also an essential requisite. Besides cultivation practices, nutrient and water management, selection of varieties, improvised propagation and plantation techniques using microbial inoculants, and postharvest management and storage have immense potential in improving the profitability of *S. cumini*.

This book on *Syzygium* provides an updated account on *S. cumini* and other little known and underutilized *Syzygium* species from a multidisciplinary perspective. The 14 chapters, contributed by specialists, cover all relevant aspects, including botany; systematics and phylogeny; life history; traditional medicinal uses; phytochemical constituents and their biological activities, including cancer immunology and uses in nanotechnology; pharmacopoeial standards; horticulture; genetic resource management; biocontrol and bioremediation values of *S. cumini*; and an enumeration of usages of eight underutilized Indo-Malaysian and Australasian species of *Syzygium*.

I wish to express my sincere gratitude to all the chapter contributors for their excellent efforts and timely inputs to this book volume. A special word of thanks is due to my senior colleague, Dr. S. K. Tewari at CSIR-National Botanical Research Institute (CSIR-NBRI), Lucknow, India, for having introduced me to Dr. Roland Hardman, who eventually entrusted me with this book project. The guidance and advice I received from Dr. Hardman as the book series editor is gratefully acknowledged. I thank Prof. S. K. Barik, the director of CSIR-NBRI for the valuable support and for writing the Foreword to this book. Jennifer Blaise, Jay Margolis, Jonathan Pennell, and John Sulzycki of CRC Press and Adel Rosario of the Manila Typesetting Company deserve special thanks for the great help and assistance they rendered at various stages of preparation of this book. It was indeed an enjoyable and rewarding experience in working with this great team. Last, but not the least, I thank my wife Rakesh Kumari and our son Aditya for the unconditional love and understanding that supported me greatly to complete the book in time. I hope the book will serve as a good reference for researchers interested in studying the genus *Syzygium* in various pursuits.

K. N. Nair

Editor

K. N. Nair, PhD, is a Senior Principal Scientist at the CSIR-National Botanical Research Institute, Lucknow, India. He is a plant taxonomist by profession who has more than 30 years of experience in systematics and diversity studies on Indian Rutaceae. His other fields of main interest include molecular systematics, genetic diversity assessment, plant conservation, and biodiversity laws and policies. From 2003 through 2007, he coordinated a plant conservation program at the national level under the stewardship of an international collaborative program, Investing in Nature–India, sponsored by the Botanic Garden Conservation International, United Kingdom, and the HSBC. He served as a member of the Expert Committee on Endemic, Endangered and Threatened Species for the National Biodiversity Authority of the Government of India in 2006–2007 and as a member of the Task Force on Agro-biodiversity Hotspots and Benefit Sharing for the Protection of Plant Varieties and Farmers Rights Authority of the Government of India in 2007–2008. He is currently involved in a project study on the molecular systematics of the *Didymocarpus–Henckelia* generic complex (Gesneriaceae) in India, funded by the Science, Engineering Research Board of the Government of India. He has authored or coauthored 47 research papers and articles and 4 books, and he is the coinventor involved in 4 patents.

Contributors

Lal Babu Chaudhary
Plant Diversity, Systematics and
    Herbarium Division
CSIR-National Botanical Research
    Institute
Lucknow, India

Rekha Chaudhury
ICAR-National Bureau of Plant Genetic
    Resources
New Delhi, India

Varughese George
Amity Institute of Phytochemistry and
    Phytomedicine
Thiruvananthapuram, Kerala, India

Roland Hardman
School of Pharmacy and Pharmacology
University of Bath
Bath, United Kingdom

K. B. Harikumar
Rajiv Gandhi Centre for Biotechnology
Thiruvananthapuram, Kerala, India

Sudhir Kumar
Medicinal and Process Chemistry
    Division
CSIR-Central Drug Research Institute
Lucknow, India

Vineet Kumar
Biotechnology Division
CSIR-Institute of Himalayan
    Bioresource Technology
Palampur, Himachal Pradesh, India

Avnesh Kumari
Biotechnology Division
CSIR-Institute of Himalayan
    Bioresource Technology
Palampur, Himachal Pradesh, India

C. Kunhikannan
Institute of Forest Genetics and Tree
    Breeding
R. S. Puram, Coimbatore, Tamil Nadu,
    India

Arun Kumar Kushwaha
Plant Diversity, Systematics and
    Herbarium Division
CSIR-National Botanical Research
    Institute
Lucknow, India

S. K. Malik
ICAR-National Bureau of Plant Genetic
    Resources
New Delhi, India

Rakesh Maurya
Medicinal and Process Chemistry
    Division
CSIR-Central Drug Research Institute
Lucknow, India

Sabira Mohammed
Rajiv Gandhi Centre for Biotechnology
Thiruvananthapuram, Kerala, India

R. C. Nainwal
Distant Research Centre
CSIR-National Botanical Research
    Institute
Lucknow, India

K. N. Nair
Plant Diversity, Systematics and
  Herbarium Division
CSIR-National Botanical Research
  Institute
Lucknow, India

Vinodkumar T. G. Nair
Jawaharlal Nehru Tropical Botanic
  Garden and Research Institute,
  Palode
Thiruvananthapuram, Kerala, India

Madan Mohan Pandey
Pharmacognosy & Ethno Pharmacology
  Division
CSIR-National Botanical Research
  Institute
Lucknow, India

Palpu Pushpangadan
Amity Institute for Herbal and Biotech
  Products Development
Thiruvananthapuram, Kerala, India

S. Rajasekharan
Jawaharlal Nehru Tropical Botanic
  Garden and Research Institute,
  Palode
Thiruvananthapuram, Kerala, India

T. S. Rana
Plant Diversity, Systematics and
  Herbarium Division
CSIR-National Botanical Research
  Institute
Lucknow, India

A. K. S. Rawat
Pharmacognosy & Ethno Pharmacology
  Division
CSIR-National Botanical Research
  Institute
Lucknow, India

Devendra Singh
Distant Research Centre
CSIR-National Botanical Research
  Institute
Lucknow, India

Sanjay Singh
Central Horticultural Experiment
  Station and Regional Station
ICAR-Central Institute of Arid
  Horticulture
Panchmahals, Gujarat, India

Wuu Kuang Soh
Botany Department
School of Natural Sciences
Trinity College
Dublin, Republic of Ireland

Vartika Srivastava
ICAR-National Bureau of Plant Genetic
  Resources
New Delhi, India

P. Sujanapal
Kerala Forest Research Institute
Peechi, Thrissur, Kerala, India

S. K. Tewari
Distant Research Centre
CSIR-National Botanical Research
  Institute
Lucknow, India

Sudesh Kumar Yadav
Center of Innovative and Applied
  Bioprocessing (CIAB)
Mohali, Punjab, India

1 Taxonomy of *Syzygium*

Wuu Kuang Soh

CONTENTS

INTRODUCTION TO *SYZYGIUM*: SIZE, DISTRIBUTION, ECOLOGY, AND USAGES

Syzygium Gaertn. is the largest genus of Myrtaceae, with an estimated 1200 species in the world (Parnell et al. 2007; Govaerts et al. 2008). *Syzygium* ranked 16th among the 57 largest genera of flowering plants (Frodin 2004) or possibly higher, within the top 10 according to Parnell et al. (2007), on the basis that many novel species are yet to be discovered and many validly described species await transfer to *Syzygium*. This is a paleotropical genus with a wide range of occurrence mainly in southern and southeastern Asia, southern China, Australia, Malesia, and New Caledonia. Some species occur in east Africa, Madagascar, the Mascarenes, southwestern Pacific Islands, Taiwan, and southern Japan. The center of diversity is in Malesia, but its basic evolutionary diversity is in the Melanesian–Australian region.

Syzygium is principally found in tropical or subtropical vegetation, including lowland to montane rainforest, swamp, ultramafic forest, savannah, and limestone forest. Some species occur in specialized habitats, such as along rivers or on ultramafic or limestone soil. Van Steenis (1981) recorded 33 rheophytic species in *Syzygium* that are morphologically characterized by a narrow leaf, short petiole, and flexible twig and leaves crowded at twig ends. In the tropical rainforest, *Syzygium* usually blooms in masses. As one of the most common tree genera in the forest ecosystem, *Syzygium* is important as a food resource for birds, insects, and small and large mammals (Parnell et al. 2007).

The breeding biology of *Syzygium* has been understudied. Several studies showed that *Syzygium* are pollinated by birds, bats, and insects (bees, wasps, moths, ants, and spiders) and have low to high self-compatibility (Nic Lughadha and Proenca 1996; Parnell et al. 2007). Chantaranothai and Parnell (1994) showed that in *S. jambos* (L.) Alston, *S. megacarpum* (Craib) Rathakr. et N. C. Nair, and *S. samarangense* (Blume) Merr. et L. M. Perry, three breeding systems occur: apomictic, inbreeding (excluding apomixis), and outbreeding. Some species display inconsistent incidents of polyembrony, for example, *S. cumini* (L.) Skeels (Nic Lughadha and Proenca

1996). In short, there is no conclusive evidence indicating a general breeding trend in *Syzygium*, especially in regards to inbreeding or outbreeding.

Despite its high species diversity, only a fraction of the *Syzygium* species are commercially cultivated or used for their fruit, timber, or medicinal properties or as spices. *S. aromaticum* (L.) Merr. et L. M. Perry is widely cultivated in the tropics for cloves and essential oil. Clove oil is an important essential oil that is commonly used in the food, perfumery, and pharmaceutical industries (De Guzman and Siemonsma 1999). Other species, for instance, *S. aqueum* (Burm. f.) Alston, *S. cumini*, *S. jambos*, and *S. malaccense* (L.) Merr. et L. M. Perry are important and commonly cultivated fruit trees in Southeast Asia and India. *Syzygium* is not regarded as a major source of timber, but some species are locally used for house building and light construction (Lemmens et al. 1995). Other forest species are locally used in traditional medicine, fabric dye, and cooking.

TAXONOMIC HISTORY OF *SYZYGIUM*

The name *Syzygium* was adopted by Gaertner (1788) from Browne's (1756) genus *Suzygium*, but with a corrected spelling. Its etymology is from the Latin *syzygia* and from the Greek *syzygos*, which means "yoked together." Browne (1756) probably coined the new generic name in allusion to the paired arrangement of the leaves and branchlets (Craven and Biffin 2010). In *Species Plantarum*, Linnaeus (1753) treated five species that are currently accepted as *Syzygium* in three separate genera: *Caryophyllus* L., *Eugenia* L., and *Myrtus* L. Although the oldest name representing *Syzygium* is *Caryophyllus*, the former name was conserved against the latter because it was more widely used (Farr and Zijlstra 2014).

The works of De Candolle (1828) and Wight (1841) are influential in breeding two schools of thought about the classification of *Syzygium*. The first school favors De Candolle's concept of multiple genera, accepts *Eugenia* for predominantly the New World species, and recognizes several genera for the Old World syzygioid species. The other, following Wight, advocated a broader concept of *Eugenia* encompassing all species of *Syzygium sensu lato* (*s.l.*).

De Candolle (1828) recognized five genera occurring in the Old World, *Acmena* DC., *Caryophyllus*, *Eugenia* (small number of species), *Jambosa* Adans., and *Syzygium*, while the New World species were placed in *Eugenia*. Other botanists in the nineteenth and twentieth centuries followed this concept, but with some modification by reduction or addition of new genera to accommodate newly discovered species that exhibited novel morphological characters (e.g., *Acicalyptus* A. Gray, *Aphanomyrtus* Miq., *Clavimyrtus* Blume, *Cleistocalyx* Blume, *Cupheanthus* Seem., *Paraeugenia* Turrill, and *Piliocalyx* Brongn. and Gris.). Following De Candolle (1828), Niedenzu (1893) proposed one of the most comprehensive reclassifications of *Syzygium*/*Eugenia*. He established several segregate genera, namely, *Acicalyptus*, *Jambosa*, *Piliocalyx*, and *Syzygium*, which were based mainly on floral characters. His system was used by many workers, but later due to the lack of distinct generic limits (e.g., between *Jambosa* and *Syzygium*), many other botanists were tempted to consider *Eugenia* in a broader sense. Because of the immense number of species described in *Syzygium s.l.*, the group became unmanageable. This prompted Merrill and Perry (1937, 1938a, 1938b, 1939a, 1939b) to redefine some of the earlier proposed segregates of the genera by using fruit, seed (seed coat adhesion and cotyledon fusion), and flower characters. In their new classification, they retained species from the New World in *Eugenia* and recognized *Acmena, Aphanomyrtus, Caryophyllus, Cleistocalyx, Paraeugenia, Syzygium, Tetraeugenia* Merrill, and tentatively *Jambosa* for the Old World species. For the main genera, *Eugenia* and *Acmena* were separated from *Syzygium* in having undivided cotyledons; therein *Acmena* was differentiated from *Eugenia* in having the testa adhering to the pericarp and also in having divaricate anther locules, while *Cleistocalyx* stands out from the rest due to its calyptrate calyx (Merrill and Perry 1937, 1938a, 1938b, 1939a, 1939b).

Wight (1841) recommended merging the five genera recognized by De Candolle (1828) into *Eugenia* but maintaining the overall structure of the classification in having five subgeneric groups by using the same epithets. The justifications for this were that first, there were no discrete variations in the flowering and fruiting characters, and second, for stabilizing the nomenclature. Bentham and Hooker (1865) adopted the same principle in their influential *Genera Plantarum*. Others, mostly, if not all, British or British-associated botanists, followed this classification in their floristic work in the South and Southeast Asian region. Among them is Henderson (1949), who revised *Syzygium* in Malaya and vehemently rejected Merrill and Perry's classification and advocation of the taxonomic importance of testa and cotyledon fusion. Merrill and Perry (1939a), based on their observations on herbarium specimens, suggested that the testa in *Syzygium* is adherent to the pericarp, while in *Eugenia* it is free from the pericarp and adherent to the cotyledons. Henderson (1949) found that in live specimens, the degree of adherence of testa to the cotyledons varies greatly in *Syzygium*. Within *Eugenia*, he recognized four sections, including *Acmena, Cleistocalyx, Fissicalyx*, and *Syzygium* (including *Caryophyllus* and *Jambosa*); he divided the *Syzygium* section into five groups, simply named groups I, II, III, IV, and V.

Schmid (1972a, 1972b) provided an extensive review of the status of *Eugenia* and *Syzygium* and its segregate genera, including *Acmena, Cleistocalyx, Caryophyllus*, and *Jambosa*. He showed that *Eugenia* and *Syzygium s.l.* are not closely related on the basis that the pathway of vascular supply to ovules in *Eugenia* is transeptal and that of *Syzygium s.l.* is axile. He further suggested that both genera have separate lines of evolution, and neither was directly ancestral to the other. His hypothesis was later supported by phylogenetic studies from DNA sequence data (Gadek et al. 1996; Wilson et al. 2001, 2005; Harrington and Gadek 2004).

Although the exclusion of *Eugenia* from *Syzygium s.l.* is now widely accepted by contemporary botanists, it does not solve the conceptual problem in the classification within *Syzygium s.l.* in the Old World. Presently, there are two schools of thought as to the approach for a classification scheme for *Syzygium*. Craven and Biffin (Craven 2001; Craven et al. 2006; Craven and Biffin 2010), based on DNA sequence phylogeny and morphological evidence, are in favor of an all-inclusive generic concept of *Syzygium*, but with an infrageneric classification to reflect evolutionary relationships among the clades. According to them, until enough molecular evidence is gathered to erect phylogenetically robust genera, a wide-ranging concept of *Syzygium* will tentatively eliminate the need to create more new genera to accommodate novel morphological characters. On the contrary, Parnell et al. (2007) advocated having multiple and smaller genera within *Syzygium s.l.* on the basis that an expanded concept of *Syzygium* will make it more polyphyletic and diminish any degrees of predictability. Indeed, this scenario is analogous to the *Eugenia sensu stricto (s.s.)–Syzygium s.l.* problem faced by nineteenth- and twentieth-century botanists, but differs from the latter on scale.

NOTES ON *SYZYGIUM CUMINI*

S. cumini is a widespread species and is found in the wild and in cultivation. The consumption of the fruits by the local people in India was documented in the Western literature as early as the seventeenth century in Van Rheede's *Hortus Malabaricus* in 1685, and could possibly imply that the species was already in cultivation before that. At present, *S. cumini* is widely cultivated throughout its range in the tropics and sub-tropics, and to date, many cultivars have been created (Verheij and Coronel 1992). Like most cultivated plants, there is wide morphological variation with gradation in the leaves and fruits of *S. cumini* in terms of their size, color, and shape. As a result of this, many new species have been described by various authors based on these subtle and overlapping morphological variations of the leaf and fruit. For example, Roxburgh (1832) recognized four distinct *Eugenia* species: *E. caryophylifolia* Lam., *E. jambolana* Lam., *E. obtusifolia* Roxb., and *E. fruticosa* Roxb. Duthie (1879) later relegated two of the species (*E. caryophylifolia* Lam. and *E. obtusifolia* Roxb.) as a variety of *E. jambolana* Lam. and recognized an additional new species, *Eugenia tenuis* Wall ex Duthie. Hence, currently there are at least 30 synonyms for *S. cumini* (Soh and Parnell 2015).

REFERENCES

Bentham, G., and J. D. Hooker. 1865. Myrtaceae. In *Genera Plantarum*. Vol. 1. London: Reeve and Co.

Browne, P. 1756. *Suzygium*. In *The Civil and Natural History of Jamaica*, 240. London: T. Osborne and J. Shipton in Gray's-Inn.

Chantaranothai, P., and J. A. N. Parnell. 1994. The breeding biology of some Thai *Syzygium* species. *Trop. Ecol.* 35: 199–208.

Craven, L. A. 2001. Unravelling knots of plaiting rope: What are the major taxonomic strands in *Syzygium sens. lat.* (Myrtaceae) and what should be done with them? In *Proceedings of the 4th Flora Malesiana Symposium*, ed. L. G. Saw, L. S. L. Chua, and K. C. Choo, 75–85. Kuala Lumpur, Malaysia: Forest Research Institute.

Craven, L. A., and E. Biffin. 2010. An infrageneric classification of *Syzygium* (Myrtaceae). *Blumea* 55: 94–99.

Craven, L. A., E. Biffin, and P. S. Ashton. 2006. *Acmena, Acmenosperma, Cleistocalyx, Piliocalyx* and *Waterhousea* formally transferred to *Syzygium* (Myrtaceae). *Blumea* 51: 131–142.

De Candolle, A. P. 1828. Myrtaceae. In *Prodromus systematis naturalis regni vegetabilis*, ed. A. P. De Candolle, 207–296. Vol. 3. Paris: Truettel et Würtz.

De Guzman, C. C., and J. S. Siemonsma (eds.). 1999. *Plant Resources of South-East Asia*, No. 13, *Spices*. Leiden: Backhuys Publishers.

Duthie, J. F. 1879. Myrtaceae. In *Flora of British India*, ed. J. D. Hooker, 497–506. Vol. 2. London: Reeve and Co.

Farr, E. R., and G. Zijlstra (eds.). 2014. Index Nominum Genericorum (Plantarum). Washington, DC: Smithsonian Institution. http://botany.si.edu/ing/ (accessed July 4, 2014).

Frodin, D. G. 2004. History and concepts of big plant genera. *Taxon* 54: 753–776.

Gadek, P. A., P. G. Wilson, and C. J. Quinn. 1996. Phylogenetic reconstruction in Myrtaceae using *matK*, with particular reference to the position of *Psiloxylon* and *Heteropyxis*. *Aust. Syst. Bot.* 9: 283–290.

Gaertner, J. 1788. *De Fructibus et Seminibus Plantarum*. Vol. 1. Stuttgart: Typis Academie Carolinae.

Govaerts, R., M. Sobral, P. Ashton, and F. Barrie. 2008. *World Checklist of Myrtaceae*. London: Royal Botanic Gardens, Kew.

Harrington, M. G., and P. A. Gadek. 2004. Molecular systematics of the *Acmena* alliance (Myrtaceae): Phylogenetic analyses and evolutionary implications with reference to Australian taxa. *Aust. Syst. Bot.* 17: 63–72.

Henderson, M. R. 1949. The genus *Eugenia* (Myrtaceae) in Malaya. *Gard. Bull. Singapore* 12: 1–293.

Lemmens, R. H. M. J., I. Soerianegara, and W. C. Wong (eds.). 1995. *Plant Resources of South-East Asia*, No. 5(2), *Timber Trees: Minor Commercial Timbers*. Leiden: Backhuys Publishers.

Linnaeus, C. 1753. *Species Plantarum*. Stockholm: Laurentii Salvii.

Merrill, E. D., and L. M. Perry. 1937. Reinstatement and revision of *Cleistocalyx* Blume (including *Acicalyptus* A. Gray), a valid genus of the Myrtaceae. *J. Arnold Arbor.* 18: 322–343.

Merrill, E. D., and L. M. Perry. 1938a. A synopsis of *Acmena* DC., a valid genus of the Myrtaceae. *J. Arnold Arbor.* 19: 1–20.

Merrill, E. D., and L. M. Perry. 1938b. On the Indo-Chinese species of *Syzygium* Gaertner. *J. Arnold Arbor.* 19: 99–116.

Merrill, E. D., and L. M. Perry, 1939a. The Myrtaceae genus *Syzygium* Gaertner in Borneo. *Mem. Am. Acad. Arts Sci.* 18: 135–202.

Merrill, E. D., and L. M. Perry. 1939b. Additional notes on Chinese Myrtaceae. *J. Arnold Arbor.* 20: 102–103.

Nic Lughadha, E., and C. Proenca. 1996. A survey of the reproductive biology of the Myrtoideae (Myrtaceae). *Ann. Mo. Bot. Gard.* 83: 480–487.

Niedenzu, F. 1893. Myrtaceae. In *Die Naturlichen Planzenfamilien*, ed. A. Engler and K. Prantl, 57–105. Vol. 3. Leipzig: Wilhelm Engelmaan.

Parnell, J. A. N., L. Craven, and E. Biffin. 2007. Matters of scale: Dealing with one of the largest genera of Angiosperms. In *Towards the Tree of Life: Systematics of Species Rich Groups*, ed. T. Hodkinson and J. Parnell, 251–274. Special Series Vol. 72. London: Systematics Association.

Roxburgh, W. 1832. *Flora Indica*. Vol. 2. London: Parbury, Allen and Co.

Schmid, R. 1972a. A resolution of the *Eugenia-Syzygium* controversy (Myrtaceae). *Am. J. Bot.* 59: 423–436.

Schmid, R. 1972b. Floral anatomy of Myrtaceae. I. *Syzygium*. *Bot. Jahrb. Syst.* 92: 433–489.

Soh, W. K., and J. A. N. Parnell. 2015. A revision of *Syzygium* Gaertn. (Myrtaceae) in Indochina (Cambodia, Laos and Vietnam). *Adansonia* 37: 179–275.

Van Steenis, C. G. G. J. 1981. *Rheophytes of the World*. Alphen aan den Rijn, Netherlands: Springer.

Verheij, W. M., and R. E. Coronel (eds.). 1992. *Plant Resources of South-East Asia*, No. 2, *Edible Fruit*. Wageningen: Pudoc.

Wight, R. 1841. *Illustrations of Indian Botany*. Vol. 2. Madras: Pharoah.

Wilson, P. G., M. M. O'Brien, P. A. Gadek, and C. J. Quinn. 2005. Relationships within Myrtaceae *sensu lato* based on *matK* phylogeny. *Plant Syst. Evol.* 251: 3–19.

Wilson, P. G., M. O. Marcelle, P. A. Gadek, and C. J. Quinn. 2001. Myrtaceae revisited: A reassessment of infrafamilial groups. *Am. J. Bot.* 88: 2013–2025.

2 Phylogeny of *Syzygium*

K. N. Nair and T. S. Rana

CONTENTS

INTRODUCTION: THE *SYZYGIUM–EUGENIA* COMPLEX

The taxonomy of *Syzygium* Gaertn. has been confusing, primarily due to the apparent lack of "good diagnostic characters" (Parnell et al. 2007) to demarcate the generic and species boundaries in the genus and its alliances. The question as to the proper generic delimitation of *Syzygium* and *Eugenia* L. has been in debate for many years, with the taxonomists divided into three major schools of thoughts. The first school supported a unified single-genus concept of *Eugenia* comprising all syzygioid genera of Myrtaceae, including *Syzygium* (Bentham and Hooker 1865; Henderson 1949; Kochummen 1995; Lemmens 1995). The second school upheld the segregation of *Syzygium* and *Eugenia* as two distinct genera and placed the largely New World species in *Eugenia* and the strictly Old World species in *Syzygium* (Schmid 1972; Johnson and Briggs 1984). The third concept adopted the recognition of a series of generic and subgeneric segregates in the *Syzygium–Eugenia* group, such as *Acmena* DC., *Cleistocalyx* Blume, *Jambosa* Adans., *Acicalyptus* A. Gray, *Acmenosperma* Kausel, *Cupheanthus* Seem., and *Piliocalyx* Brong. & Gris (De Candolle 1828; Wight 1841; Niedenzu 1893; Merrill and Perry 1938, 1939; Kostermans 1981; Hyland 1983; Chantaranothai and Parnell 1994).

The historical developments surrounding the *Syzygium–Eugenia* generic complex have been adequately reviewed in earlier works (Schmid 1972; Parnell 1999; Parnell et al. 2007; Craven and Biffin 2010; Soh in Chapter 1 of this volume). Schmid's (1972) contention on separating *Syzygium* and *Eugenia* was based mainly on the substitution of "transeptal" vascular supply to the ovules in *Eugenia sensu stricto* (*s.s.*) to the "axile" ones in *Syzygium sensu lato* (*s.l.*). Additionally, Schmid (1972) found three organographic criteria—nature of bracteoles, presence or absence of pubescence, and presence or absence of pseudopedicels—as supporting evidence to distinguish between *Eugenia s.s.* and *Syzygium s.l.* More evidence from wood anatomy showing the presence of vasicentric tracheids and fibers with conspicuous bordered pits in *Eugenia* and the absence of vasicentric tracheids and fibers with simple or indistinctly bordered pits in *Syzygium* (Ingle and Dadswell 1953), morphological and anatomical data (Johnson and Briggs 1984), and differences in leaf surface sculpturing, outer

TABLE 2.1
Summary of Phenetic and Phylogenetic Studies on *Syzygium–Eugenia* Generic Complex

Author(s)	Summary
Linnaeus (1753)	Included the then known species of *Syzygium* under three genera: *Caryophyllus* L., *Eugenia* L., and *Myrtus* L.
De Candolle (1828)	All New World species in *Eugenia* and the Old World species in four genera: *Acmena* DC., *Caryophyllus*, *Jambosa* Adans., and *Syzygium* Gaertn.
Wight (1841)	Recognized *Eugenia* with *Acmena*, *Caryophyllus*, *Jambosa*, and *Syzygium* as its subgenera
Bentham and Hooker (1865)	Accepted *Eugenia* as a unified genus, for *Eugenia*, *Jambosa*, and *Syzygium*
Niedenzu (1893)	Retained *Eugenia* to include all the New World and a few Old World species, and other Old World species in four genera: *Acicalyptus* A. Gray, *Jambosa* (including *Cleistocalyx* Blume and others), *Piliocalyx* Brongn. & Gris, and *Syzygium* (including *Acmena* and others)
Henderson (1949)	Recognized Malayan Peninsula species under *Eugenia* with four sections, *Acmena*, *Cleistocalyx*, *Fissicalyx* Bentham, and *Syzygium*, and section *Syzygium* again split into five groups
Merrill (1951)	Considered syzigioid species in separate genera: *Acmena*, *Aphanomyrtus* Miq., *Caryophyllus*, *Pareugenia* Turrill, *Syzygium*, *Tetraeugenia* Merr., and *Jambosa*
Ingle and Dadswell (1953)	The southwest Pacific species of *Eugenia s.l.* in two groups: (1) *Eugenia s.s.* and (2) *Acmena*, *Cleistocalyx*, and *Syzygium*
Schmid (1972)	Segregated *Eugenia s.s.* and *Syzygium s.l.* as separate genera based on floral anatomical evidence
Johnson and Briggs (1984)	*Eugenia* and *Syzygium* were recognized as separate genera in two distinct and remotely placed clades
Khatijah et al. (1992)	Leaf anatomical characters supported recognition of *Eugenia* with retention of two sections, *Acmena* and *Syzygium*, but not the other sections, *Cleistocalyx* and *Fissicalyx*
Parnell (1999)	Numerical analyses of morphological characters supported reinstatement of *Jambosa* as a section under *Syzygium* and did not support recognition of other sections, *Acmena* and *Cleistocalyx*
Harrington and Gadek (2004)	nrDNA ITS and ETS analyses did not support recognition of subgeneric segregates within *Syzygium*, such as *Acmena*, *Acmenosperma* Kausel, *Cleistocalyx*, *Jambosa*, *Syzygium*, *Waterhousea* B. Hyland, and *Anetholea* Peter G. Wilson
Wilson et al. (2005)	cpDNA sequence analyses of *matK* and *ndhF* genes and *rpl16* intron revealed no support for the conventional segregation of subgeneric taxa within *Syzygium*

(Continued)

TABLE 2.1 (CONTINUED)
Summary of Phenetic and Phylogenetic Studies on *Syzygium–Eugenia* Generic Complex

Author(s)	Summary
Biffin et al. (2006)	*matK* sequence data analyzed against a strong morphological and anatomical background classified the Myrtaceae to comprise two subfamilies and 17 tribes; accorded generic status to *Syzygium*, *Acmena*, *Acmenosperma*, *Anetholea*, and *Waterhousea* in Syzygeae tribe, and *Eugenia* in a separate tribe, Myrteae
Craven and Biffin (2010)	Based on molecular and morphological data, proposed an infrageneric classification for *Syzygium*, with recognition of six subgenera: *Acmena* (DC.) Craven & Biffin, *Sequestratum* Craven & Biffin, *Perikion* Craven & Biffin, *Anetholea* (Peter G. Wilson) Craven & Biffin, and *Wesa* Craven & Biffin; the subgenus *Acmena* was subdivided into seven sections: *Acmena* (DC.) Craven & Biffin, *Piliocalyx* (Brongn. & Gris) Craven & Biffin, *Agaricoides* Craven & Biffin, *Waterhousea* (B. Hyland) Craven & Biffin, *Glenum* Craven & Biffin, *Monimioides* Craven & Biffin, and *Gustavioides* Craven & Biffin
Soh and Parnell (2011)	Leaf anatomical and macromorphological data strongly supported recognition of the subgenera *Perikion*, *Sequestratum*, and *Syzygium*, but not the subgenus *Acmena*

stomatal rims, and stomatal types (Noorma and Moore 1996) came in support of differentiating the two genera. There is a growing consensus on the recognition of *Syzygium s.s.* for the Old World species and *Eugenia s.s.* for the largely New World species. A summary of the historical developments in systematic and phylogenetic studies on the *Syzygium–Eugenia* generic complex is presented in Table 2.1.

PHYLOGENY OF *SYZYGIUM*: THE BOTTLENECKS

One of the major constraints in understanding the phylogeny of *Syzygium* (including its alliances) is the large size of the genus, with an estimated 1200 (Govaerts et al. 2008) or 1800 (SYZWG 2016) species distributed across the Old World tropics of Africa, Asia, Malaysia, Australia, New Zealand, and southwest Pacific. The lack of systematic revision and monographic studies on the *Syzygium–Eugenia* group in its major centers of diversity is another major lacuna. Although floristic or revisionary accounts of *Syzygium* are available from a few floristic regions (see Parnell et al. 2007; Ashton 2011; Soh and Parnell 2015; Byng et al. 2016), undercollecting *Syzygium* in many parts of its geographical range, for example, in Thailand, Malesia, and India, is one of the main impediments to undertaking systematic and phylogenetic studies in the genus (Parnell et al. 2007). Deciphering good diagnostic characters for species- and supra-species-level evolution is critically important in resolving the existing taxonomic bottlenecks in *Syzygium* and its close alliances. Considering the urgent need for a taxonomic update on the regional treatments and also in view of the emerging generic and infrageneric classification, a genus-wide taxonomic update of *Syzygium* through a series of 22 regional revisions, including 9 in the *Flora Malesiana* region, has been launched recently by the *Syzygium* Working Group (SYZWG 2016).

Phylogenetic inferences in the *Syzygium–Eugenia* complex were attempted earlier using morphological and anatomical characters by Johnson and Briggs (1984), who proposed an informal system of "alliances" and "suballiances" within the Myrtaceae. According to Johnson and Briggs (1984), *Eugenia* formed a clade separate from *Syzygium* with other Myrtoideae, whereas *Syzygium* formed another clade with *Acmena* and other Old World species remote from the *Eugenia* clade. Leaf anatomical studies on peninsular Malayan species of *Eugenia* sects. *Acmena*, *Cleistocalyx*, *Fissicalyx* Bentham, and *Syzygium* by Khatijah et al. (1992) supported the recognition of sects. *Acmena* and *Syzygium*, but not the other sections (*Cleistocalyx* and *Fissicalyx*) that were found to be similar to sect. *Syzygium*. Phenetic and phylogenetic studies based on morphological and anatomical data were used for defining the generic and subgeneric boundaries of *Syzygium* and associated genera (Johnson and Briggs 1984; Parnell 1999).

MOLECULAR PHYLOGENY OF *SYZYGIUM* AND ALLIANCES

Harrington and Gadek (2004) examined phylogeny of the *Syzygium* group (66 species comprising *Acmena*, *Acmenosperma*, *Anetholea*, *Cleistocalyx*, *Syzygium*, and *Waterhousea* plus 6 unnamed species) by analyzing sequence variations in internal transcribed spacer (ITS) and external transcribed spacer (ETS) regions of the nuclear ribosomal DNA. The above study did not support segregation of the above Old World syzygioid genera, because all of these *Syzygium* suballiances were found nested within *Syzygium*. In another study with an expanded taxon coverage, Biffin et al. (2006) inferred phylogeny in as many as 87 species of the syzygioid genera using sequence analyses of the chloroplast DNA genes *matK*, *ndhF*, and the *rpl16* intron. The results of this study supported Harrington and Gadek (2004) in recognizing all the conventional *Syzygium* suballiances in one single genus, *Syzygium*, but with a few major cryptic clades. Using the RNA secondary structure partitioning and RNA-specific evolutionary models (paired sites) of the nrDNA ITS sequences from 76 taxon *Syzygium* and 45 taxon Myrtaceae data sets, Biffin et al. (2007) demonstrated that the phylogeny of *Syzygium* was found to be consistent with the topologies drawn from the standard four-state models of ITS analysis by Harrington and Gadek (2004) and the cpDNA analysis by Biffin et al. (2006). However, the RNA-specific approach of Biffin et al. (2007) showed several topological differences in the phylogeny of the Myrtaceae from the conventional tribal classifications of Wilson et al. (2005). This indicates that RNA-specific models, which account for the mutational dynamics of ITS, may have an impact on the accuracy of phylogenies estimated from these regions compared with the standard four-state models.

Craven and Biffin (2010), based on the molecular phylogenetic analyses of Biffin et al. (2006), proposed an infrageneric classification for *Syzygium*, which was also supported by the available morphological evidences. Accordingly, Craven and Biffin (2010) divided *Syzygium* into six subgenera: *Syzygium*, *Acmena* (DC.) Craven & Biffin, *Sequestratum* Craven & Biffin, *Perikion* Craven & Biffin, *Anetholea* (Peter G. Wilson) Craven & Biffin, and *Wesa* Craven & Biffin. The subgenus *Acmena* was subdivided into seven sections: *Acmena* (DC.) Craven & Biffin, *Piliocalyx* (Brongn. & Gris)

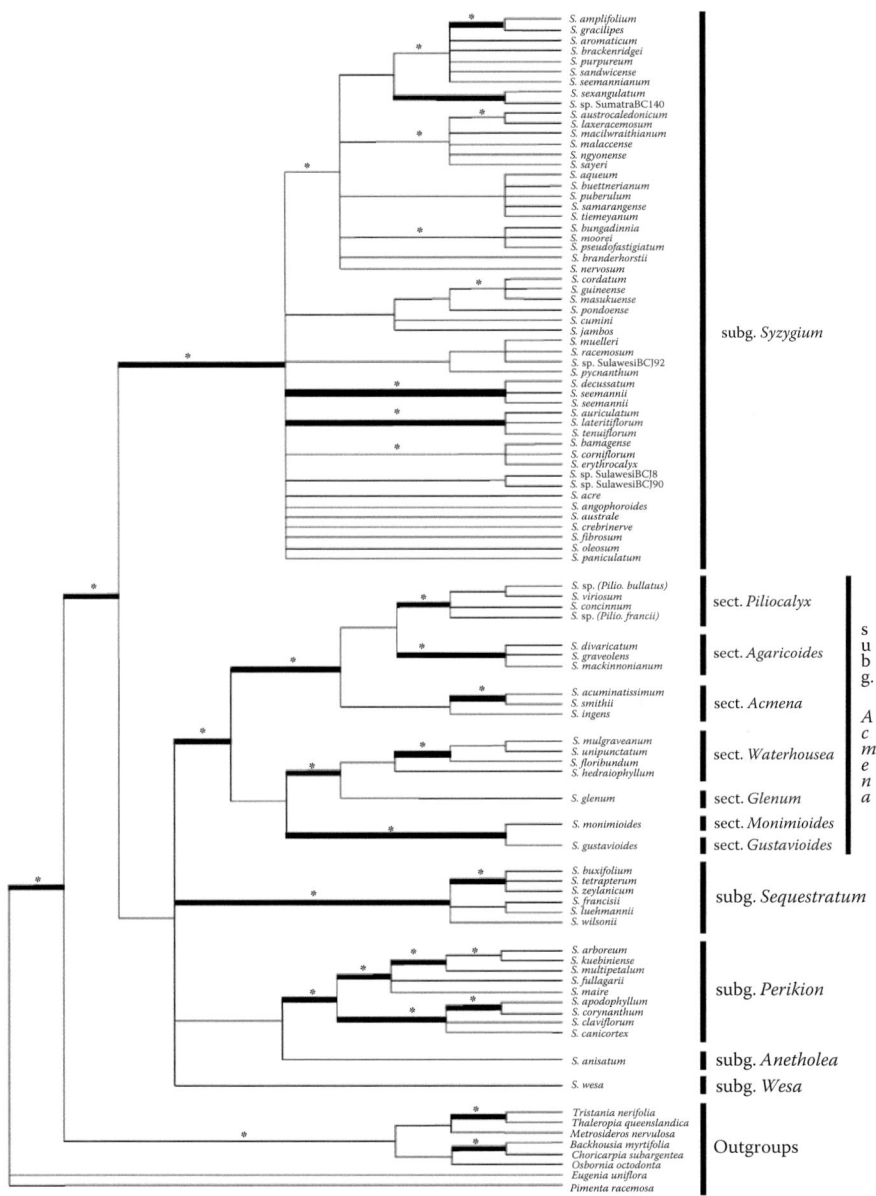

FIGURE 2.1 Strict consensus tree drawn from chloroplast DNA sequence data (*matK*, *ndhF*, and *rpl16*) showing the infrageneric and sectional grouping in *Syzygium*. (Reproduced with permission from Craven, L. A., and Biffin, E., *Blumea*, 55, 94–99, 2010.)

Craven & Biffin, *Agaricoides* Craven & Biffin, *Waterhousea* (B. Hyland) Craven & Biffin, *Glenum* Craven & Biffin, *Monimioides* Craven & Biffin, and *Gustavioides* Craven & Biffin (Figure 2.1). Craven and Biffin (2010) considered the following morphological characters as important in segregating the subgenera and sections recognized under *Syzygium*: (1) presence and absence of numerous fiber bundles in the hypanthium wall, (2) position of the placenta, (3) orientation and arrangement of the ovules on the placenta, (4) presence and absence of a testa, and (5) nature of the tissues developed in fruit from the chalaza.

In order to infer the intertribal relationships, character evolution, and lineage divergence times in Myrtaceae, Biffin et al. (2010) analyzed the nuclear (rDNA ITS, 91 taxa) and plastid (*matK*, 96 taxa; *ndhF*, 84 taxa) DNA sequences of a data set representing 14 of the 17 tribes (cf. Wilson et al. 2005), with focus on the fleshy fruited Myrtaceae. This study concluded that the fleshy fruits evolved independently in the tribes Syzygieae and the Myrteae and inferred a recent Oligocene–Miocene radiation of the two tribes, with an exceptionally high diversification rate. This was contrary to the earlier hypothesis (Sytsma et al. 2004; Wilson et al. 2005), according to which the fleshy fruit of Myrtaceae had multiple origins, arising separately within Syzygieae, Myrteae, and elsewhere in the family. Biffin et al. (2010) suggested that the evolution of fleshy fruit was a key innovation for rainforest Myrtaceae.

Soh and Parnell (2011) carried out separate and combined analyses of nrDNA ETS and ITS sequences and 28 qualitative morphological (leaf anatomical and macromorphological) characters in 33 species (representing four of the six subgenera of *Syzygium* recognized by Craven and Biffin 2010; i.e., *Acmena, Perikion, Sequestratum,* and *Syzygium*). The study strongly supported the recognition of the subgenera *Perikion, Sequestratum,* and *Syzygium*, but subg. *Acmena* was only moderately supported. Soh and Parnell (2011) also studied the leaf anatomy of 81 species of the above four subgenera and identified four stomatal types (anisocytic, anomocytic, cyclostaurocytic, and paracytic) and two major vascular systems differentiated by the presence or absence of adaxial phloem partition. No unique leaf anatomical characters allowed delimitation of the four subgenera in *Syzygium*, whereas the combinations of nonunique leaf anatomical characters (including stomatal types, crystal types, and frequency, and midrib vascular system) were diagnostic for the subgeneric grouping within *Syzygium* (Soh and Parnell 2011).

SYZYGIUM PHYLOGENY: FUTURE PROSPECTS

The studies so far carried out on the phylogeny of *Syzygium* and its alliances have been based on limited taxon and geographical coverage. Both phenetic and phylogenetic analyses, including molecular phylogenetic studies, need to be carried out in each floristic province or country, with special attention to identifying key characters and tracing their evolution at species and supraspecific levels. Testing the predictability of the emerging generic and infrageneric classifications of *Syzygium* and its alliances is also an important aspect to be carried out within a phylogenetic framework. More collection of materials and a comprehensive taxonomic update of regional floristic treatments as envisioned by SYZWG (2016) will result in a robust taxonomy of the genus, which is a prerequisite for testing the many complex questions about evolution and ecology that *Syzygium* could help address.

REFERENCES

Ashton, P. S. 2011. Myrtaceae. In *Tree Flora of Sabah and Sarawak* 7, ed. E. Soepadmo, L. G. Saw, R. C. K. Chung, and R. Kiew. Kuala Lumpur, Malaysia: Forest Research Institute.

Bentham, G., and J. D. Hooker. 1865. Myrtaceae. In *Genera Plantarum*. Vol. 1. London: Reeve and Co.

Biffin, E., Craven, L. A., Crisp, M. D., and P. A. Gadek. 2006. Molecular systematics of *Syzygium* and allied genera (Myrtaceae): Evidence from the chloroplast genome. *Taxon* 55: 79–94.

Biffin, E., Harrington, M. G., Crisp, M. D., Craven, L. A., and P. A. Gadek. 2007. Structural partitioning, paired-sites models and the evolution of the rDNA ITS transcripts in *Syzygium* and Myrtaceae. *Mol. Phylogenet. Evol.* 43: 134–139.

Biffin, E., Lucas, E. J., Craven, L. A., Costa, I. R., Harrington, M. G., and M. D. Crisp. 2010. Evolution of exceptional species richness among lineages of fleshy-fruited Myrtaceae. *Ann. Bot.* 106: 79–93.

Byng, J. W., Barthelat, F., Snow, N., and B. Bernardini. 2016. Revision of *Eugenia* and *Syzygium* (Myrtaceae) from the Comoros Archipelago. *Phytotaxa* 252: 163–184.

Chantaranothai, P., and J. Parnell. 1994. A revision of *Acmena, Cleistocalyx, Eugenia s.s.* and *Syzygium* (Myrtaceae) in Thailand. *Thai Forest Bull.* 21: 1–123.

Craven, L. A., and E. Biffin. 2010. An infrageneric classification of *Syzygium* (Myrtaceae). *Blumea* 55: 94–99.

De Candolle, A. P. 1828. Myrtaceae. In *Prodromus systematis naturalis regni vegetabilis*, ed. A. P. De Candolle, 207–296. Vol. 3. Paris: Truettel et Würtz.

Govaerts, R., Sobral, M., Ashton, P. et al. (2008). World checklist of Myrtaceae. London: Royal Botanic Gardens, Kew. http://www.kew.org/wcsp/ (accessed July 2015).

Harrington, M. G., and P. A. Gadek. 2004. Molecular systematics of the *Acmena* alliance (Myrtaceae): Phylogenetic analyses and evolutionary implications with reference to Australian taxa. *Aust. Syst. Bot.* 17: 63–72.

Henderson, M. R. 1949. The genus *Eugenia* (Myrtaceae) in Malaya. *Gard. Bull. Singapore* 12: 1–293.

Hyland, B. P. M. 1983. A revision of *Syzygium* and allied genera (Myrtaceae) in Australia. *Aust. J. Bot. Suppl. Ser.* 9: 1–164.

Ingle, H. D., and H. E. Dadswell. 1953. The anatomy of the timbers of the south-west Pacific area III. Myrtaceae. *Aust. J. Bot.* 1: 353–401.

Johnson, L. A. S., and B. G. Briggs. 1984. Myrtales and Myrtaceae—A phylogenetic analysis. *Ann. Mo. Bot. Gard.* 71: 700–756.

Khatijah, H. H., Cutler, D. F., and D. M. Moore. 1992. Leaf anatomical studies of *Eugenia* L. (Myrtaceae) species from the Malay Peninsula. *Bot. J. Linn. Soc.* 110: 137–156.

Kochummen, K. M. 1995. *Eugenia* in *Tree Flora of Malaya*, ed. F. S. P. Ng, 172–247. Vol. 3. London: Longman. Reprint of the 1978 edition.

Kostermans, A. J. G. H. 1981. *Eugenia, Syzygium* and *Cleistocalyx* (Myrtaceae) in Ceylon. *Quart. J. Taiwan Mus.* 34: 117–188.

Lemmens, R. H. M. J. 1995. *Syzygium* Gaertner. In *Plant Resources of South-East Asia*, No. 5, *Timber Trees: Minor Commercial Timbers*, ed. R. H. M. J. Lemmens, I. Soerianegara, and W. C. Wong, 441–474. Leiden: Backhuys Publishers.

Linnaeus, C. 1753. *Species Plantarum*. Stockholm: Laurentii Salvii.

Merrill, E. D. 1951. Readjustments in the nomenclature of Philippine *Eugenia* species. *Phil. J. Sci.* 79: 351–430.

Merrill, E. D., and L. M. Perry. 1938. A synopsis of *Acmena* DC., a valid genus of the Myrtaceae. *J. Arnold Arbor.* 19: 1–20.

Merrill, E. D., and L. M. Perry. 1939. The Myrtaceae genus *Syzygium* Gaertner in Borneo. *Mem. Am. Acad. Arts Sci.* 18: 135–202.

Niedenzu, F. 1893. Myrtaceae. In *Die Naturlichen Planzenfamilien*, ed. A. Engler and K. Prantl, 57–105. Vol. 3. Leipzig: Wilhelm Engelmaan.

Noorma, W. H., and Moore, D. M. 1996. The taxonomic significance of leaf micromorphology in the genus *Eugenia* L. (Myrtaceae). *Bot. J. Linn. Soc.* 120: 265–277.

Parnell, J. 1999. Numerical analysis of Thai members of the *Eugenia-Syzygium* group (Myrtaceae). *Blumea* 44: 351–379.

Parnell, J. A. N., L. Craven, and E. Biffin. 2007. Matters of scale: Dealing with one of the largest genera of Angiosperms. In *Towards the Tree of Life: Systematics of Species Rich Groups*, ed. T. Hodkinson and J. Parnell, 251–274. Special Series Vol. 72. London: Systematics Association.

Schmid, R. 1972. A resolution of the *Eugenia-Syzygium* controversy (Myrtaceae). *Am. J. Bot.* 59: 423–436.

Soh, W. K., and J. Parnell. 2011. Comparative leaf anatomy and phylogeny of *Syzygium* Gaertn. *Pl. Syst. Evol.* 297: 1–32.

Soh, W. K., and J. A. N. Parnell. 2015. A revision of *Syzygium* Gaertn. (Myrtaceae) in Indochina (Cambodia, Laos and Vietnam). *Adansonia* 37: 179–275.

Sytsma, K. J., Litt, A., Zjhra, M. L. et al. 2004. Clades, clocks, and continents: Historical and biogeographical analysis of Myrtaceae, Vochysiaceae, and relatives in the southern hemisphere. *Int. J. Plant Sci.* 165: S85–S105.

SYZWG (*Syzygium* Working Group). 2016. *Syzygium* (Myrtaceae): Monographing a taxonomic giant via 22 coordinated regional revisions. PeerJ Preprints. https://doi.org/10.7287/peerj.preprints.1930v1.

Wight, R. 1841. *Illustrations of Indian Botany*. Vol. 2. Madras: Pharoah.

Wilson, P. G., M. M. O'Brien, P. A. Gadek, and C. J. Quinn. 2005. Relationships within Myrtaceae *sensu lato* based on *mat*K phylogeny. *Plant Syst. Evol.* 251: 3–19.

3 The Genus *Syzygium* in Western Ghats

P. Sujanapal and C. Kunhikannan

CONTENTS

INTRODUCTION

Western Ghats is an unbroken mountainous stretch (with the exception of the "Palakkad Gap" of an average 30 km), running parallel to the Arabian seacoast, in a north–south direction, for about 1500 km from the river Tapti (about 21° 16′ N) in Gujarat to Kanyakumari (about 8° 19′ N) in Tamil Nadu, at the tip of the Indian peninsula. This mountain range is phytogeographically categorized into southern Western Ghats, central Western Ghats, and northern Western Ghats. It has luxuriant vegetation because of its position and different microclimatic conditions prevailing in different parts of the mountain ranges (WGEEP 2011). It has an area of approximately 129,037 km², and it stretches to a width of 210 km in Tamil Nadu and narrows to as small as 48 km in Maharashtra. The Western Ghats is second to the Eastern Himalaya in richness of biological diversity in India, and with its geographical extension in the wet zone of Sri Lanka, it is considered one of the eight "hottest hotspots" of biodiversity (Nayar 1996; Myers et al. 2000).

It is estimated that about 7% of the area of the Western Ghats is presently under primary vegetation cover, although a much larger area is under secondary forest or some form of tree cover. Nearly 15% of the Ghats is under the protected area system. The great topographic heterogeneity, from sea level to the highest point; the Anaimudi peak (2695 m); and a strong rainfall gradient from <500 mm in valleys in the east to >7000 mm along west-facing slopes combine to give rise to a rich diversity of life forms and vegetation types, including tropical wet evergreen forest, montane wet temperate forest (shola) and grassland, lateritic plateaus, moist and dry deciduous forest, dry thorn forests, and grasslands.

The importance of the Western Ghats in terms of its biodiversity is evident from the known document of its plant and animal diversity and the high endemism made to declare this region as a biodiversity hotspot, along with Sri Lanka (Mittermeier et al. 2005; Gunawardene et al. 2007). Nearly 5800 species of flowering plants are reported from the Western Ghats (Rao 2012). But a recent compilation (Nayar et al. 2014) gave a precise number, that is, 8080 taxa, including 7402 species, 117 subspecies, and 476 varieties. Among these, 5588 species are indigenous, 376 are naturalized exotics, and 1438 are cultivated. Out of the known species, 645 species are evergreen trees, with 56% endemic to the Western Ghats (Nayar et al. 2014). Among the lower plant groups, there are 850–1000 species of bryophytes, out of which 682 species are mosses, with 28% endemic, and 280 species are liverworts, with 43% endemic. The other group consists of pteridophytes, with 277 species, and lichens, with 949 species, in 150 genera and 54 families, with 253 endemic species (26.7%). Similar to the floristic diversity, fauna of the Western Ghats is also very rich. The faunal diversity of the Ghats includes 120 species of mammals; more than 500 species of birds; 225 species of reptiles, with 62% endemic, including primitively burrowing snakes of the family Uropeltidae, restricted mostly to the southern hills of the Western Ghats; and 220 species of amphibians, with the highest level of endemism (78%). The recent discovery of a new genus and species of frog, *Nasikabactrachus sahyadrensis*, with Indo-Madagascan affinity in the southern Western Ghats, affirms the importance of the region in harboring these ancient Gondwanan lineages. Similarly, the Ghats is unique in its caecilian diversity, harboring 16 of the 20 species known in India, with all 16 species being endemic. The fish fauna of the Ghats is known to comprise 318 species, with 43% of these endemic to this region. The invertebrate species described from the Western Ghats include 350 (20% endemic) species of ants, 330 (11% endemic) species of butterflies, 174 (40% endemic) species of odonates (dragonflies and damselflies), and 269 (76% endemic) species of mollusks (land snails) (WGEEP 2011).

GENUS *SYZYGIUM*: ECONOMIC VALUE

The family Myrtaceae, especially the genera *Syzygium* Gaertn. and *Eugenia* L., has received much attention since ancient times due to its multidimensional uses for mankind. Cloves (*S. aromaticum* [L.] Merr. & L. M. Perry), the universal spice, were one of the first spices to be traded. They were imported into Alexandria in 176 CE. Used in Southeast Asia for thousands of years, they were regarded as a panacea for almost all ills. It is recorded that Chinese officials, in 266 BCE, would chew on cloves to sweeten their breath before audiences with the emperor (Herbal Encyclopedia 2016). In many countries, the astringent bark is a mouthwash for thrush. Clove is the most economically important known representative of this family. Clove oil is used for the treatment of inflammatory diseases and also has antioxidant properties. Clove oil shows significant antimicrobial activity against a collection of 25 different genera of test bacteria. Fruits of many species, such as *S. aqueum* (Burm. f.) Alston, *S. cumini* (L.) Skeels, *S. jambos* (L.) Alston, and *S. malaccense* (L.) Merr. &

L. M. Perry, used in the vine industry, are also edible. *S. cumini* has been regarded as sacred and protected. In popular medicine, the fruit is used for the treatment of insulin-dependent diabetes mellitus (DMID). The flowers are astringent and used in Taiwan to treat fever and halt diarrhea, and they also show weak antibiotic action. A root bark decoction of *S. malaccense* is used for treating dysentery and amenorrhea. Powdered leaves are used for cracked tongues. Root bark is used as an abortifacient. Decoctions of the bark, leaf, and root of *S. cordatum* Hochst. ex Krauss are used to treat tuberculosis, respiratory problems, diarrhea, and stomach complaints, and also as an emetic. *S. alternifolium* (Wight) Walp. is used to treat diabetes and dry cough. Leaves are fried in cow ghee and used as a curry. In Malaya, the greenish fruits of many *Syzygium* are eaten raw with salt or may be cooked as a sauce. They are also stewed with true apples (Quattrocchi 2012). Large flowered *Syzygium* are commonly known as lillypillies, which are one of the most beautiful plant groups for landscaping and gardens. Although this genus has a wide range of utilitarian importance, many of the species in the forests of the Western Ghats are underexplored for their economic importance.

SYSTEMATICS OF *SYZYGIUM* IN INDIA

Syzygium is represented by more than 1200 species all over the world, distributed in tropical regions of Asia, Africa, and Australia and in southwestern Pacific regions (Parnell et al. 2007; Govaerts et al. 2008). The genus *Syzygium*, so richly distributed in the Indian subcontinent, has received only very little attention compared with other angiosperm families and genera. The forests, especially the evergreen and shola forests in the high ranges of the Western Ghats and northeastern region, are ideal habitats for *Syzygium* with a high rate of endemism. Even though *Syzygium* forms very significant parts of the endemic flora of India, it has not gained due attention from taxonomic, phytogeographic, and conservation perspectives. Duthie (1878–1879), in Hooker's *Flora of British India*, treated the Indian specimens of this complex under the generic limit of *Eugenia*, with 131 species in three sections (*Jambosa* with 31 species, *Syzygium* with 77 species, and *Eugenia* with 22 species). Of these 131 species, 21 were originally described from south India by Wight (1842, 1846), Beddome (1864, 1872, 1874), and Duthie (1878–1879). Later, workers like Gamble (1919) treated the south Indian specimens of *Eugenia* in three separate genera, such as *Eugenia*, *Jambosa* DC., and *Syzygium*, with 10 new species. Later still, a few more species were added to the peninsular Indian flora (Sheeba et al. 2003; Murugan and Manickam 2004; Viswanathan and Manikandan 2008; Shareef et al. 2010, 2012a,b; Sujanapal et al. 2013). Many of the species are known only by their type collections, and some of these have been relocated during the floristic studies in some protected areas (Sasidharan 2013; Nayar et al. 2014). Andaman and Nicobar Islands are also found to be important centre of *Syzygium* with 17 species (Pandey and Divakar 2008) with latest addition of *S. hookeri* M.V. Ramana, Chorghae & Venu and *S. sanjappaiana* M.V. Ramana (Ramana et al. 2014).

Among the known 645 trees species in the Western Ghats (Nayar et al. 2014), *Syzygium* forms the single largest genus, with 48 taxa, of which 27 are endemic to the region. Nayar et al. (2006) and Sasidharan (2013) reported that 39 taxa occur in the Kerala part of the Western Ghats, with 19 endemic species. Chithra (1987) reported 30 species and three varieties of *Syzygium* in Tamil Nadu, whereas Viswanathan and Manikandan (2008) recorded 33 species and four varieties of the genus in the state. The following are the new species and varieties of *Syzygium* described recently from various parts of the Western Ghats: *S. parameswaranii* Mohanan & Henry (Mohanan and Henry 1987); *S. zeylanicum* (L.) DC. var. *ellipticum* Henry, *S. zeylanicum* var. *megamalayanum* Ravikumar & Lakshmanan, and *S. sriganesanii* Ravikumar & Lakshmanan (Ravikumar 1999); *S. periyarensis* Jomy & Sasidharan (Sasidharan and Jomy 1999); *S. agastyamalayanum* M. B. Viswan. & Manik. (Viswanathan and Manikandan 2008); *S. fergusonii* (Shareef et al. 2012a); *S. palodense* Shareef, E. S. S. Kumar & Shaju (Shareef et al. 2012b); *S. chemunjianum* Shareef, E. S. S. Kumar & Roy (Shareef et al. 2013a); *S. munnarensis* Shareef, Roy & Krishnaraj (Shareef et al. 2013b); *S. sasidharanii* Sujanapal (Sujanapal et al. 2013); *S. dhaneshiana* Ratheesh, Shareef, & Nandakumar (Ratheesh Narayanan et al. 2014); and *S. sahyadricum* Sujanapal (Sujanapal et al. 2014). Taxonomists used various characters of the inflorescence and flower, particularly the calyx tube and perianth, for species delimitation in *Syzygium*. However, there are so many species with intermediate character states, and therefore one or few character differences should not be expected to separate the taxa within this genus. Also, in the present state of ignorance about the world's *Syzygium* and *Eugenia*, unbridled speculation by various trivial characters can only leads to errors. These ambiguities were reflected in the identity of some of the recently described taxa. Hence, these taxa have been reduced by various workers while revising a species or a section due to insufficient character sets. The present chapter illustrates 48 systematically confirmed taxa so far known from the Western Ghats. Detailed citations and other taxonomic details of all the species are provided in the species enumeration section.

DISTRIBUTION PATTERN OF *SYZYGIUM* IN THE WESTERN GHATS

Generally, the species of *Syzygium* occur at an altitude between 100 and 2400 m asl, mainly on the windward side of the Western Ghats. The forests, especially the evergreen and shola forests in the high ranges of the Western Ghats, are ideal habitats for *Syzygium* species with a high rate of endemism. Among the 48 species, 5 species, *S. aqueum*, *S. aromaticum*, *S. jambos*, *S. malaccense*, and *S. samarangense* (Blume) Merr. & L. M. Perry, are exotics and cultivated commonly in homesteads. Although *S. cumini* is an indigenous species, it is widely planted throughout the region due to its multifarious uses. State-wise diversity of *Syzygium* in relation to areas of the Western Ghats is provided in Figure 3.1. Five states share the geographic area of the Western Ghats, and among them, Kerala represents the highest number of *Syzygium* (39), followed by Tamil Nadu (35), Karnataka (23), Goa (10), and Maharashtra (9).

The diversity and endemism of *Syzygium* are high in the southern Western Ghats, especially in the Kerala part, when compared with central and the northern Western Ghats. This is mainly due to the peculiarity of tropical evergreen forests and the edaphic types on the windward side of the southern Western Ghats region. About 50% Western Ghats endemism in this group demonstrates the microhabitat peculiarity of tropical forests in these mountain ranges. Species such as *S. chavaran*

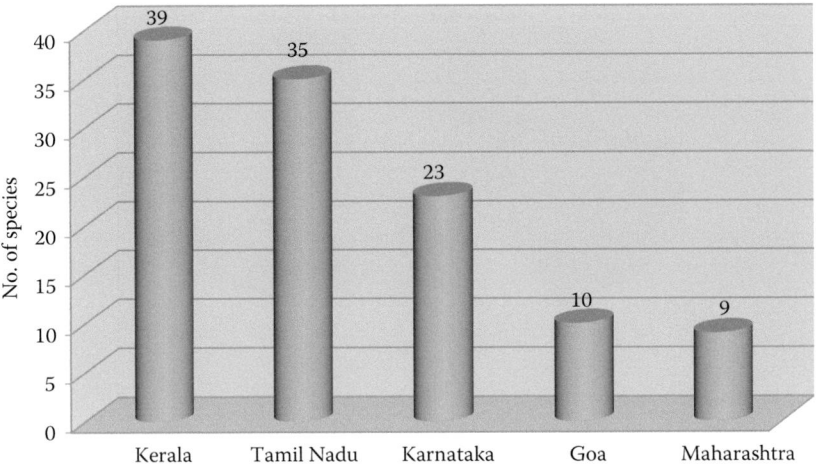

FIGURE 3.1 State-wise distribution of *Syzygium* in the Western Ghats.

(Bourd.) Gamble, *S. makul* Gaertn., *S. neesianum* Arn., *S. palghatense* Gamble, *S. palodense*, *S. periyarense*, *S. sasidharanii*, and *S. sahyadricum* are known only from the Kerala part of the Western Ghats, while *S. agasthyamalayanum* and *S. beddomei* (Duthie) Chithra are restricted to Tamil Nadu region, whereas *S. kanarense* (Talbot) Raizada is strictly endemic to Karnataka. A total of eight species, *S. calophyllifolium* (Wight) Walp., *S. courtallense* (Gamble) Alston, *S. fergusonii*, *S. tamilnadensis* Rathkr. & Chithra, *S. microphyllum* Gamble, *S. mundagam* (Bourd.) Chithra, *S. parameswaranii*, and *S. rama-varmae* (Bourd.) Chithra, are found in both Kerala and Tamil Nadu.

ENDEMISM AND RARITY

More than 50% of the *Syzygium* in the Western Ghats are endemic to this region, and some of the nonendemic species have shared their distribution with tropical forests of Sri Lanka (Figure 3.2). In the Western Ghats, 27 species are endemic: *S. agastyamalayanum*; *S. beddomei*; *S. benthamianum* (Wight ex Duthie) Gamble; *S. bourdillonii* (Gamble) Rathakr. & N. C. Nair; *S. courtallense*; *S. densiflorum* Wall. ex Wight & Arn.; *S. kanarense*; *S. laetum* (Buch.-Ham.) Gandhi; *S. malabaricum* (Bedd.) Gamble; *S. microphyllum*; *S. mundagam*; *S. munroi* (Wight) Chandrabose; *S. myhendrae* (Bedd. ex Brandis) Gamble; *S. occidentale* (Bourd.) Gandhi; *S. palghatense* Gamble; *S. palodense*; *S. parameswaranii*; *S. periyarensis*; *S. rama-varmae*; *S. sahyadricum*; *S. salicifolium* (Wight) Graham; *S. sasidharanii*; *S. sriganesanii*; *S. stocksii* (Duthie) Gamble; *S. tamilnadensis*; *S. travancoricum* Gamble; and *S. utilis* (Talbot) Rathakr. & N. C. Nair. Most of these species are narrowly distributed in certain forest patches of Agasthyamala, Anamalai, and Nilgiris in the southern Western Ghats.

Although most of the endemic species are threatened at regional and global levels for a variety of reasons, only a few have been assessed by the International Union for Conservation of Nature and Natural Resources (IUCN) criteria. Only 19 species of

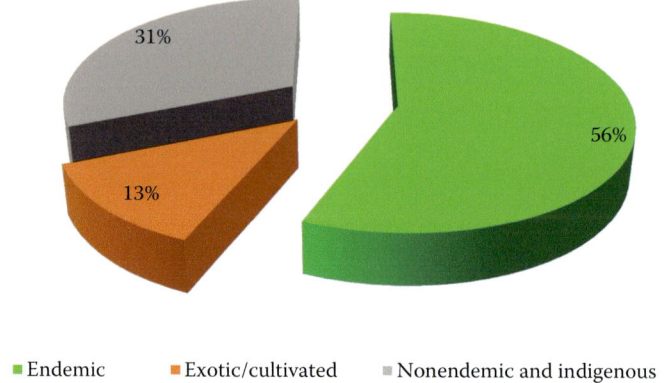

■ Endemic   ■ Exotic/cultivated   ■ Nonendemic and indigenous

FIGURE 3.2   Endemism of *Syzygium* in the Western Ghats.

Syzygium have been placed under various threat categories of IUCN (2010). These include *S. beddomei*, *S. bourdillonii*, S. *caryophyllatum* (L.) Alston, *S. chavaran*, *S. fergusonii*, *S. microphyllum*, *S. myhendrae*, *S. parameswaranii*, and *S. stocksii* under the Endangered B1+2c (ver 2.3) category; *S. courtallense*, *S. kanarense*, and *S. travancoricum* under the Critically Endangered B1+2abcde (ver 2.3) category; *S. benthamianum*, *S. densiflorum*, *S. occidentale*, *S. makul*, *S. neesianum*, and *S. rama-varmae* under the Vulnerable A1d (ver 2.3.) category; and *S. utilis* under the Data Deficient category (Figure 3.3). The threat assessments made by other workers are mentioned at appropriate places under species enumeration. Some species are known only from a single locality, like *S. palghatense* in Kerala. *S. beddomei* and *S. rubicundum* Wight & Arn. are known only by their type collections. There are no recent records of these species. A thorough assessment and reinvestigation of the

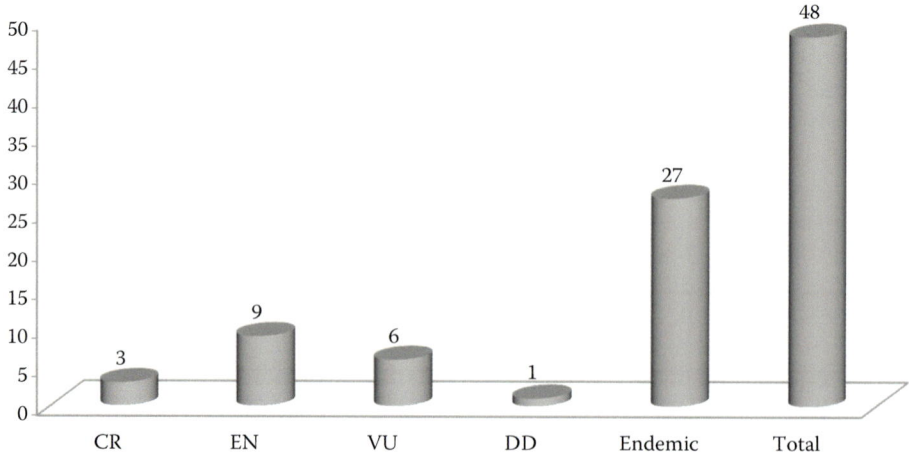

FIGURE 3.3   Threatened species of *Syzygium* of the Western Ghats assessed as per IUCN (2010) criteria. CR: Critically Endangered; DD: Data Deficient; EN: Endangered; VU: Vulnerable.

endemic and endangered species of *Syzygium* are essential for finding the correct conservation status and developing appropriate conservation measures before the species vanishes from the isolated forest patches in the Western Ghats.

SYSTEMATICS OF *SYZYGIUM* OF THE WESTERN GHATS

Syzygium (Gaertn., Fruct. 1: 166, 1788, nom. cons.)

Trees, bark thick, granular; twigs usually glabrous. Leaves opposite, entire, penninerved, usually gland dotted; lateral nerves united, forming a clear or faint intramarginal vein. Flowers bisexual, in terminal or axillary corymbose cymes or panicles; calyx tube hemispherical, globose, or turbinate, tube produced above the summit of ovary, lobes 4 or 5, ovate to suborbicular, imbricate; petals 4 or 5, orbicular, pellucidglandular; stamens numerous, filaments inflexed in bud; staminal disc broad or absent; anthers globose; ovary inferior, 2 celled; ovules few to several in each cell; style 1, subulate, stigma simple. Fruit a berry, 1 celled; seeds few.

Key to the Species

1. Calyx tube funneliform	2
1. Calyx tube turbinate, globose, or campanulate	11
2. Leaves small, 0.6–3 × 0.2–1 cm; calyx tube resinous-scaly	*S. microphyllum*
2. Leaves large; calyx tube without resinous scales	3
3. Flowers >1 cm across with a prominent staminal disc	*S. munroi*
3. Flowers <1 cm across without prominent staminal disc	4
4. Leaves sessile, base cordate	5
4. Leaves petiolate or subsessile, base acute or rounded	6
5. Secondary vein arching, obscure below; inflorescence dense	*S. parameswaranii*
5. Secondary vein horizontal, prominent below; inflorescence lax	*S. fergusonii*
6. Flowers in terminal panicle of umbellules	7
6. Flowers in terminal or axillary cymes	8
7. Leaves prominently glandular, highly aromatic on crushing; fruit white	*S. zeylanicum*
7. Leaves with few or without glands and aroma; fruit pink to dark black	*S. agastyamalayanum*
8. Branchlets distinctly quadrangular, leaves narrow; stamens pale pink	*S. sriganesanii*
8. Branchlets terete, leaves broad; stamens white or slightly yellowish	9
9. Leaves lanceolate or elliptic with few or without glands	*S. lanceolatum*
9. Leaves obovate or elliptic-oblanceolate, prominently glandular	10
10. Lateral nerves few, distant; flower bud highly aromatic	*S. aromaticum*
10. Lateral nerves close, parallel; flower bud not aromatic	*S. palghatense*
11. Inflorescence exclusively on old wood or defoliated mature twigs	12

11. Inflorescence terminal, axillary, or subterminal, not on old wood   13
12. Leaves secondary nerves canaliculated above;
    flowers white   *S. rama-varmae*
12. Leaves secondary nerves not canaliculated above;
    flowers mostly red   *S. malaccense*
13. Flowers >1 cm across; calyx tube with a thickened
    staminal tube at mouth   14
13. Flowers <0.5 cm across; calyx tube without a thickened
    staminal disc at mouth   26
14. Leaves cordate or rounded at base   15
14. Leaves acute or narrowed at base   19
15. Leaf nerves and intramarginal nerve not
    conspicuous, irregular   *S. courtallensis*
15. Leaf nerves and intramarginal nerve conspicuous, regular   16
16. Leaves chartaceous; fruits pyriform   *S. samarangense*
16. Leaves coriaceous; fruits globose or ovoid   17
17. Leaves usually <10 cm long, obtusely long acute at
    apex; flowers sessile   *S. beddomei*
17. Leaves usually >15 cm long, acute or small acuminate
    at apex; flowers pedicellate   18
18. Leaves glandular-punctate; cymes many flowered   *S. mundagam*
18. Leaves without conspicuous glands; cymes few
    flowered (usually 2–6)   *S. megacarpum*
19. Leaves narrow; linear-lanceolate, narrowly elliptic, or oblong   20
19. Leaves broad; lanceolate, elliptic, oblong, or obovate   21
20. Secondary nerves indistinct; pedicel flexuous; fruit
    densely glandular-punctate   *S. occidentale*
20. Secondary nerves prominent, pedicel stout; fruit not
    glandular-punctate   *S. jambos*
21. Pedicel 2–4 mm long   22
21. Pedicel 7–20 mm long   23
22. Leaves and flower bud densely glandular-punctate;
    fruit globose or ovoid   *S. bourdillonii*
22. Leaves and flower bud not glandular-punctate; fruit pyriform   *S. aqueum*
23. Leaves chartaceous; pedicel flexuous; stamens pink,
    red, or yellow   *S. laetum*
23. Leaves thickly coriaceous; pedicel stout; stamens white   24
24. Inflorescence mostly axillary; flowers 1–2 cm across   *S. grande*
24. Inflorescence mostly terminal; flowers more than 2.5 cm across   25
25. Leaves broadly elliptic or broadly obovate; calyx
    tube 8–13 mm long   *S. periyarensis*
25. Leaves elliptic or elliptic-lanceolate; calyx
    tube 4–6 mm long   *S. hemisphericum*
26. Branchlets or young shoots distinctly quadrangular   27
26. Branchlets or young shoots terete   34
27. Leaves to 2.5 cm broad, lateral nerves many, close, parallel   28

27. Leaves more than 3 cm broad; lateral nerves few, distant   29
28. Leaves ovate or ovate-lanceolate; petals calyptrate   *S. rubicundam*
28. Leaves oblanceolate or obovate; petals free, caducous   *S. myhendrae*
29. Leaves thickly coriaceous; inflorescence congested,
     stout   *S. tamilnadensis*
29. Leaves coriaceous or chartaceous; inflorescence loose, not stout   30
30. Cymes mostly axillary; sometimes terminal or subterminal   31
30. Cymes lateral usually from the scar of fallen leaves,
     sometimes axillary   33
31. Leaves chartaceous, nerves prominent; petiole
     >1 cm; calyx turbinate   *S. travancoricum*
31. Leaves coriaceous, nerves obscure; petiole <1 cm;
     calyx globet shaped or campanulate   32
32. Intramarginal nerve obscurely 2 tiered; calyx
     globet shaped   *S. neesianum*
32. Intramarginal nerve 1 tiered; calyx campanulate   *S. palodense*
33. Inflorescence a reduced metabotryoid; peduncle
     flattened toward the apex   *S. sahyadricum*
33. Inflorescence a cyme; peduncle uniform throughout   35
34. Leaves glaucous beneath; petals free   *S. malabaricum*
34. Leaves not glaucous beneath; petals calyptrate   *S. stocksii*
35. Petiole <5 mm long, stout   36
35. Petiole >5 mm long, stout, or slender   39
36. Leaves orbicular-ovate, apex rounded or
     emarginate; petals gland dotted   *S. calophyllifolium*
36. Leaves elliptic, oblong, or lanceolate; apex acute or
     acuminate; petals not gland dotted   37
37. Leaves narrowly elliptic; flowers mostly in cymes on
     defoliated mature twigs   *S. salicifolium*
37. Leaves broader; flowers mostly in terminal or
     subterminal corymbose cymes   38
38. Leaf base cuneate, sometimes attenuate; peduncle
     long, slender   *S. revolutum*
38. Leaf base cordate or obtuse; peduncle small, stout   *S. benthamianum*
39. Inflorescence mostly from the defoliated mature twigs   40
39. Inflorescence strictly terminal, subterminal, or axillary   41
40. Lateral nerves close and parallel; inflorescence dense;
     hypanthium turbinate   *S. cumini*
40. Lateral nerves distant, arching; inflorescence lax;
     hypanthium hemispheric   *S. nervosum*
41. Leaves chartaceous; branches of cymes slender, pedicel slender   42
41. Leaves coriaceous; branches of cymes stout; pedicel stout   43
42. Intramarginal nerves in 2 tiers; mature fruit with a
     terminal ring of calyx   *S. kanarense*
42. Intramarginal nerve 1 tiered; mature fruit without a
     terminal ring of calyx   *S. gardneri*

43. Leaves obovate, apex rounded or subacute; mature
    fruit yellow, >2.5 cm diameter   *S. sasidharanii*
43. Leaves oblong or elliptic; apex acuminate acute; mature
    fruit purple or pink, <1.5 cm diameter   44
44. Leaves oblong or elliptic-oblong; berry obliquely ventricose   *S. chavaran*
44. Leaves elliptic; berry globose or obovoid   45
45. Leaf apex obtuse or rounded; flowers in terminal
    spreading paniculate cymes   *S. caryophyllatum*
45. Leaf apex acute or acuminate; flowers in small axillary
    and terminal cymes   46
46. Lateral nerves of leaves distant, arching; branches of
    cymes terete   *S. makul*
46. Lateral nerves of leaves close, parallel; branches of
    cymes quadrangular   47
47. Cymes congested, stalk and pedicel short; fruit
    oblong-ovoid   *S. densiflorum*
47. Cymes elongate or lax; stalk and pedicel long; fruit globose   *S. utilis*

SPECIES ENUMERATION

The botanical name with author citation, a few important synonyms, a brief description, and the distribution, threat status, and flowering and fruiting seasons (Fl. & fr.) for all 48 species of *Syzygium* in the Western Ghats are provided under each species.

1. ***Syzygium agastyamalayanum*** M. B. Viswan. & Manik., Adansonia 30 (1):
   113–118, 2008 (Figures 3.4a and 3.5a).
       A medium-sized tree, up to 15 m tall; branchlets 4 angled, later becoming subterete. Leaves up to 6.6 × 2.9 cm, obovate to oblanceolate or elliptic, base acute, apex obtuse or subacute or retuse, margin recurved, coriaceous, glabrous; midrib grooved above, raised beneath; secondary nerves 14–20 pairs, with prominent intramarginal nerve; petiole up to 4 mm long. Inflorescence in terminal and axillary umbellate panicles, up to 4.5 cm. Flowers about 20 in each umbellule; sepals 4 or 5, glabrous, up to 1 × 2.5 mm, with entire margin and rounded apex; petals 4 or 5, white, concave, reniform-orbicular, 2–2.8 × 2.4–2.6 mm, cordate or rounded at base, obtuse at apex, glabrous, gland dotted; stamens many, varying in length, 2.7–7.5 mm, glabrous; filaments yellow and incurved; anthers yellow, reniform, 1.1 × 0.9 mm; ovary obovoid, fleshy, 3.8 × 2.5 mm; ovules many; style yellow, glabrous; stigma minute. Berries pink to dark black, subglobose or globose, 15–17 × 14.5–15 mm. Seed pale brown, solitary, subglobose, 10 × 8 mm, glabrous.
       Distribution: Endemic to southern Western Ghats (Kalakkad-Mundanthurai Tiger Reserve, Tamil Nadu).
       Fl. & fr.: January–April.
       Threat status: Critically Endangered (Viswanathan and Manikandan 2008).

FIGURE 3.4   (a) *Syzygium agastyamalayanum*. (b) *Syzygium benthamianum*. (c) *Syzygium bourdillonii*. (d) *Syzygium calophyllifolium*.

2. **Syzygium aqueum** (Burm. f.) Alston, Ann. Roy. Bot. Gard. (Peradeniya) 11: 204, 1929. *Eugenia aquea* Burm. f., Fl. Ind. 114, 1768; Duthie in Hook. f., Fl. Brit. India 2: 473, 1878.

   Small trees, 3–10 m tall. Leaves 7–25 × 2.5–16 cm, elliptic-cordate to obovate-oblong; petiole 0.5–1.5 mm long. Inflorescences terminal and axillary, 3–7 flowered. Flowers 2.5–3.5 cm in diameter; petals yellow-white petals; stamens numerous. Berry turbinate, 1.5–2 × 2.5–3.5 cm, white, pinkish to red, glossy, flesh very juicy. Seeds 1–6, rounded, small.

   Distribution: Native to Malaysia; cultivated in other tropical countries, including the Indian states of Kerala, Tamil Nadu, Karnataka, Goa, and Maharashtra.

   Fl. & fr.: December–March.

   Threat status: No threat.

   Note: Mostly, the tree is planted in the courtyards or gardens for its edible fruits. The fruit tastes sweet-sour.

3. **Syzygium aromaticum** (L.) Merr. & L. M. Perry, Mem. Am. Acad. Arts 18: 196, 1939. *Caryophyllus aromaticus* L., Sp. Pl. 515, 1753. *Eugenia caryophyllata* Thunb., Diss. 1, 1788. *Myrtus caryophyllus* Spreng., Syst. 2: 485, 1825. *Jambosa caryophyllus* (Spreng.) Neidz., Pflanzenfam. 3 (7): 84, 1893. *Eugenia caryophyllus* (Spreng.) Bullock & Harrison, Kew Bull. 13: 52, 1958.

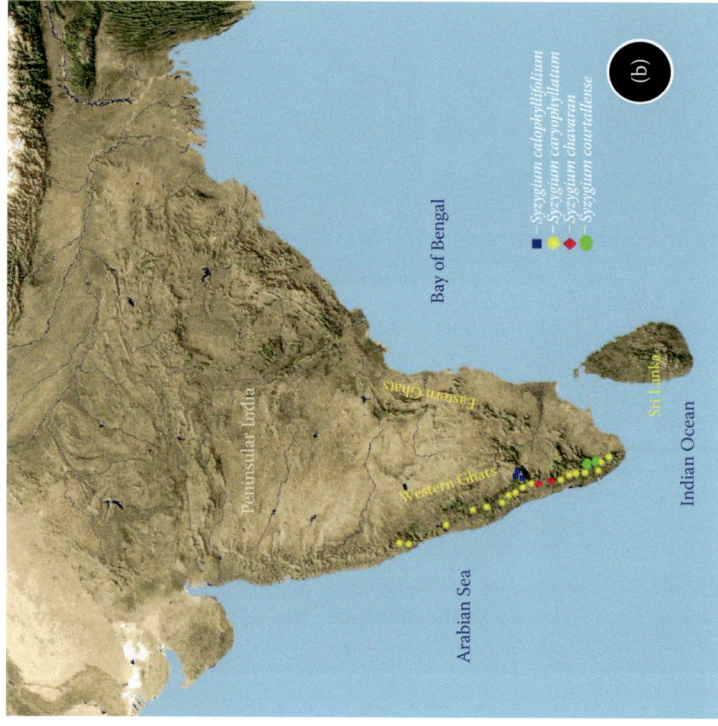

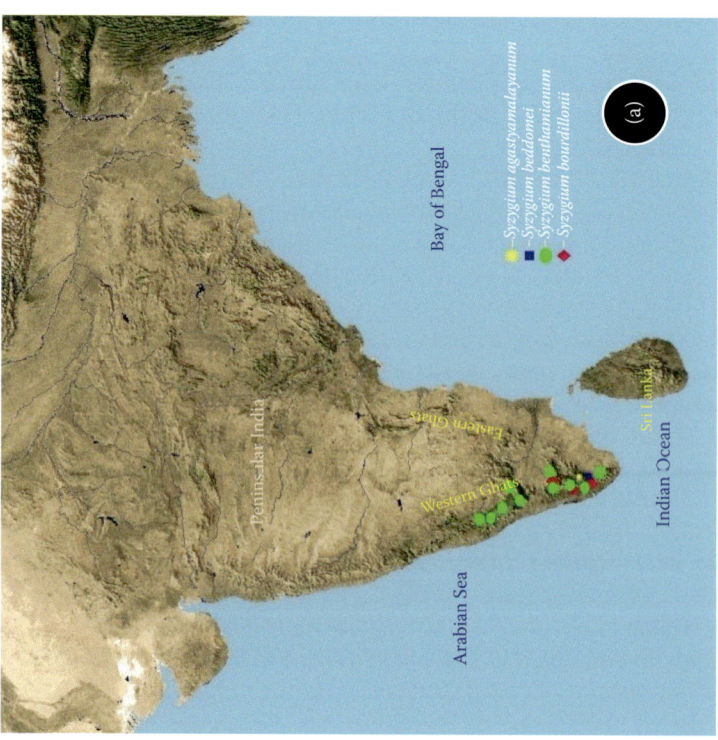

FIGURE 3.5 Map showing geographic distribution of *Syzygium* in the Western Ghats. (a) *S. agastyamalayanum*, *S. beddomei*, *S. benthamianum*, and *S. bourdillonii*. (b) *S. calophyllifolium*, *S. caryophyllatum*, *S. chavaran*, and *S. courtallense*.

A small evergreen tree, up to 6 m tall with conical crown and pale brown bark; branchlets slender, terete at base, subround at apex. Leaves 7–12 × 3–5 cm, elliptic or oblanceolate, base attenuate or cuneate, apex acuminate, margin entire, glabrous, coriaceous, punctate beneath; lateral nerves many, parallel, obscure, with intramarginal nerve; petiole 10–20 mm long, slender, glabrous. Inflorescence terminal or axillary cymes. Flowers pinkish-white, fragrant, in 4 cm long cymes; calyx 1.5 × 6 mm, tubular, verrucose with 4 hook-like involute ascending segments; petals up to 10 × 5 mm, elliptic, calyptrate; stamens many, inflexed in the bud; ovary inferior. Berry globoid or ellipsoid, dark purple with persistent calyx ring on the top.

Distribution: Native of Molucanna Islands (Indonesia), now widely cultivated in many tropical regions, including India, mainly in the states of Kerala, Tamil Nadu, Karnataka, Goa, and Maharashtra for the clove of commerce.

Fl. & fr.: December–July.

Threat status: No threat.

Note: Flower buds are the clove of commerce, used as a flavoring agent and spice. Leaves and buds are potential source of antioxidant and acetylcholinesterase inhibitor, which can be used in Alzheimer's disease (Darusman et al. 2013).

4. *Syzygium beddomei* (Duthie) Chithra in N. C. Nair & A. N. Henry, Fl. Tamil Nadu 1: 155, 1983. *Eugenia beddomei* Duthie in Hook. f., Fl. Brit. India 2: 476, 1878. *Jambosa beddomei* (Duthie) Gamble, Fl. Madras 1: 475, 1919 (Figure 3.5a).

Large trees, to 30 m tall. Leaves 10–13 × 6–8 cm, broadly ovate-elliptic, base rounded, apex rounded, or emarginate, coriaceous, polished above; midrib and lateral nerves prominent, stout, midrib channeled above, lateral nerves 9–13 pairs; petiole 1–2 cm. Inflorescence elongate, terminal corymbose panicles, branches acutely tetragonous. Flowers; petals free, ovate. calyx tube short. Berry ovoid, small; seed 1, globose.

Distribution: Endemic to southern Western Ghats; so far known from the Tirunelvelly hills of Tamil Nadu.

Fl. & fr.: August–December.

Threat status: Endangered B1+2c ver. 2.3 (IUCN 2015).

5. *Syzygium benthamianum* (Wight ex Duthie) Gamble, Fl. Madras 1: 478, 1919. *Eugenia arnottiana* (Walp.) Wight var. *benthamiana* Wight ex Duthie in Hook. f., Fl. Brit. India 2: 484, 1878 (Figures 3.4b and 3.5a).

Small to medium-sized trees; branchlets tetragonous. Leaves 3–6 × 2.5–3 cm, elliptic-oblong, apex acuminate, folded, base cordate or obtuse, coriaceous, nearly black when dry; lateral nerves many, close and parallel, with intramarginal nerve. Cymes terminal and axillary, rarely from leafless axils, flowers in corymbs of umbellules; calyx tube short, turbinate; petals usually free. Berry oblong-ovoid, dark purple, fleshy; seed 1, globose.

Distribution: Endemic to southern Western Ghats. Reported from Kerala, Karnataka, and Tamil Nadu.

Fl. & fr.: November–February.

Threat status: Vulnerable B1+2c ver. 2.3 (IUCN 2015).

Note: *Syzygium benthamianum* is reported to possess chemical constituents with antimicrobial (Kiruthiga et al. 2011; Deepika et al. 2013), antioxidant and anticancer activities (Deepika et al. 2013).

6. **Syzygium bourdillonii** (Gamble) Rathakr. & N. C. Nair, J. Econ. Taxon. Bot. 4: 287. 1983. *Jambosa bourdillonii* Gamble, Bull. Misc. Inform. Kew 1918: 239, 1918; Gamble, Fl. Madras 1: 414, 1919 (Figures 3.4c and 3.5a).

Medium-sized tree, up to 10 m tall; branchlets terete, woody, and glabrous. Leaves elliptic, lanceolate, or oblanceolate, base cuneate or acute, apex acuminate, margin entire, glandular-punctate; lateral nerves 8–10 pairs, parallel, with intramarginal nerve; petiole slender glabrous, grooved above. Flowers white, in terminal few flowered cymes; pedicel 5 mm long; calyx tube bell shaped, 1 cm long, lobes 4, rounded, 3 mm long, recurved, persistent, furnished with thickened staminal disc; petals 4, orbicular, gland dotted; many free stamens, longer than the petals; ovary inferior, 2 loculed, many ovuled; style slender, hairy; stigma slightly acute. Berry small, crowned by the calyx tube and thickened disc.

Distribution: Endemic to southern Western Ghats; in Thiruvanthapuram, Quilon, and Idukki districts of Kerala, including places like Shenduruny Wildlife Sanctuary (Sasidharan 1997), Periyar Tiger Reserve, and Agasthyamala Hills.

Fl. & fr.: February–April.

Threat status: Endangered B1+2c ver. 2.3 (IUCN 2015).

7. **Syzygium calophyllifolium** (Wight) Walp., Rep. 2: 180, 1843; Gamble, Fl. Madras 1: 480, 1919. *Eugenia calophyllifolia* Wight, Ic. t. 1000, 1845; Duthie in Hook. f., Fl. Brit. India 2: 494, 1878 (Figures 3.4d and 3.5b).

Medium-sized trees, up to 20 m; branchlets terete. Leaves up to 5 × 3 cm, obovate or suborbicular, base obtuse or rounded, apex slightly emarginate, margin entire, glabrous and coriaceous with pellucid glands; lateral nerves many, parallel, slender, very close, prominent, with intramarginal nerve; petiole 2–3 mm, stout, glabrous. Flowers white, in dense terminal corymbs; peduncle 4 angled; calyx tube 3 mm long, ovate, lobes 4, minute; petals 4, calyptrate; stamens many, bent inward in bud; ovary inferior, 2 loculed; ovules many; style up to 6 mm in length. Berry up to 12 mm long, dark purple.

Distribution: Western Ghats and Sri Lanka. Reported from Nilgiri phytogeographical region in the Kerala and Tamil Nadu parts of southern Western Ghats.

Fl. & fr.: February–May.

Threat status: No threat.

Note: This tree form is a characteristic vegetation cover of the shola ecosystem (Nair et al. 2001). The wood is strong and used for building purposes (Vinod Kumar 2003). Fruits are edible (Thomas et al. 2012). Leaves contain compounds that are antioxidant and antimicrobial in nature and have potential medicinal values (Vignesh et al. 2013; Saranya et al. 2014).

8. **Syzygium caryophyllatum** (L.) Alston in Trimen, Handb. Fl. Ceylon 6 (Suppl.): 116, 1931. *Myrtus caryophyllata* L., Sp. Pl. 472, 1753. *Syzygium caryophyllaeum* sensu Gamble, Fl. Madras 1: 480, 1919; *non* Gaertn., 1788.

FIGURE 3.6 (a) *Syzygium caryophyllatum*. (b) *Syzygium chavaran*. (c) *Syzygium densiflorum*. (d) *Syzygium laetum*.

Eugenia caryophyllaea Wight, Ic. Pl. Ind. Or. 2: t. 540, 1842; Duthie in Hook. f., Fl. Brit. India 2: 490, 1878 (Figures 3.5b and 3.6a).

Small trees reaching up to 6 m; bark thick, reddish-brown; branchlets terete. Leaves 11 × 5.2 cm, obovate to obovate-oblong, base attenuate or acute, apex obtuse, obtusely acute or emarginate, margin entire, glabrous, coriaceous, brown on drying, gland dotted; lateral nerves many, close, prominent forming intramarginal nerve; petiole up to 3 mm long, stout, glabrous. Flowers white, small, 5 mm across, in terminal corymbose cymes; calyx tube 2–2.5 mm long, turbinate; petals calyptrate; stamens numerous, bent inward at the middle in bud, 2.5–3.5 mm long; ovary inferior, 2 loculed, ovules many; style 1; stigma simple. Berry globose, purple.

Distribution: South India and Sri Lanka. Common in lower lateritic hillocks and forest boundaries in Kerala, Tamil Nadu, Karnataka, Goa, and Maharashtra.

Fl. & fr.: February–August.

Threat status: Endangered B1+2c ver. 2.3 (IUCN 2015).

Note: The fruits are edible and the leaves possess medicinal properties and are used as blood purifier, appetizer (Vinod Kumar 2003), and remedy for skin infections (Gayathri et al. 2012) and exhibit antimicrobial activity against pathogens like *Staphylococcus aureus* and scavenging activity against free radicals (Gayathri et al. 2012).

9. ***Syzygium chavaran*** (Bourd.) Gamble, Fl. Madras 1: 480, 1919. *Eugenia chavaran* Bourd., For. Trees Travancore 188, 1908 (Figures 3.5b and 3.6b).

Medium-sized tree up to 25 m tall; bark greyish-brown; branchlets terete. Leaves 10–21 × 4–9 cm, elliptic, oblong, or elliptic-oblong, base attenuate or acute, apex obtuse or acuminate, margin entire, glossy above, glabrous, coriaceous, gland dotted; lateral nerves many, parallel, close, slender, with prominent intramarginal nerve; petiole 13–30 mm, grooved above, glabrous. Flowers, 1–1.5 cm across, white, in axillary and terminal compound cymes, peduncle branches divaricating at right angles; calyx tube turbinate, less than 5 mm long, mouth 5–6 mm across, lobes 4; petals calyptrate; stamens many, free, bent inward in bud; ovary inferior, 2 loculed, ovules many in each locule; style 1; stigma simple. Berry bluish-purple, 3–4 × 2–2.5 cm, crowned with 3–5 mm long calyx limb.

Distribution: Western Ghats (Thrissur and Palakkad district of Kerala) and Thailand.

Fl. & fr.: December–April.

Threat status: Endangered B1+2c ver. 2.3 (IUCN 2015).

10. **Syzygium courtallense** (Gamble) Alston in Trimen, Handb. Fl. Ceylon 6 (Suppl.): 115, 1931. *Jambosa courtallensis* Gamble, Bull. Misc. Inform. Kew 1918: 239, 1918; Gamble, Fl. Madras 1: 475, 1919 (Figure 3.5b).

Small trees, 7–12 m tall. Leaves 7–11 × 3–5.5 cm, elliptic, base rounded, apex obtusely acute; intramarginal nerves double. Flowers 2–3.2 cm across, pedicellate, white, in terminal panicled cymes; calyx with a staminal disc, tube thick, subcylindric, 1.2 cm long, lobes rounded, persistent; petals 4, inserted on the top of the mouth of the calyx, broad, concave, obtuse; stamens many and free, longer than the petals, bent fleshy inward in bud; ovary 2 loculed, many ovuled; style slender; stigma slightly acute. Berry 1 or 2 seeded, crowned by the calyx limb and thickened disc. Seeds large, angled.

Distribution: Endemic to southern Western Ghats; reported only from southern parts of Kerala and Tamil Nadu.

Fl. & fr.: January–March.

Threat status: Critically Endangered B1+2cde ver. 2.3 (IUCN 2015).

11. **Syzygium cumini** (L.) Skeels, Bull. Bur. Pl. Industr. U.S.D.A. 248: 25, 1912. *Myrtus cumini* L., Sp. Pl. 471, 1753. *Eugenia jambolana* Lam., Encycl. 3: 198, 1789; Duthie in Hook. f., Fl. Brit. India 2: 499, 1879. *Syzygium jambolanum* (Lam.) DC., Prodr. 3: 259, 1828; Gamble, Fl. Madras 1: 481, 1919.

Medium-sized trees; bark white, light pink inside. Leaves to 18 × 8 cm, ovate, oblong, base acute, apex long-acuminate; nerves many, close, shining above; petiole 1.5–2 cm long. Panicles to 10 cm across, on leafless branchlets. Flowers 6–9 mm across, subsessile; calyx tube 3 mm broad, turbinate; filaments 7 mm long. Berry 10 × 7 mm, obovoid, deep violet-blue.

Distribution: India, Sri Lanka, south China, and Malaysia.

Fl. & fr.: April–October.

Threat status: No threat.

Note: The tree is also cultivated for its fruits. Wood is heavy and used for construction of bridges, door and window frames, and railway sleepers, as it is not affected by white ants. The bark is astringent and used in asthma, bronchitis, sore throat, and ulcer. It is also used for extraction of tannin. Fruits are edible and used in the treatment of diarrhea and bladder stones. Seeds are very good in managing diabetes (Quattrocchi 2012).

12. **Syzygium densiflorum** Wall. ex Wight & Arn., Prodr. Fl. Ind. Orient. 329, 1834; Chithra in N. C. Nair & A. N. Henry, Fl. Tamil Nadu 1: 156, 1983. *Eugenia arnottiana* Wight, Ill. Ind. Bot. 2: 17, 1841, Duthie in Hook. f., Fl. Brit. India 2: 483, 1878. *Syzygium arnottianum* (Wight) Walp., Repert. Bot. Syst. 2: 180, 1843, *nom. superfl.*; Gamble, Fl. Madras 1: 478, 1919 (Figures 3.6c and 3.7a).

Large trees, to 20 m tall; bark surface blackish-grey, rough; branchlets terete. Leaves 3.5–9 × 1.8–3.7 cm, elliptic, elliptic-lanceolate or elliptic-oblong, base attenuate or acute, apex acuminate or caudate-acuminate, margin entire, glabrous, glandular-punctate, coriaceous, olive-green when dry; petiole 3–20 mm long, slender, grooved above, glabrous; lateral nerves many, parallel, close, prominent, looped at the margin forming intramarginal nerve, intercostae reticulate, prominent. Flowers, creamy, 10–12 mm long, sessile, in dense clusters forming compact, terminal trichotomous cymes; calyx tube to 5 mm, turbinate, lobes 4, no thick disc; petals free, deciduous; stamens many free, bent inward at the middle in bud; ovary inferior, 2 loculed, ovules many; style 1; stigma simple. Berry oblong-ovoid, dark purple, fleshy.

Distribution: Southern Western Ghats (shola forests of Kerala and Tamil Nadu).

Fl. & fr.: October–February.

Threat status: Vulnerable B1+2c ver. 2.3 (IUCN 2015).

Note: Potential source of essential oils and other chemicals, with biological properties and uses such as antidiabetic (Mohan Maruga Raja et al. 2013), antimicrobial (Saranya et al. 2012), and insect repellent, which can be prospected as products for the management of mosquitoes and the treatment of skin problems and cancer.

13. **Syzygium fergusonii** (Trimen) Gamble, Bull. Misc. Inform. Kew 1920: 52, 1920. *Eugenia fergusonii* Trimen, Handb. Fl. Ceylon 1: 156, 1894; plates, Handb. Fl. Ceylon t. 38, 1895. *Eugenia aquea sensu* Bedd., Fl. Sylv. S. India t. 109, 1872; *non* Burm. f., 1768. *Syzygium chandrasekharanii* Chandrab. & V. Chandras., J. Bombay Nat. Hist. Soc. 78: 354, f. 1–8, 1981 (Figure 3.7a).

Trees up to 10 m tall with young branchlets 4 angled, later becoming terete with prominent leaf scars. Leaves broadly ovate-elliptic, base subcordate, apex obtuse, or emarginate, margin revolute, coriaceous; midrib grooved, purplish brown above; lateral nerves 8–10, reticulate, with 2-tiered intramarginal nerves; sessile to subsessile. Inflorescence a terminal corymb with 3-flowered umbels. Flowers sessile, peduncle up to 1.5 cm long; calyx tubular funnel shaped, tapering to slender stalk-like base, 1.5–2.1 cm long, mouth produced beyond the ovary, yellowish green, becoming purplish pink from the mouth, calyx lobes 4–5, obtuse, with recurved parts; petals calyptrate with 5 petals uneven and thin at margin, slightly subequal and gland dotted, fugaceous; stamens many; filaments white, up to 2 cm; ovary 2 loculed; style 2.2 cm with tapering end. Berries oblong-ellipsoid, up to 1.7 × 0.80 cm, purple, 1 seeded.

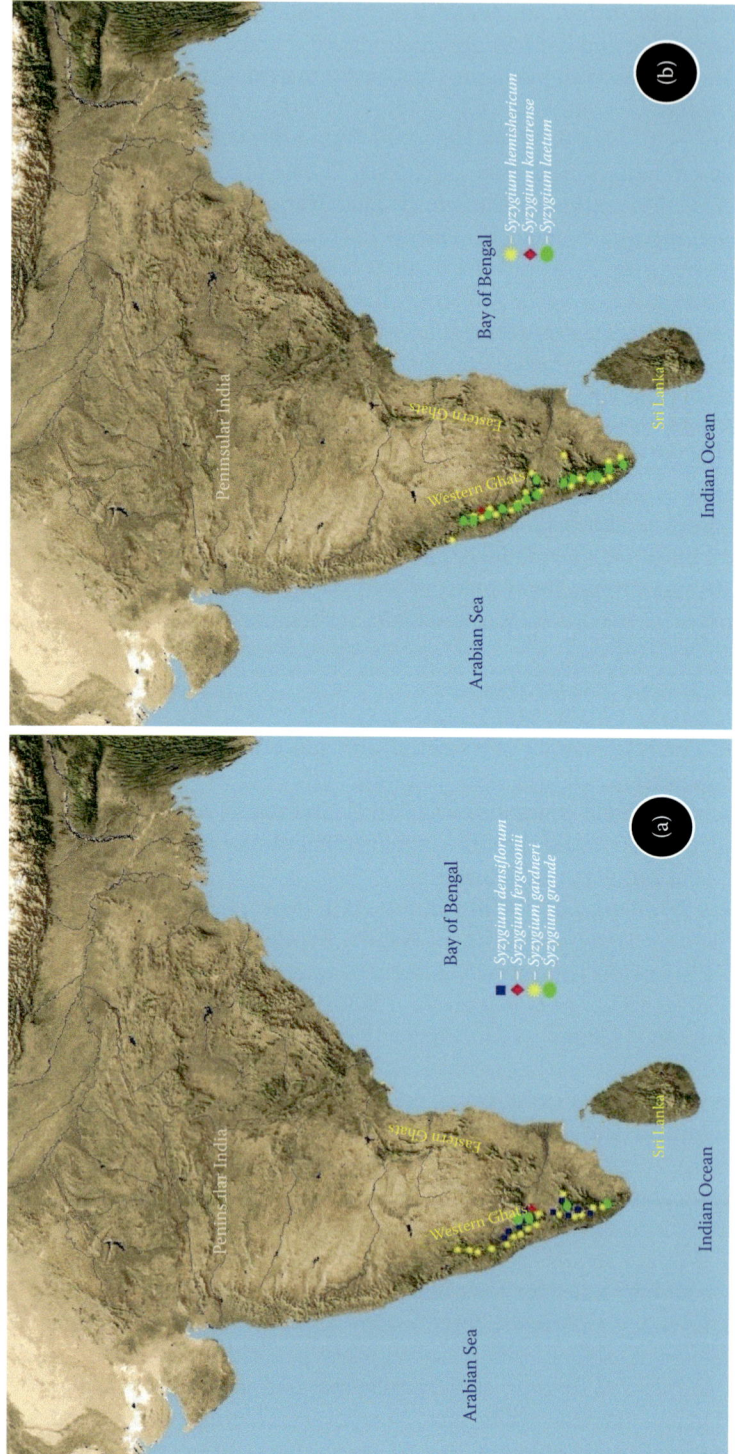

FIGURE 3.7 Map showing geographic distribution of *Syzygium* in the Western Ghats. (a) *S. densiflorum, S. fergusonii, S. gardneri,* and *S. grande.* (b) *S. hemisphericum, S. kanarense,* and *S. laetum.*

Distribution: Southern Western Ghats; in shola vegetation above 1800 m asl (Kerala and Tamil Nadu) and Sri Lanka (Shareef et al. 2012a).

Fl. & fr.: January–April.

Threat status: Endangered B1+2c ver. 2.3 (IUCN 2015).

14. **Syzygium gardneri** Thwaites, Enum. Pl. Zeyl. 117, 1859; Gamble, Fl. Madras 1: 479, 1919. *Eugenia gardneri* (Thwaites) Bedd., Fl. Sylv. S. India 108, 1872; For. Man. Bot. 108, 1874; Duthie in Hook. f., Fl. Brit. India 2: 489, 1878. *Eugenia cymosa* Lam. var. *rostrata* Duthie in Hook. f., Fl. Brit. India 2: 482, 1878. *S. munnarensis* Shareef, Roy & Krishnaraj, Webbia: J. Pl. Taxon. & Geogr. 69(1): 53–57, 2013. *S. dhaneshiana* Ratheesh, Shareef, & Nandakumar, Intl. J. Adv. Res. 2:3, 1055–1058, 2014 (Figure 3.7a).

Tall trees, up to 35 m tall; bark yellowish-white, peeling off in small flakes; branchlets slender, terete. Leaves 5–10 × 2.5–5 cm, elliptic-ovate or ovate-lanceolate, base acute, apex caudate-acuminate, margin entire, glabrous, gland dotted; lateral nerves many, parallel, close, with intramarginal nerve, prominent; petiole 10–15 mm long, slender, grooved above. Flowers, 4–5 mm across, white, in axillary and terminal trichotomous cymes; pedicels 2–5 mm long; calyx tube 2 × 3 mm, turbinate, lobes 4, obtuse, obscure; petals calyptrate; stamens many, free, bent inward in bud; filaments 4 mm long; ovary inferior, 2 loculed, ovules many; style 1; stigma simple. Berry 5–8 mm across, ovoid, rarely globose, purple.

Distribution: Western Ghats and Sri Lanka. Kerala, Karnataka, and Tamil Nadu regions of Western Ghats.

Fl. & fr.: January–April.

Threat status: No threat.

Note: The leaves contain essential oils that were tested and shown to have antimicrobial activity against bacteria, as well as the fungi *Candida albicans* and *Candida glabrata* (Gopan et al. 2008).

15. **Syzygium grande** (Wight) Walp., Repert. Bot. Syst. 2: 180, 1843; *Eugenia grandis* Wight, Ill. Ind. Bot. 2: 17, 1841; Duthie in Hook. f., Fl. Brit. India 2: 475, 1878 (Figure 3.7a).

Medium-sized trees; branchlets terate. Leaves 5–13 × 3–8 cm, elliptic, obovate, or elliptic-obovate, base cuneate, apex obtuse, or emarginate, margin entire, glabrous, coriaceous, pellucid dotted; lateral nerves parallel, slender, slightly distant with intramarginal nerve; petiole 5–16 mm, stout. Flowers white, 1–2 cm across, in axillary or terminal corymbose cymes, not exceeding the leaves; calyx tube 7 mm long; petals calyptrate; stamens many, bent inward in bud, filaments 5–10 mm long; ovary inferior, 2 loculed; ovules many; style 1; stigma simple. Berry 1 cm across, globose, crowned by calyx limb.

Distribution: Indo-Malesia. In peninsular India, it is reported from Kerala, Karnataka, and Tamil Nadu parts of Western Ghats.

Fl. & fr.: November–May.

Threat status: No threat.

Note: Govaerts et al. (2008) and Byng et al. (2015) reported its occurrence in Kerala, Tamil Nadu, and Karnataka parts of Western Ghats; however, needs confirmation for its occurrence in this region.

16. ***Syzygium hemisphericum*** (Wight) Alston in Trimen, Handb. Fl. Ceylon 6 (Suppl.): 115, 1931. *Eugenia hemispherica* Wight, Ic. t. 525, 1842; Duthie in Hook. f., Fl. Brit. India 2: 477, 1878. *Jambosa hemispherica* (Wight) Walp., Rep. 2: 191, 1843; Gamble, Fl. Madras 1: 474, 1919 (Figure 3.7b).

Trees, up to 27 m tall; bark dark brown, smooth; branchlets terete. Leaves 5.5–17.5 × 2–5 cm, elliptic, elliptic-lanceolate, elliptic-oblanceolate, or ovate-lanceolate, base cuneate or acute, apex acute, acuminate, or caudate-acuminate, margin entire, coriaceous, glabrous, gland dotted; lateral nerves 8–16 pairs, obscure, with intramarginal nerve; petiole 10–20 mm, stout, grooved above, glabrous. Flowers white or rose, densely packed, 3–4 cm across; pedicel 5 mm long; calyx tube 6 mm long, shortly and stoutly obconic, lobes 4, 3 × 6 mm, obtuse; disc thick; petals 4, 7 mm across, free; stamens many, bent inward in bud; filaments 15 mm long; ovary inferior, 2 loculed, ovules many; style slender; stigma slightly acute. Berry 23–25 mm across, purple, crowned by calyx lobes.

Distribution: South India and Sri Lanka. In the Western Ghats, Kolhapur, Ratnagiri, and Sindhudurg (Maharashtra); Chikmangalur, Hassan, North Kanara, Shimoga, and South Kanara (Karnataka); and Kannur, Kozhikkode, Wyanad, Malappuram, Thrissur, Idukki, Kollam, Palakkad, Pathanamthitta, Thiruvananthapuram (Kerala), Coimbatore, Dindigul, Nilgiri, and Tirunelveli (Tamil Nadu).

Fl. & fr.: March–June.

Threat status: No threat.

17. ***Syzygium jambos*** (L.) Alston in Trimen, Handb. Fl. Ceylon 6 (Suppl.): 115, 1931. *Eugenia jambos* L., Sp. Pl. 470, 1753; Duthie in Hook. f., Fl. Brit. India 2: 474, 1878. *Jambosa vulgaris* DC., Prodr. 3: 286, 1828; Gamble, Fl. Madras 1: 474, 1919.

Trees, up to 15 m tall; branchlets terete, glabrous. Leaves 10–18 × 2.5–5.5 cm, elliptic, elliptic-oblong, or elliptic-lanceolate, base acute, obtuse, or cuneate, apex acute or acuminate, margin entire, glabrous, coriaceous; lateral nerves 10–16 pairs, pinnate, prominent, with intramarginal nerves; petiole 7–10 mm long, slender, glabrous, grooved above. Flowers, white, to 6 cm across in terminal cymes to 10 cm; pedicel to 2 cm; calyx tube 1.5 cm, turbinate; lobes 4, 8 × 6 mm, ovate-orbicular, subequal, persistent; petals 4, 1.5 × 1.8 cm, free, concave, spreading, orbicular; disc thick, lining the calyx; stamens many; filaments exserted, basally subconnate, unequal; ovary inferior, to 8 mm long, 2 loculed, ovules many; style filiform, subulate. Berry 3 × 2.5 cm, white or pink, fleshy, oblong; seeds brown.

Distribution: Native of Malaysia, planted in tropical Asia and Australia. In India cultivated in the Western Ghats region, including Maharashtra, Goa, Karnataka, Kerala, and Tamil Nadu and also in the Eastern Ghats region.

Fl. & fr.: October–January.

Threat status: No threat.

18. **Syzygium kanarense** (Talbot) Raizada, Indian For. 74: 336, 1948. *Eugenia kanarensis* Talbot, J. Bombay Nat. Hist. Soc. 11: 236, 1897; Saldanha, Fl. Karnataka 2: 28, 1996 (Figure 3.7b).

Large trees; bark smooth, white; branchlets terete. Leaves up to 4.5 × 9 cm, elliptic, base acute, apex acuminate, margin wavy and slightly revolute, drying black, densely gland dotted; midrib reddish-brown, channeled above, lateral nerves conspicuous, numerous, and parallel; petiole 1.3 cm long. Cymes axillary and terminal, shorter than the leaves. Flowers white, 0.6–0.8 cm across, nearly sessile, buds globose, creamy white; calyx 0.4 × 0.4 cm, almost truncate, turbinate; corolla calyptrate 2.5 × 1.8 mm, thick; corolla lobes 3, very thin, gland dotted, varying in size, largest 1.5 × 1.6 mm, shortly clawed, another 1.3 × 1.5 mm, hastate with two claws, smallest 1.2 × 1.2 mm with a single claw; stamens 0.5 cm long; ovary 2 loculed; style simple. Berry to 2 × 1.2 cm, 1 seeded, purple, pulpy, crowned with the persistent calyx.

Distribution: Endemic to Western Ghats; reported only from Shimoga and North Kanara districts of Karnataka (Raizada 1948; Saldanha 1996; Shenoy et al. 2015).

Fl. & fr.: December–March.

Threat status: Critically Endangered (IUCN 2015).

19. **Syzygium laetum** (Buch.-Ham.) Gandhi in Saldanha & Nicolson, Fl. Hassan Dist. 282, 1976. *Eugenia laeta* Buch.-Ham., Mem. Wern., Nat. Hist. Soc. 5: 338, 1826; Duthie in Hook. f., Fl. Brit. India 2: 479, 1878. *Jambosa laeta* (Buch.-Ham) Blume, Mus. Bot. Lugd.-Bat. 1: 104, 1849; Gamble, Fl. Madras 1: 474, 1919. *Eugenia pauciflora* Wight, Ic. t. 526, 1842. *Jambosa pauciflora* (Wight) Wight, Illustr. 2: 14, 1850 (Figures 3.6d and 3.7b).

Medium-sized trees, up to 10 m tall; branchlets slender, terete. Leaves 6–15 × 2–6 cm, elliptic-ovate or elliptic-lanceolate, base cuneate or acute, apex acuminate, margin entire, chartaceous, gland dotted; lateral nerves 10–15 pairs, parallel, with intramarginal nerve near the margin (not at the margin); petiole 5–10 mm long, grooved above. Flowers crimson or lemon yellow, 4–5 cm across, solitary or 2–5 together in axillary or terminal cymes; pedicel 2–5 cm long; calyx tube 1.5–2 cm long, lobes 4, 8 × 8 mm, orbicular, persistent; petals 4, 10 × 10 mm, orbicular; stamens numerous, 2–3 cm long, yellow or pink, bent inward in bud; ovary 2 loculed, ovules many; style longer than the stamens. Berry oblong, crowned by calyx lobes.

Distribution: Endemic to southern Western Ghats (Kerala, Tamil Nadu, and Karnataka).

Fl. & fr.: December–July.

Threat status: No threat.

20. **Syzygium lanceolatum** (Lam.) Wight & Arn., Prodr. 330, 1834. *Eugenia lanceolata* Lam., Encycl. 3: 200, 1789. *Syzygium wightianum* Wall. ex Wight & Arn., Prodr. 330, 1834; Gamble, Fl. Madras 1: 478, 1919. *Eugenia wightiana* (Wall. ex Wight & Arn.) Wight, Ic. t. 529, 1842; Duthie in Hook. f., Fl. Brit. India 2: 485, 1878. *Syzygium claviflorum sensu* Shareef et al.,

Rheedea 20: 52, 2010; *non* (Roxb.) Wall. ex A. M. Cowan & Cowan, 1929 (Figures 3.8a and 3.9a).

Small trees, up to 12 m tall; bark greyish-brown, smooth; branchlets terete. Leaves 7.5–12.5 × 2–3.7 cm, elliptic, or elliptic-lanceolate, base cuneate or acute, apex acuminate or acute, margin entire, chartaceous, gland dotted; lateral nerves many, parallel, obscure, with intramarginal nerve; petiole grooved above. Flowers white, 5 mm across, in axillary and terminal cymes; calyx tube 12 × 3 mm, funnel shaped, lobes 4, short; petal 4, often many, 3 × 3 mm, suborbicular; stamens many, bent inward in bud; ovary 2 loculed; ovules many; style 1; stigma simple. Berry 10–12 mm long, ovoid-turbinate, bright scarlet.

Distribution: South India and Sri Lanka. In the Western Ghats, frequent in south and central Western Ghats (Kerala, Tamil Nadu, and Karnataka) and rare in Maharashtra.

Fl. & fr.: March–April.

Threat status: No threat.

Note: Byng et al. (2015) resurrected *S. wightianum* as a separate species. *S. lanceolatum* and *S. wightianum* are closely related and distributed in southern Western Ghats. Field observation shows that these taxa show a wide range of morphological variation with respect to altitude, topography, climate, and so forth. Hence, detailed field studies in various populations are essential for

FIGURE 3.8   (a) *Syzygium lanceolatum*. (b) *Syzygium malabaricum*. (c) *Syzygium mundagam*. (d) *Syzygium munroi*.

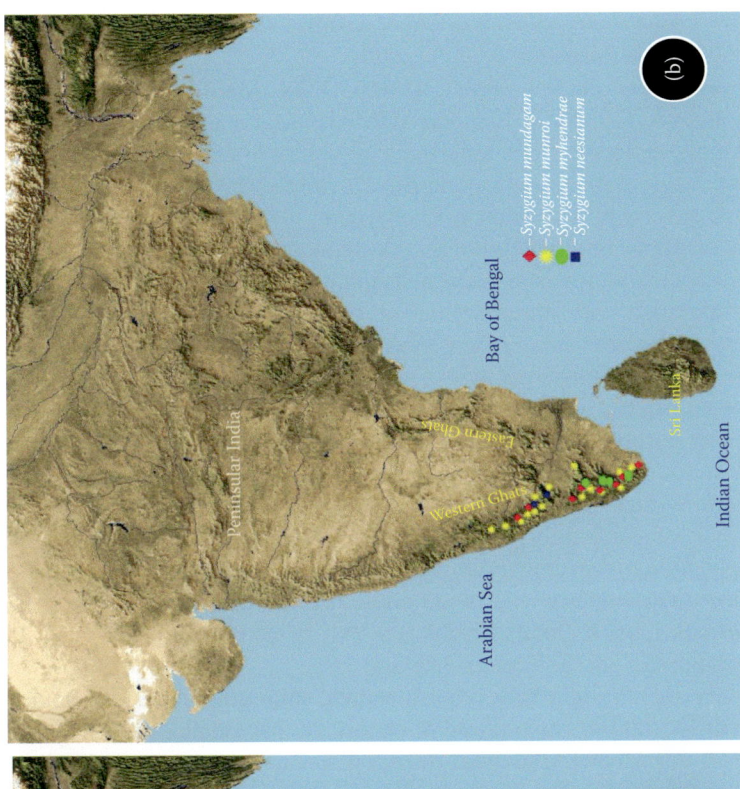

FIGURE 3.9 Map showing geographic distribution of *Syzygium* in the Western Ghats. (a) *S. lanceolatum, S. makul, S. malabaricum,* and *S. microphyllum.* (b) *S. mundagam, S. munroi, S. myhendrae,* and *S. neesianum.*

fixing the status of *S. wightianum*. Here, we are following the earlier concept for the systematic treatment.

21. ***Syzygium makul*** Gaertn., Fruct. 1: 166, 1778; Manilal et al., J. Econ. Taxon. Bot. 5: 419, 1984. *Eugenia sylvestris* Moon ex Wight, Ic. t. 532, 1843; Duthie in Hook. f., Fl. Brit. India 2: 493, 1879 (Figure 3.9a).

Small trees; bark smooth, pale brown, thinly flaked; branches terete, glabrous. Leaves bright crimson when young, 9–17 × 4.5–7 cm, narrowly obovate or elliptic or elliptic-obovate, base cuneate, apex acuminate, acumen twisted, chocolate-brown beneath on drying, coriaceous; lateral nerves many, slender, parallel, with intramarginal nerve, prominent; petiole 8–15 mm, long, stout. Flowers white, in dense terminal or axillary cymes; calyx up to 3 × 2 mm, campanulate; petals 4–5, up to 4 mm long, concave, fugacious; stamens numerous, up to 4 mm long; ovary 2 loculed, ovules many. Berry 1 × 0.8 cm, purplish, subglobose, with a terminal unlobed crown.

Distribution: South India and Sri Lanka. It is reported from the Kerala part of the Western Ghats.

Fl. & fr.: January–July.

Threat status: Vulnerable A1c ver. 2.3 (IUCN 2015).

22. ***Syzygium malabaricum*** (Bedd.) Gamble, Fl. Madras 1: 481 (340), 1919. *Eugenia malabarica* Bedd., Fl. Sylv. t. 199, 1872; Duthie in Hook. f., Fl. Brit. India 2: 497, 1879 (Figures 3.8b and 3.9a).

Tree, up to 12 m tall; bark greyish-brown, smooth; branchlets 4 angled. Leaves 6–13 × 2.2–8 cm, obovate or obcordate, base cuneate, apex obtuse, or retuse, coriaceous, glabrous, nearly glaucous beneath, gland dotted; lateral nerves 8–10 pairs, faint, meeting in loops only; petiole 6–25 mm, grooved above. Flowers white, in short axillary or lateral cyme, 2.5 mm across; pedicel short; calyx tube turbinate, lobes 4, minute, triangular; petals 4, free, orbicular; stamens numerous, bent inward in bud; ovary 2 loculed, ovules many; style 1; stigma simple. Berry 5–6 mm across, globose.

Distribution: Endemic to semievergreen forests in southern Western Ghats (Kerala, Tamil Nadu, and Karnataka).

Fl. & fr.: July–August.

Threat status: Data Deficient.

23. ***Syzygium malaccense*** (L.) Merr. & L. M. Perry, J. Arnold Arbor. 19: 215, 1938. *Eugenia malaccensis* L., Sp. Pl. 470, 1753; Duthie in Hook. f., Fl. Brit. India 2: 471, 1878.

Tree, up to 10 m tall; bark grey-brown and smooth; branchlets terete compressed. Leaves 16–34 × 5–13 cm, elliptic, base cuneate, coriaceous, glossy, gland dotted; lateral nerves 10–15 pairs, prominent, with intramarginal nerves; petiole 8–15 mm, stout, grooved above. Flowers large; calyx tube 1.5 cm; lobes round and unequal; petals large, glandular, suborbicular; stamens 2 cm, many, bent inward in bud; ovary 2 loculed, ovules many; style long. Berry large, to 5 × 3 cm.

Distribution: Native of Malaysia; cultivated in other parts of tropical Asia, including India in Kerala, Tamil Nadu, Karnataka, Goa, and Maharashtra.

Fl. & fr.: February–June.

Threat status: No threat.

Note: The fruits can be used for making frozen pulp, juice, jam, and ice cream, which act as good sources of vitamin C, dietary fiber, and antioxidant (Augusta et al. 2010). The aroma present in leaves and fruits of this species has important properties (Wong and Lai 1996; Pino et al. 2004; Karioti et al. 2007; Ismail et al. 2010).

24. **Syzygium megacarpum** (Craib) Rathakr. & N. C. Nair, J. Econ. Taxon. Bot. 4: 287, 1983. *Eugenia megacarpa* Craib, Fl. Siam. 1: 652, 1931. *Eugenia latilimba* Merr., Lingnan Sci. J. 13: 64, 1934. *Syzygium latilimbum* (Merr.) Merr. & L. M. Perry, J. Arnold Arbor. 39: 216, 1938.

Medium-sized trees, up to 20 m tall; branchlets green when dried, slightly compressed. Leaves 14–30 × 6–13 cm, narrowly long elliptic to elliptic, base rounded to sometimes cordate, apex acuminate, leathery, abaxially pale green when dry, adaxially green when dry, both surfaces without conspicuous glands; lateral nerves 15–22 on each side of midrib and 1–1.3 cm apart, reticulate veins conspicuous, intramarginal veins 4–5 mm from margin and an additional inconspicuous intramarginal vein ca. 1.5 mm from margin; petiole 5–10 mm. Inflorescences terminal, cymes, 2–6 flowered; peduncle very short. Flowers white, large; hypanthium long obconic, 1.5–2 × ca. 1.5 cm; calyx lobes 4, rounded, 6–7 × 8–9 mm; petals distinct, rounded, ca. 2 cm; stamens numerous, 2.5–3 cm; style ca. 4 cm. Berry ovoid-globose, ca. 5 cm.

Distribution: Southern Western Ghats (Tamil Nadu), Bangladesh, Myanmar, Thailand, and Vietnam. Cultivated in different parts of China, Myanmar, and Thailand for edible fruits.

Fl. & fr.: November–June.

Threat status: No threat.

Note: The authors believe that detailed studies are required for any taxonomic assessment on the species. Byng et al. (2015) considered *S. coarctatum* (Blume) Byng and *S. megacarpum* as conspecific, and adopted the former name for the taxon. They, however, did not mention any localities of *S. coarctatum* in peninsular Indian region. Therefore, pending further detailed studies, we followed here the concept of Rathakrishnan and Nair (1983) in describing this taxon under *Syzygium megacarpum*.

25. **Syzygium microphyllum** Gamble, Fl. Madras 1: 479, 1919. *Eugenia microphylla* Bedd., For. Man. Bot. 110, 1874. *Syzygium gambleanum* Rathakr. & V. Chithra in Nair & Henry, Fl. Tamil Nadu 1: 157, *nom. illeg.* (Figure 3.9a).

Small trees, up to 10 m tall; bark greyish brown with white patches; branchlets 4 angled in young, later terete. Leaves opposite, decussate, small, 0.6–3 × 0.2–1 cm, lanceolate, elliptic to obovate, base acute or rounded, apex obtuse, margin revolute, gland dotted, coriaceous, glabrous; midrib grooved; lateral nerves and tertiary nerves obscure; petiole 0.1 cm long, canaliculate, glabrous. Inflorescence terminal and axillary umbels, rarely solitary. Flowers white; pedicel 0.2 cm long. Berry 0.8 × 5 cm, globose to oblong, crowned by persistent calyx, purple, 1 seeded.

Distribution: Known only from the type locality, the species is endemic to a small area in the Agasthyamala Hills of Kerala and Tamil Nadu.

Fl. & fr.: Not available.

Threat status: Endangered B1+2c ver. 2.3 (IUCN 2015).

26. **Syzygium mundagam** (Bourd.) Chithra in Nair & Henry, Fl. Tamil Nadu 1: 157, 1983. *Eugenia mundagam* Bourd., For. Trees Travancore 182, 1908. *Jambosa mundagam* (Bourd.) Gamble, Fl. Madras 1: 473, 1919 (Figures 3.8c and 3.9b).

Small trees, up to 15 m tall; bark brown and smooth; branchlets 4 angled. Leaves 12–30 × 5–12.5 cm, oblong or elliptic-oblong, base cordate, apex acute or obtuse, margin entire, coriaceous, glandular; lateral nerves 18–22 pairs, pinnate, with intramarginal nerve; petiole 3–7 mm, glabrous. Flowers white, 4 cm across, in terminal many flowered corymbs; pedicel 5 mm long; calyx tube stout, funnel shaped, lobes 4, 6 mm across, suborbicular; petals 4, 1.2 × 1.5 cm, orbicular; stamens many, free, bent inward in bud; ovary inferior, 2 loculed, many ovuled; style slender, longer than stamens; stigma slightly acute. Berry, 2.5 cm across, ovoid, greenish-pink.

Distribution: Endemic to the southern Western Ghats (Kerala and Tamil Nadu).

Fl. & fr.: February–March.

Threat status: Data Deficient.

27. **Syzygium munroi** (Wight) Chandrabose in Sharma et al., Biol. Mem. 2: 58, 1977 as "Munronii." *Eugenia munronii* Wight, Ill. Ind. Bot. 2: 14, 1841; Icon. Pl. Ind. Orient. t. 546, 1842; Duthie in Hook. f., Fl. Brit. India 2: 472, 1878. *Jambosa munronii* (Wight) Walp., Rep. 2: 191, 1843; Gamble, Fl. Madras 473, 1919. *Syzygium munroi* (Wight) N. P. Balakrishnan, Bull. Bot. Surv. India 22 (1–4): 174, 1982 [1980], *nom. superfl.* (Figures 3.8d and 3.9b).

Small trees; branchlets quadrangular. Leaves to 22 × 7 cm, lanceolate, acuminate, glabrous; lateral nerves and intramarginal veins prominent; subsessile. Cymes 5–10 cm broad, terminal. Flowers few, 4 cm across; pedicels 1 cm long; calyx 2.5 cm long, funnel shaped, lobes 1 × 1 cm, orbicular; petals larger, 15 × 15 mm, yellowish white; filaments 2.5 cm long, white. Berry 25 × 15 mm, ellipsoid, glabrous, crowned with calyx lobes.

Distribution: Endemic to the southern Western Ghats, found as an understory tree in evergreen forests of Kerala, Tamil Nadu, and Karnataka, especially in Coorg region.

Fl. & fr.: December–May.

Threat status: Data Deficient.

28. **Syzygium myhendrae** (Bedd. ex Brandis) Gamble, Fl. Madras 1: 478, 1919. *Eugenia myhendrae* Bedd. ex Brandis, Indian Trees 325, 1906 (Figure 3.9b).

Small trees, 12 m tall; bark greyish; branchlets 4 angled. Leaves 3–7 × 2–2.5 cm, oblanceolate or obovate, base cuneate, apex acuminate, margin entire, coriaceous; lateral nerves many, close, parallel, obscure, with intramarginal nerves; petiole 2–5 mm long. Flowers small, white, sessile in terminal corymbose cymes of umbellules, branches 4 angled; calyx tube 3 mm, turbinate, lobes 4; petals 4, caducous; stamens many, incurved in bud; ovary

2 loculed, ovules many; style filiform, shorter than the stamens; stigma simple, acute. Berry sessile, 7–8 mm across, globose, pink-purple, crowned by persistent calyx.

Distribution: Endemic to the southern Western Ghats region of Karnataka, Tamil Nadu, and Kerala; reported from Agasthyamalai Hills, Pandimotta of Thiruvananthapuram, and Shenduruny Wildlife Sanctuary in Kollam district and Periyar Tiger Reserve in Idukki district of Kerala.

Fl. & fr.: January–June.

Threat status: Endangered B1+2c ver. 2.3 (IUCN 2015).

Note: The tree is highly attractive and can be introduced as an ornamental. The fruits are edible (Shareef and Beegam 2015).

29. **Syzygium neesianum** Arn., L. Nova Acta Phys. Med. Acad. Caes. Leop. Carol Nat. Cur. 18: 335, 1836; Ashton in Dassanayake & Fosberg, Rev. Handb. Fl. Ceylon 2: 442, 1981. *Eugenia neesiana* Wight, Ic. t. 533, 1843; Ill. 2: 15, 1850; Duthie in Hook. f., Fl. Brit. India 2: 493, 1879 (Figure 3.9b).

Large trees up to 25 m tall; bark pale brown, shallowly flaking; branches pendulous; twigs rather slender, at first quadrangular, becoming terete, smooth, pale brown. Leaves 5.5–10 × 1.8–5 cm, elliptic, base obtuse or subcordate, apex caudate-acuminate, margin shallowly subrevolute, undulate, frequently subplicate, chartaceous, or thinly coriaceous; lateral nerves many, with intramarginal nerve obscurely 2 tiered, the inner ca. 1–2 mm within margin, rather straight; midrib slender but prominent beneath; petiole 1–3 mm long, slender. Cymes to 8 cm long, terminal or axillary, lax, spreading. Flowers white; calyx 4 × 3 mm, globet shaped, lobes 4 or 5, obscure, obtuse; petals 3 × 2 mm, elliptic, concave; stamens many, 4 mm long. Berry 1 cm across, subglobose, dark purple, with prominent, 3 mm diameter unlobed calyx limb.

Distribution: South India and Sri Lanka. In southern Western Ghats (reported from Silent Valley National Park; Manilal and Sabu 1984; Manilal 1988) and Vellarimala of Kerala (Pradeep 2000).

Fl. & fr.: January–May.

Threat status: Vulnerable A1c ver. 2.3 (IUCN 2015).

30. **Syzygium nervosum** DC., Prodr. 3: 260, 1828. *Cleisticalyx cerasoides* (Roxb.) I. M. Turner, Gard. Bull. Singapore 57: 26, 2005. *Syzygium cerasoides* (Roxb.) Raizada, Indian For. 84: 478, 1958. *Eugenia holtzei* F. Muell., Aust. J. Pharm. 1: 199, 1886. *Syzygium operculatum* (Roxb.) Nied., Die Naturlichen Pflanzenfam. 3 (7): 85, 1893. *Cleistocalyx operculatus* (Roxb.) Merr. & L. M. Perry, J. Arnold Arbor. 18: 337, 1937. *Eugenia operculata* Roxb., Fl. Ind. (ed. 2) 2: 486, 1832. *Eugenia cerasoides* Roxb., Fl. Ind. (ed. 2) 2: 488, 1832.

Medium-sized trees, up to 15 m tall; bark greyish brown; branchlets flattened and furrowed; Leaves 11–17 × 4.5–7 cm, oblong to elliptic, base broadly cuneate to rounded, apex acute to acuminate, thinly leathery, gland dotted, emit mango-like odor when crushed; midrib grooved or depressed on the upper surface; lateral nerves 9–13 pairs, close with intramarginal nerve; petiole 1–2 cm. Inflorescence a panicle on lateral leafless branches, 6–12 cm; buds oval, 5 × 3.5 mm; calyx tube hemispheric, 3 mm; corolla calyptrate, 2–3 mm, apex beaked; petals obsolete; stamens 5–8 mm; style 3–5 mm, equal or shorter than the stamens. Berry violet to black, broadly ovoid, 1–1.2 × 1–1.4 cm. Seed solitary, 8 mm.

Distribution: China, India, Myanmar, Thailand, Vietnam, Malaysia, Indonesia, and Philippines to Australia. In India, it is reported from Karnataka and Delhi.

Fl. & fr.: Data not available.

Threat status: No threat.

31. **Syzygium occidentale** (Bourd.) Gandhi in Saldanha & Nicolson, Fl. Hassan Dist. 282, 1976. *Eugenia occidentalis* Bourd., Indian For. 30: 195, t. 3, 1904. *Jambosa occidentalis* (Bourd.) Gamble, Fl. Madras 1: 474, 1919 (Figures 3.10a and 3.11a).

Small trees 3–6 m tall. Leaves 10–18 × 3–4 cm, linear-lanceolate, acuminate at both ends. Cymes 10 cm broad, terminal. Flowers 5 cm in diameter, white; calyx tube 15 mm long, funnel shaped; lobes ovate, obtuse, 6 mm across; petals 4, 8 mm broad, orbicular; filaments 25 mm long, yellow; style 3.5 cm long. Berry greenish-pink, 2 × 1.5 cm, obovoid to globose.

Distribution: Endemic to southern Western Ghats in evergreen forests and along the streams in Kerala and Tamil Nadu.

Fl. & fr.: March–April.

Threat status: Vulnerable (IUCN 2015).

FIGURE 3.10 (a) *Syzygium occidentale*. (b) *Syzygium palghatense*. (c) *Syzygium parameswaranii*.

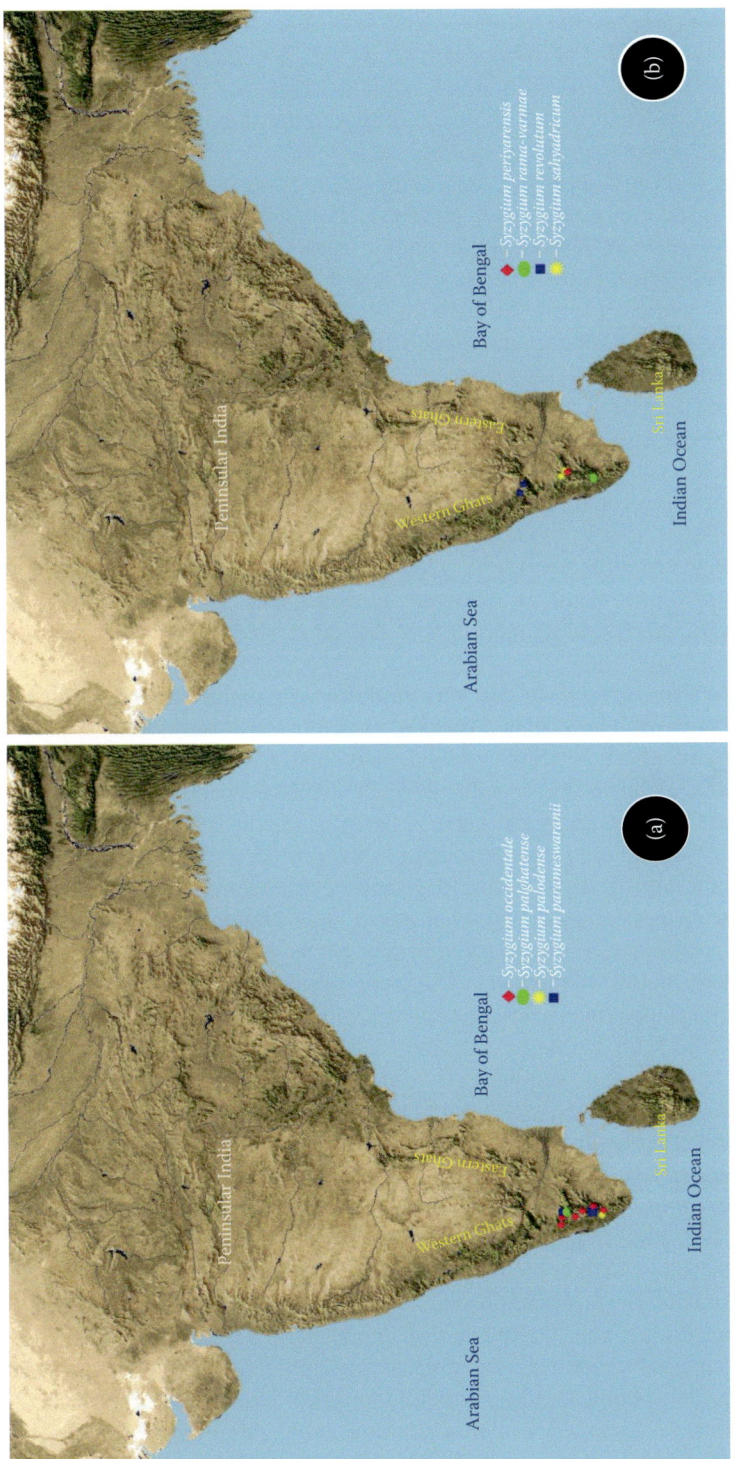

FIGURE 3.11 Map showing geographic distribution of *Syzygium* in the Western Ghats. (a) *S. occidentale, S. palghatense, S. palodense,* and *S. parameswaranii.* (b) *S. periyarensis, S. rama-varmae, S. revolutum,* and *S. sahyadricum.*

32. **Syzygium palghatense** Gamble, Kew Bull. 240, 1918; Fl. Madras 1: 480, 1919; Sasidh., Fl. Parambikulam WLS 122, 2002; Sujanapal & Sasidh., Rheedea 12: 189, 2002 (Figures 3.10b and 3.11a).

Trees, up to 15 m tall; branchlets subtetragonous, becoming terete on maturity. Leaves opposite or subopposite 3.5–8.3 × 2–3.5 cm, elliptic, elliptic-oblong, or elliptic-obovate, base cuneate or acute, apex obtusely acuminate or acute, margin reflexed, glabrous, chartaceous, sparsely gland dotted on lower surface; lateral nerves many, parallel, slightly distant, slender, faint, with intramarginal nerve; petiole 3–5 mm, grooved above, glabrous. Flowers, white, in terminal or axillary cymes of 2–3 cm long; pedicel to 2 mm long; calyx tube funnel shaped, lobes 4; petals 2.5 mm across, creamy white, calyptrate, cauducous at early stage; stamens many, bent inward in bud; ovary inferior, 2 loculed, ovules many; style 1; stigma simple. Berry 10–13 × 4–5 mm, crowned by the calyx limb.

Distribution: Rare and endemic to the southern Western Ghats in Kerala. It was first described in 1918 from the Palakkad Hills, and a few trees were spotted in the forests of Parambikulam Wildlife Sanctuary (Sasidharan 2002).

Fl. & fr.: March–April.

Threat status: Critically Endangered (Sujanapal and Sasidharan 2002).

33. **Syzygium palodense** Shareef, E. S. S. Kumar & Shaju, *Phytotaxa* 71: 28–33 (Figure 3.11a).

Large trees up to 18 m tall; bark smooth and pale brown; twigs 4 angled, winged. Leaves opposite or alternate, crimson red when young, 5.5–9.5 × 2.4–5.4 cm, elliptic to elliptic-oblong, base cuneate, apex caudate-acuminate, coriaceous, gland dotted beneath; midrib channeled above; lateral nerves 24–35, close, parallel, intramarginal nerve 1 tiered, 2 mm from the margin; petiole 2–5 mm long. Inflorescence terminal or subterminal, 9 cm long; peduncle and branches 4 angled, slightly winged. Flowers calyptrate, sessile, creamy white; calyx campanulate, outer yellowish green, inner creamy white, 3.5 × 4 mm; lobes 4, persistent, deltoid to suborbicular; petals 4, free, creamy white, 2.5 × 3 mm, orbicular to suborbicular, calyptrate; stamens many, filaments pointed; ovary 2 loculed, ovules many; style 5.5 mm, long, pointed. Berry subglobose to obovoid, 2.2 × 1.8 cm, fleshy, dark purple. Seed 1, subglobose, 1.6 × 1.3 cm.

Distribution: In southern Western Ghats, Palode forest areas in Thiruvananthapuram district, Kerala.

Fl. & fr.: April–July.

Threat status: Data Deficient.

34. **Syzygium parameswaranii** Mohanan & Henry, J. Bombay Nat. Hist. Soc. 84 (2): 408, 1987 (Figures 3.10c and 3.11a).

Trees, to 6 m tall; branchlets tetragonous. Leaves 2–3 × 2–3.5 cm, ovate, base round or obtuse, apex subacute or acute obtuse, margin entire, recurved, glabrous, coriaceous; lateral nerves many, parallel, slender, rather close, faint, looped at the margin forming intramarginal nerve, intercostae reticulate, obscure; petiole 1–3 mm, stout, glabrous. Flowers, greenish-white, 8 × 2.8 mm, funnel shaped, in dense terminal umbellate cymes; panicles sub-sessile; pedicel 3–18 mm long, slender, glabrous; calyx tube 1–1.2 cm long, lobes 4, 1 × 1–1.5 mm, ovate, obtuse; petals 4, 3 × 2.5 mm, suborbicular, calyptrate, obtuse, gland dotted along the main nerve; stamens many, filaments 2.5–3 mm long, dilated at base; ovary inferior, 2 loculed, ovules many; style 9 mm long; stigma simple. Berry 1–1.5 × 0.5–0.6 cm, top shaped, 1 or 2 seeded with calyx tube at the top.

Distribution: Endemic to southern Western Ghats; so far reported from Idukki and Thiruvananthapuram districts in Kerala.

Fl. & fr.: January–April.

Threat status: Endangered (IUCN 2015).

35. ***Syzygium periyarensis*** Augustine & Sasidh., Rheedea 9: 155, 1999 (Figure 3.11b).

Trees, up to 15 m tall; bark greyish-white, smooth; branchlets terete. Leaves 11–15 × 7–9 cm, obovate or broadly elliptic, base acute or obtuse, apex obtusely acute, acumen ca. 0.5 cm long, margin entire, glabrous, coriaceous, sparingly black punctate below; lateral nerves 7–14 pairs, parallel with intramarginal nerve, prominent; petiole 7–15 mm long, stout, dark brown, glabrous. Flowers white, in terminal corymbose cymes, 5–8 cm across, few flowered; pedicel 4.5 mm long; pseudopedicel 3 mm long; calyx tube 12 × 12 mm, tube above the ovary 3–4 mm high; lobes 4 ca. 6 × 12 mm, broadly ovate, obtuse; petals 4, white, 13 mm across, orbicular, concave; stamens numerous, many seriate, filaments 11–18 mm long, inflexed in bud, anthers 1.5 mm long, ovate, obtuse; disc prominent, 2–3 mm thick, shortly creneate; ovary inferior, conical, 2 loculed, ovules many; style 22 mm long; stigma indistinct. Fruit a berry; 1–2 cm diameter globose or ovoid; seed 1.

Distribution: Endemic to southern Western Ghats; so far known only from the Periyar Tiger Reserve, Kerala.

Fl. & fr.: March–April.

Threat status: Data Deficient.

36. ***Syzygium rama-varmae*** (Bourd.) Chithra in N. C. Nair & Henry, Fl. Tamil Nadu I: 157, 1983; Gopalan & Henry, Endemic Pl. Agasthiyamala 403, 2000; Mohanan & Sivad., Fl. Agasthyamala 264, 2002. *Eugenia rama-varmae* Bourd., Indian For. 30: 147, t. 2, 1904. *Jambosa rama-varmae* (Bourd.) Gamble, Fl. Madras 1: 474, 1919 (Figures 3.11b and 3.12a).

Trees, up to 15 m tall; bark brown. Leaves 11.5–26 × 4.2–8 cm, elliptic-oblong, elliptic-ovate, or elliptic-lanceolate, base rounded, apex acute to acuminate; margin entire, coriaceous; lateral nerves 15–20 pairs, parallel, prominent, with intramarginal nerves; petiole stout, 0.65 mm long, glabrous. Flowers white, 4–5 cm across, in few flowered lateral cymes; calyx tube short, truncate, lobes 4; petals 4, 1–1.5 cm orbicular; stamens many; filaments 1–1.3 cm long, bent inward in middle when in bud; ovary inferior, 2 loculed, ovules many; style 1.5 cm long; stigma slightly acute. Berry greenish-pink, spherical, 1 or 2 seeded.

Distribution: Endemic to the southern Western Ghats; reported only from Agasthiamalai region in Thiruvanathapuram district, Kerala.

Fl. & fr.: January–June.

Threat status: Vulnerable (IUCN 2015).

FIGURE 3.12   (a) *Syzygium rama-varmae.* (b) *Syzygium sahyadricum.* (c) *Syzygium salici-folium.* (d) *Syzygium sasidharanii.*

37. ***Syzygium revolutum*** (Wight) Walp., Repert. Bot. Syst. 2: 180, 1843; Chithra in N. C. Nair & A. N. Henry, Fl. Tamil Nadu 1: 157, 1983. *Eugenia revoluta* Wight, Icon. Pl. Ind. Orient. 2: t. 534, 1842; Duthie in Hook. f., Fl. Brit. India 2: 492, 1878; Fischer, Rec. Bot. Surv. India 9: 80, 1921 (Figure 3.11b).

Medium-sized trees; branchlets terete or obscurely 4 angled. Leaves lanceolate to ovate or broadly obovate, with a small obtuse point, margins often revolute, coriaceous. Inflorescence terminal long peduncled cymes. Flowers sessile congested at the ends of the branchlets; calyx 4–5 toothed; petals free, or cohering and falling off as a lid. Berry spherical, reddish, size of a small cherry.

Distribution: South India (southern Western Ghats: Coimbatore and Nilgiri in Tamil Nadu) and Sri Lanka.

Fl. & fr.: January–April.

Threat status: Data Deficient.

38. ***Syzygium rubicundum*** Wight & Arn., Prodr. 330, 1834; Gamble, Fl. Madras 1; 479, 1919. *Eugenia rubicunda* (Wight & Arn.) Wight, Ic. t. 538, 1842; Duthie in Hook. f., Fl. Brit. India 2: 495, 1878. *Eugenia lissophylla* (Thw.) Bedd., For. Man. Bot. 108, 1874; Duthie in Hook. f., Fl. Brit. India 2: 488, 1878. *Syzygium lissophylla* Thw., Enum. Pl. Zeyl. 117, 1859 (Figure 3.13a).

Trees up to 15 m tall; bark pale brown, thin, smooth; branchlets 4 angled. Leaves 2.5–10 × 1.5–5.5 cm, ovate-lanceolate or elliptic-obovate, base attenuate or cuneate, apex caudate-acuminate or obtusely acuminate,

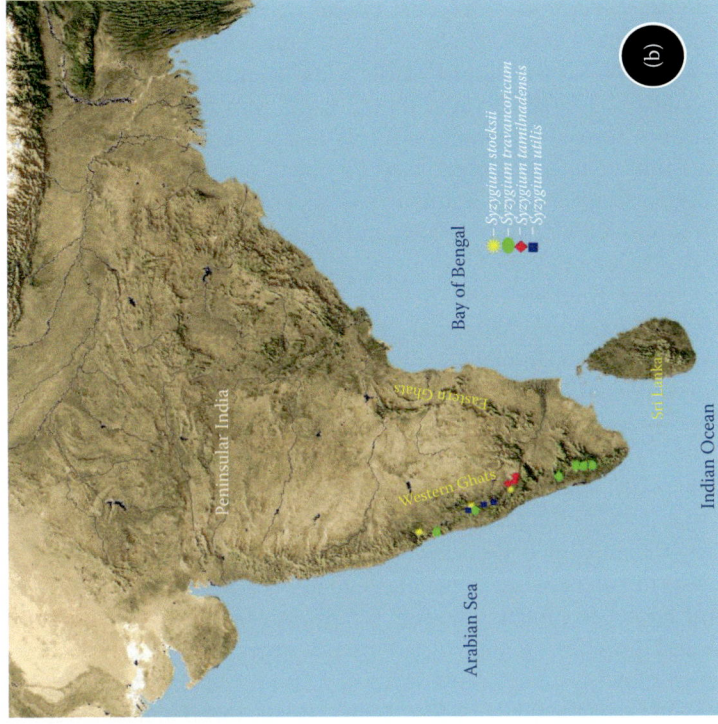

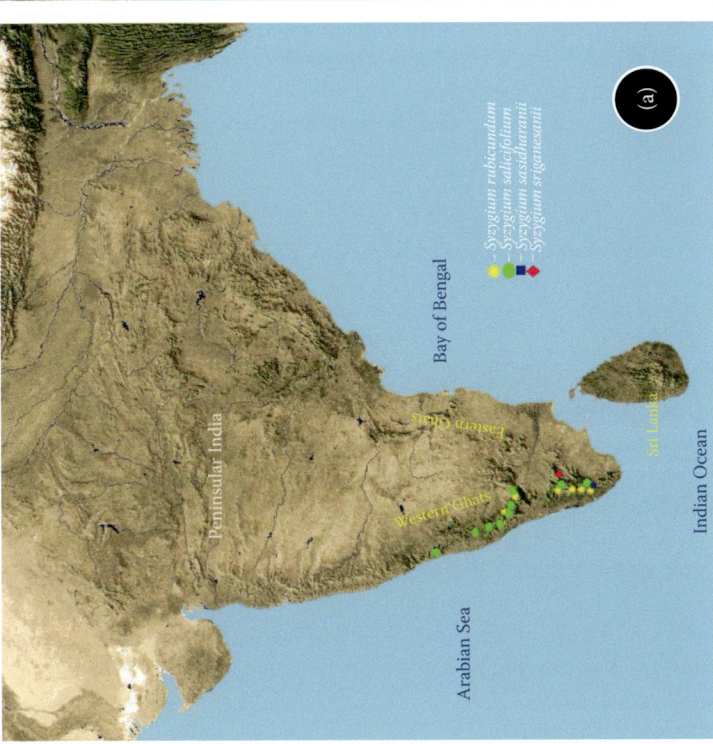

FIGURE 3.13 Map showing geographic distribution of *Syzygium* in the Western Ghats. (a) *S. rubicundum, S. salicifolium, S. sasidharanii,* and *S. sriganesanii.* (b) *S. stocksii, S. travancoricum, S. tamilnadensis,* and *S. utilis.*

margin entire, coriaceous, glabrous; lateral nerves many, parallel, close, with intramarginal nerve; petiole 5–6 mm long, slender, grooved above. Flowers small, pinkish-white, in axillary and terminal, densely flowered, corymb, shortly pedicelled; calyx tube 2 × 2.5 mm long, turbinate, lobes 4; petals 3 mm across, calyptrate; stamens many, free, bent inward in bud; filaments 4 mm long, spreading; ovary 2 loculed, many ovules in each cell; style 1; stigma simple. Berry 6 mm across, globose, purplish-black.

Distribution: Kerala, Tamil Nadu, and Sri Lanka.

Fl. & fr.: December–April.

Threat status: Data Deficient.

39. **Syzygium sahyadricum** Sujanapal et al., Phytotaxa 174 (5): 285, 2014. *Syzygium olivifolium* (Duthie) Gamble, Kew Bull. 1920: 52, 1920. *Syzygium spathulatum sensu* Chithra in N. C. Nair & Henry, Fl. Tamil Nadu 1: 158, 1983; *non* Thwaites, 1859 (Figures 3.11b and 3.12b).

Small trees up to 12 m tall; bark pale brown, smooth; branchlets 4 angled. Leaves opposite, decussate, yellowish when young, 4.5–6.5 × 2–3 cm, obovate to elliptic-oblong, base acute or obtuse, apex obtusely acute, margin slightly revolute, glossy, coriaceous, gland dotted; lateral nerves 12–16 pairs with intramarginal nerve; petiole 7–10 mm long. Inflorescences lateral or axillary branched umbels. Flowers 5 × 4 mm, sessile, 5–7 in an umbel; peduncle and branches 4 angled; calyptrate, greenish-white; hypanthium 3 × 3 mm, yellowish green; calyx lobes 4, deltoid, 1 × 1 mm, basally connate; petals 4, creamy white, 1.5 × 1 mm, ovate to orbicular, calyptrate; stamens many, filaments 0.5–2 mm long, pointed at apex; ovary 2 loculed, ovules many; style yellow, 1 mm long; stigma globose, white. Berry obovoid, 4–8 × 4 mm, dark purple when ripe, 1 seeded.

Distribution: Endemic to Kerala and Tamil Nadu part of southern Western Ghats.

Fl. & fr.: December–March.

Threat status: Data Deficient.

40. **Syzygium salicifolium** (Wight) Graham, Cat. Pl. Bombay 73, 1839. *Eugenia salicifolia* Wight, Ill. Ind. Bot. 2: 16, 1841. *Eugenia heyneana* Duthie in Hook. f., Fl. Brit. India 2: 500, 1879. *Syzygium heyneanum* (Duthie) Wall. ex Gamble, Fl. Madras 1; 482, 1919 (Figures 3.12c and 3.13a).

Small trees up to 6 m tall; branchlets obscurely 4 angled. Leaves 6–10 × 1.3–3 cm, oblong or elliptic, base acute, apex obtuse or retuse, margin entire, gland dotted; lateral nerves many, very slender, with intramarginal nerve; petiole 4 5 mm long, grooved above. Flowers small, sessile, white, in lateral or terminal cymes; calyx tube 2 × 3.5 mm, turbinate; petals calyptrate, 3 mm across; stamens many, free, bent inward at the middle when in bud, filaments 3 mm long; ovary 2 loculed; ovules many; style 1; stigma simple. Berry obovoid, 1–1.5 × 0.5–0.7 cm, crowned with the cuplike calyx limb.

Distribution: Endemic to India; Kerala, Tamil Nadu, Karnataka, and Goa regions of Western Ghats.

Fl. & fr.: April–May.

Threat status: No threat.

41. ***Syzygium samarangense*** (Blume) Merr. & L. M. Perry, J. Arnold Arbor. 19: 115, 1938. *Myrtus samarangensis* Blume, Bijdr. 1084, 1826. *Eugenia javanica* Lam., Encycl. 3: 200, 1789. *Eugenia samarangensis* (Blume) O. Berg, Fl. Bras. 14: 646, 1859. *Jambosa samarangensis* DC., Prod. 3: 286, 1828. *Eugenia javanica* Lam., Enc. 3: 200, 1789; *non S. javanicum* Miq., Fl. Ind. Bat. 1: 461, 1855.

Medium-sized trees, up to 15 m tall. Leaves 10–25 × 5–12 cm, elliptic to elliptic-oblong, coriaceous, gland dotted; petiole thick, 3–5 mm long. Inflorescences terminal and in axils of fallen leaves, 3–30 flowered. Flowers 3–4 cm across; calyx tube 1.5 cm long, lobes 3–5 mm long; petals 4, orbicular to spathulate, 10–15 mm long, light yellow; stamens numerous, up to 3 cm long; style up to 3 cm long. Berry, broadly pyriform, crowned by the fleshy calyx lobes, up to 5.5 × 5.5 cm, light red to white, pulp white spongy, juicy, aromatic, sweet-sour in taste. Seeds 0–2, subglobose, up to 8 mm across.

Distribution: Native of Malaysia. Cultivated throughout Malaysia, Thailand, Indonesia, Taiwan, and South America. In India, cultivated throughout peninsular India.

Fl. & fr.: Generally flowers during dry season, but varies with location and as per the climatic conditions.

Threat status: No threat.

Note: The tree has antidiarrheal, antimicrobial, antidiabetic, immunostimulant, antihyperglycemic, cytotoxic, anticholinesterase, analgesic, and anti-inflammatory properties (Ratnam and Raju 2008). Different parts of the plant have different uses. Leaves used as astringent, to treat fever and diarrhea. It is also used in diabetes, cough, and headaches. Fruits edible and characterized by high levels of total phenolics, thus a significant source of phenolic antioxidants; useful in diabetes and as diuretic, emmenagogue, abortifacient, and febrifuge. Bark juice is used to treat wounds, and the bark is used as a stringent in mouthwash (Tina et al. 2011).

42. ***Syzygium sasidharanii*** Sujanapal et al., Int. J. Advan. Res. 1 (5): 44, 2013. *S. chemunjianum* Shareef, E. S. S. Kumar & Roy, Phytotaxa 129 (1): 34, 2013 (Figures 3.12d and 3.13a).

Small trees, up to 7 m tall; young shoots reddish-brown colored, branchlets slightly 4 angled when young, later becoming terete. Leaves opposite, decussate, 4–9 × 2–4 cm, obovate or obovate-elliptic, base cuneate, apex obtusely acute, subacute, rounded, or rarely retuse, margin recurved, glabrous; midrib canaliculate; lateral nerves 20–30 pairs, brochidodromous with intramarginal nerve; petiole 0.8–1 cm long. Flowers in terminal cymose panicles; peduncles 4 angled, branchlets opposite, basal ones long reaching the apex of the main axis, noncorymbiform. Flowers lax, sessile, up to 1 cm across; hypanthium funneliform, up to 7 mm across; sepals 4, broadly ovate, 3 × 3 mm; petals 4, calyptrate, white, orbicular, 3 × 4 mm, membranous, margin undulate, gland dotted; stamens many, yellowish, of different lengths, 4–8 mm long, filaments to 7 mm long, incurved; ovary obovoid, ovules many; style yellow, stigma minute. Berry globose, yellow, with spongy pulp, up to 3 cm diameter. Seed 1, greyish, up to 1.5 cm diameter.

Distribution: Endemic to the southern Western Ghats in Pongalpara, Chemmunji Hills, and Pandimotta areas in Agasthyamala region of Thiruvananthapuram district, Kerala (Sujanapal et al. 2013).

Fl. & fr.: October–March.

Threat status: Critically Endangered (Sujanapal et al. 2013).

43. ***Syzygium sriganesanii*** K. Ravik. & V. Lakshm., *Rheedea* 9: 59, 1999 (Figure 3.13a).

Medium-sized trees up to 20 m tall; Branchlets 4 angled. Leaves 5–8 × 1.5–3 cm, oblong-elliptic to ovate or suborbicular, 3 obtuse to acute at apex, entire to deflexed along margins, shiny; leaf scars distinct; petioles 1 mm long, twisted; midrib depressed above; nerves many, obscure and close. Cymes terminal, trichasial, 3–5 cm long, 1–3 flowered. Flowers 1 cm across; calyx tube obconical, 1–2 cm long, grooved and truncate at apex, green, calyx teeth 5, 1 × 1 mm; petals suborbicular, 8 mm across, white; stamens many, pale pink to white; styles 1 cm long, simple, pale pink; ovary oblong, 1 mm long. Berries obconical, 2 × 1 cm, cuneate at base, with hollow depression at apex, orange-yellow when mature. Seed 1, 1.8 cm long, white.

Distribution: Endemic to southern Western Ghats; so far known only from Pulney hills (Megamalai) in Tamil Nadu (Ravikumar 1999).

Fl. & fr.: March–April.

Threat status: Data Deficient.

44. ***Syzygium stocksii*** (Duthie) Gamble, Fl. Madras 1; 481, 1919; Ratheesh Narayanan, Fl. Stud. Wayanad Dist. 363, 2009. *Eugenia stocksii* Duthie in Hook. f., Fl. Brit. India 2: 498, 1878 (Figures 3.13b and 3.14a).

Medium-sized trees, up to 15 m tall; branchlets 4 angled. Leaves 8.5–17 × 4–7.5 cm, elliptic-oblong or elliptic-obovate, base narrowed and decurrent on petiole, apex rounded or acuminate, margin entire, chartaceous; lateral nerves 9–12 pairs, parallel, distant, prominent, with intramarginal nerve; petiole 10–20 mm long, grooved above. Flowers small in axillary cymes or from the leafless axils; peduncle 2.5–5 cm, branches slender angled; calyx lobes 4, rounded or subacute; petals calyptrate; stamens many, bent inward in bud; ovary 2 loculed, ovules many; style 1; stigma simple. Berry pink-purple, 0.8 × 6 cm.

Distribution: Endemic to the southern Western Ghats; narrowly distributed in Wayanad (Kerala) and South Kanara (Karnataka).

Fl. & fr.: March–June.

Threat status: Endangered (IUCN 2015).

45. ***Syzygium tamilnadensis*** Rathkr. & Chithra in N. C. Nair & Henry, Fl. Tamil Nadu 1: 158, 1983. *Eugenia montana* Wight, Ic. t. 1060, 1846; *nom. illeg.*, non Aubl., 1775; Hook. f., Fl. Brit. India 2: 488, 1878. *Syzygium montanum* (Wight) Gamble, Fl. Pres. Madras 479, 1919; non Thw. (Figures 3.13b and 3.14b).

FIGURE 3.14 (a) *Syzygium stocksii.* (b) *Syzygium tamilnadensis.* (c) *Syzygium travancoricum.*

Medium-sized trees, up to 20 m tall; bark dark brown, peeling off in small flakes; branchlets tetragonous. Leaves 3–12.5 × 2.5–7.5 cm, elliptic, obovate, or elliptic-obovate, base cuneate, apex obtuse or emarginate, margin entire, glabrous, coriaceous, pellucid dotted; lateral nerves many, parallel, slender, slightly distant, prominent, with intramarginal nerve; petiole 5–16 mm, stout, grooved above, glabrous. Flowers white, 5 mm across, in axillary or terminal corymbose panicles with stout branches; calyx tube 3 mm long, lobes 4; petals calyptrate; stamens many, bent inward in bud, filaments 5–6 mm long; ovary inferior, 2 loculed; ovules many; style 1; stigma simple. Berry 6–8 mm across, globose, purple, succulent, crowned by calyx limb.

Distribution: Southern Western Ghats.

Fl. & fr.: February–May.

Threat status: Data Deficient.

46. ***Syzygium travancoricum*** Gamble, Bull. Misc. Inform. Kew 1918: 240, 1918; Gamble, Fl. Madras 1: 480, 1919 (Figures 3.13b and 3.14c).

Large trees, up to 25 m tall; bark longitudinally fissured, peeling off in thin irregular flakes; branchlets 4 angled. Leaves 8–16.5 × 5–8.5 cm, ovate or ovate-oblong, base narrowed and decurrent on petiole, apex acuminate, chartaceous; lateral nerves 10–15 pairs, parallel, distant, with intramarginal nerve; petiole 10–20 mm long, grooved above. Flowers white, in axillary lax corymbose cyme; peduncle 4.5–5 (8) cm long, their branches also long, ascending; calyx tube short, 1 mm across, lobes 4, very short; petals white, calyptrate; stamens many, free, bent inward in bud; ovary 2 loculed, ovules many; style 1; stigma simple. Berry oblong-obtuse on both sides, 1 × 0.5 cm, deep violet, juicy, 1 seeded.

Distribution: Endemic to the southern Western Ghats; reported from Kerala (Gamble 1919; Sasidharan 2004), Tamil Nadu (Thomas and Ramachandran 2014), Karnataka (Chandran et al. 2008), and Goa (Prabhugaonkar et al. 2014).

Fl. & fr.: April–June.

Threat status: Critically Endangered (IUCN 2015).

Note: Byng et al. (2015) synonymized this species under *S. stocksii*. However, detailed studies based on molecular and population data are essential for fixing the status. Hence, we are treating it as a separate species in the present enumeration.

47. ***Syzygium utilis*** (Talbot) Rathakr. & N. C. Nair, J. Econ. Taxon. Bot. 4: 288, 1983. *Eugenia utilis* Talbot, J. Bombay Nat. Hist. Soc. 11: 235, 1897; Sharma et al., Fl. Karnataka 102, 1984 (Figure 3.13b).

Large trees; branches terete. Leaves 5–7 × 2.5–3 cm, elliptic, base acute, apex abruptly and shortly acuminate, conspicuously gland dotted, coriaceous; lateral nerves close, rather inconspicuous; petiole 1 cm long. Flowers small in axillaty or terminal corymbose, panicled cymes, shorter than the leaves. Flowers sessile, usually fascicled in 3; buds globose or pear shaped; calyx shortly turbinate, scarcely lobed, or truncate; petals separate or slightly cohering. Berry small, black, succulent.

Distribution: Central Western Ghats, in Chikamangalur, Mysore, Hassan, and North Kanara.

Fl. & fr.: April–July.

Threat status: Data Deficient (IUCN 2015).

48. ***Syzygium zeylanicum*** (L.) DC., Prodr. 3: 260, 1828; Gamble, Fl. Madras 1: 479, 1919. *Myrtus zeylanica* L., Sp. Pl. 472, 1753. *Eugenia zeylanica* (L.) Wight, Illustr. 2: 15, 1841; Duthie in Hook. f., Fl. Brit. India 2: 485, 1878. *Syzygium zeylanicum* (L.) DC. var. *lineare* (Wall.) Alston in Trimen, Handb. Fl. Ceylon 6 (Suppl.): 115, 1931. *Syzygium zeylanicum* var. *ellipticum* A. N. Henry, Chandrab. & N. C. Nair, Bot. Hist. Hort. Malab. 161, 1980. *Syzygium zeylanicum* (L.) DC. var. *magamalayanum* K. Ravik. & V. Lakshm., Rheedea 9 (1): 61, 1999.

Small trees, up to 10 m tall; bark black. Leaves 2.4–9.5 × 0.8–5.5 cm, elliptic-lanceolate, linear-lanceolate, ovate, ovate-lanceolate, or oblong, base obtuse, round, or acute, apex acute or acuminate, margin entire, gland dotted, glabrous, coriaceous; lateral nerves many, parallel, with intramarginal nerve; petiole 3–8 mm long, grooved above. Flowers in axillary and terminal cymes, small, white; peduncle 4–6 mm long; pedicel 3–4 mm long; calyx tube campanulate, 25–30 mm long; lobes 5, 1 mm long, ovate; petals 5,

1.5–2 mm long, deciduous, white; stamens many, free, curved inward in bud, filaments 4–6 mm long; no thickened staminal disc; ovary inferior, 2 loculed, ovules many; style 1, 4–6 mm long; stigma simple. Berry, 6–12 mm long, white when ripe, crowned with calyx.

Distribution: East Madagascar, through India, south China, to Southeast Asia. In India, it is distributed in Kerala, Tamil Nadu, Karnataka, Orissa, and Goa.

Fl. & fr.: January–April.

Threat status: No threat.

The leaves showed potent antioxidant activity, and it can be a possible source of natural antioxidants (Nomi et al. 2012).

REFERENCES

Augusta, I. M., J. M. Resende, S. V. Borges, M. C. A. Maia, and M. A. P. G. Couto. 2010. Physical and chemical characterization of Malay red-apple (*Syzygium malaccensis* (L.) Merryl & Perry) skin and pulp. *Cienc. Tecnol. Aliment.* 30: 928–932.

Beddome, R. H. 1864. Contributions to the botany of southern India. *Madras J. Lit. Sci, ser. 3*, 1: 37–59.

Beddome, R. H. 1872. *Forester's Manual of Botany* [included in *Flora Sylvatica for Southern India*] Part 17. Madras: Gantz Brothers

Beddome, R. H. 1874. *Icones Plantarum Indiae Orientalis*. Part 14. Madras: Gantz Brothers & London: Van Voorst.

Byng, W. J., P. G. Wilson, and N. Snow. 2015. Typifications and nomenclatural notes on Indian Myrtaceae. Phytotaxa. Available from http://biotaxa.org/Phytotaxa/article/view /phytotaxa.217.2.1.

Chandran, M. D. S., D. K. Mesta, G. R. Rao, S. Ali, K. V. Gururaja, and T. V. Ramachandra. 2008. Discovery of two critically endangered tree species and issues related to relic forests of the Western Ghats. *Conser. Biol. J.* 2: 1–8.

Chithra, V. 1987. *Syzygium*. In *Flora of Tamil Nadu Series I: Analysis*, ed. A. N. Henry, R. Kumari, and V. Chithra, 152–158. Vol. 2. Coimbatore: Botanical Survey of India.

Darusman, L., K. W. T. Wahyuni, and F. Alwi. 2013. Acetylcholinesterase inhibition and anti-oxidant activity of *Syzygium cumini*, *S. aromaticum* and *S. polyanthum* from Indonesia. *J. Biol. Sci.* 13: 412–416.

Deepika, N., P. Eganathan, P. Sujanapal, and P. Ajay. 2013. Chemical composition of *Syzygium benthamianum* (Wt. ex Duthie) Gamble essential oil—An endemic and vulnerable tree species. *J. Essential Oil Bearing Plants* 16: 289–293.

Duthie, J. F. 1878–1879. Order LIX Myrtaceae. In *Flora of British India*, ed. J. D. Hooker, 462–506. Vol. 2. London: Reeve & Co.

Gamble, J. S. 1919. Myrtaceae. In *Flora of the Presidency of Madras*, 391–577. Vol. 1, Part 3. London: Adlard & Son Ltd.

Gayathri, A., R. P. M Benish, J. Saranya, P. Eganathan, P. Sujanapal, and A. K. Parida. 2012. Antimicrobial, antioxidant, anticancer activities of *Syzygium caryophyllatum* (L.) Alston. *Int. J. Green Pharmacy* 6: 285–288.

Gopan, R., G. Varughese, N. S. Pradeep, and M. G. Sethuraman. 2008. Chemical composition and antimicrobial activity of the leaf oil from *Syzygium gardneri* Thw. *J. Essential Oil Res.* 20: 72–74.

Govaerts, R., M. Sobral, P. Ashton, F. Barrie, B. K. Holst, L. R. Landrum et al. 2008. *World Checklist of Myrtaceae*. London: Royal Botanic Gardens, Kew. Available from http:// apps.kew.org/wcsp/ (accessed July 22, 2012).

Gunawardene, N. R., A. E. D. Daniels, I. A. U. N. Gunatilleke, C. V. S. Gunatilleke, P. V. Karunakaran, K. G. Nayak et al. 2007. A brief overview of the Western Ghats—Sri Lanka biodiversity hotspot. *Current Sci.* 93: 1567–1572.

Herbal Encyclopedia. Common Medicinal Herbs for Natural Health. Retrieved, January 10, 2016. http://www.cloverleaffarmherbs.com/cloves/.

Ismail, I. S., N. Ismail, and N. Lajis. 2010. Ichthyotoxic properties and essential oils of *Syzygium malaccense* (Myrtaceae). *Pertanika J. Sci Technol.* 18: 1–6.

IUCN (International Union for Conservation of Nature and Natural Resources). 2010. The IUCN Red List of threatened species. Version 2012.2. Gland, Switzerland: IUCN. Available from http://www.iucnredlist.org/.

IUCN (International Union for Conservation of Nature and Natural Resources). 2015. The IUCN Red List of threatened species. Version 2015.3. Gland, Switzerland: IUCN. Available from www.iucnredlist.org (accessed September 25, 2015).

Karioti, A., H. Skaltsa, and A. A. Gbolade. 2007. Analysis of the leaf oil of *Syzygium malaccense* Merr. et Perry from Nigeria. *J. Essential Oil Res.* 19 (4): 313–315.

Kiruthiga, K., J. Saranya, P. Eganathan, P. Sujanapal, and A. K. Parida. 2011. Chemical composition, antimicrobial, antioxidant and anticancer activity of leaves of *Syzygium benthamianum* (Wight ex Duthie) Gamble. *J. Biol. Active Prod. Nat.* 1: 273–278.

Manilal, K. S. 1988. *Flora of Silent Valley Tropical Rain Forests of India.* Calicut: Calicut University.

Manilal, K. S., and T. Sabu. 1984. Discovery of two species of *Syzygium* Gaertn, hitherto endemic to Sri Lanka, from Silent Valley, India. *J. Econ. Taxon. Bot.* 5: 418–420.

Mittermeier, R. A., R. P. Gil, M. Hoffman, J. Pilgrim, T. Brooks, C. G. Mittermeier et al. 2005. *Hotspots Revisited: Earth's Biologically Richest and Most Endangered Terrestrial Ecoregions.* Boston: University of Chicago Press.

Mohan Maruga Raja, M. K., D. Agilandeswari, and S. P. Dhanabal. 2013. Pharmacognostical, antidiabetic and antioxidant studies on *Syzygium densiflorum* leaves. *Contemp. Invest. Observ. Pharmacy* 2: 43–51.

Mohanan, M., and A. N. Henry. 1987. *Syzygium parameswaranii* (Myrtaceae)—A species from southern India. *J. Bombay Nat. Hist. Soc.* 84: 408–409.

Murugan, C., and V. S. Manickam. 2004. Two new additions to Myrtaceae of India. *J. Econ. Tax. Bot.* 28: 523–526.

Myers, N., R. A. Mittermeier, C. G. Mittermeier, G. A. B. da Fonseca, and J. Kent. 2000. Biodiversity hotspots for conservation priorities. *Nature* 403: 853–858.

Nair, K. K. N., S. K. Khanduri, and K. Balasubramanyan, eds. 2001. *Shola Forests of Kerala: Environment and Biodiversity.* Thrissur: Kerala Forest Research Institute.

Nayar, M. P. 1996. *Hotspots of Endemic Plants of India, Nepal and Bhutan.* Trivandrum: Tropical Botanic Garden and Research Institute.

Nayar, T. S., A. R. Beegam, N. Mohanan, and G. Rajkumar. 2006. *Flowering Plants of Kerala—A Handbook.* Thiruvananthapuram: Tropical Botanic Garden and Research Institute.

Nayar, T. S., A. R. Beegam, and M. Sibi. 2014. *Flowering Plants of the Western Ghats, India,* 669–687. Thiruvanathapuram: Jawaharlal Nehru Tropical Botanical Garden and Research Institute.

Nomi, Y., S. Shimizu, Y. Sone, M. T. Tuyet, T. P. Gia, M. Kamiyama et al. 2012. Isolation and antioxidant activity of zeylaniin A, a new macrocyclic ellagitannin from *Syzygium zeylanicum* leaves. *J. Agric. Food Chem.* 60: 10263–10269.

Pandey, R. P., and P. G. Divakar. 2008. An integrated checklist of Andaman and Nicobar Islands, India. *J. Econ. Taxon. Bot.* 32: 403–500.

Parnell, J. A. N., L. A. Craven, and E. Biffin. 2007. Matters of scale: Dealing with one of the largest genera of angiosperms. In *Reconstructing the Tree of Life, Taxonomy and Systematics of Species Rich Taxa,* ed. T. Hodkinson and J. Parnell, 251–273. Boca Raton, FL: Taylor & Francis.

Pino, J. A., R. Marbot, A. Rosado, and C. Vázquez. 2004. Volatile constituents of Malay rose apple [*Syzygium malaccense* (L.) Merr. & Perry]. *Flavour Fragr. J.* 19: 32–35.

Prabhugaonkar, A., D. K. Mesta, and M. K. Janarthanam. 2014. First report of three red-listed tree species from swampy relics of Goa State, India. *J. Threat. Taxa* 6: 5503–5506.

Pradeep, A. K. 2000. *Floristic Studies on Vellarimala on the Western Ghats of Kerala.* Calicut: Department of Botany, University of Calicut.

Quattrocchi, U. 2012. *CRC World Dictionary of Medicinal and Poisonous Plants: Common Names, Scientific Names, Eponyms, Synonyms, and Etymology.* 5 vols. Boca Raton, FL: CRC Press/Taylor & Francis Group.

Raizada, M. B. 1948. *Syzygium kanarense* (Talbot) Raizada (Myrtaceae). *Indian For.* 74: 336.

Ramana, M. V., A. Chorghe, and P. Venu. 2014. Two new species of *Syzygium* (Myrtaceae) from Saddle Peak National Park, Andaman and Nicobar Islands India, *Blumea* 59: 42–48.

Rao, R. R. 2012. Floristic diversity in Western Ghats: Documentation, conservation and bioprospection—A priority agenda for action. *Sahyadri E-News*, Western Ghats Biodiversity Information System. Bangalore: Indian Institute of Sciences.

Rathakrishnan, N. C., and N. C. Nair. 1983. Nomenclatural changes in some myrtaceous plants. *J. Econ. Tax. Bot.* 4: 287–288.

Ratheesh Narayanan, M. K., S. M. Shareef, T. Shaju, A. R. Sivu, K. A. Sujana, M. K. Nandakumar, and K. T. Satheesh. 2014. A new species of *Syzygium* (Myrtaceae) from the southern Western Ghats of Kerala, India. *Int. J. Adv. Res.* 2 (3): 1055–1058.

Ratnam, K. V., and R. R. V. Raju. 2008. *In vitro* antimicrobial screening of the fruit extracts of two *Syzygium* species (Myrtaceae). *Adv. Biol. Res.* 2: 17–20.

Ravikumar, K. 1999. Novelties from high wavy mountains, southern Western Ghats, Theni District, Tamil Nadu, India. *Rheedea* 9: 55–75.

Saldanha, C. J. 1996. *Flora of Karnataka*. Vol. II. New Delhi: Oxford & IBH Publishers.

Saranya, J., P. Eganathan, P. Sujanapal, and A. K. Parida. 2012. Chemical composition of leaf essential oil of *Syzygium densiflorum* Wall. ex Wt. & Arn.—A vulnerable tree species. *J. Essential Oil Bearing Plants* 15: 283–287.

Saranya, J., E. Palanisami, and P. Sujanapal. 2014. Antioxidant activity of the leaf essential oil of *Syzygium calophyllifolium*, *Syzygium makul*, *Syzygium grande* and *Eugenia cotinifolia* ssp. *codyensis*. *J. Biol. Active Prod. Nat.* 4: 12–18.

Sasidharan, N. 1997. *Studies on the Flora of Shenduruny Wildlife Sanctuary with Emphasis on Endemic Species*, 125–126. Kerala Forest Research Institute Research Report No. 128. Thrissur: Kerala Forest Research Institute.

Sasidharan, N. 2002. *Floristic Studies in Parambikulam Wildlife Sanctuary.* Kerala Forest Research Institute Research Report No. 246. Thrissur: Kerala Forest Research Institute.

Sasidharan, N. 2004. *Biodiversity Documentation for Kerala.* Part 6. Thrissur: Kerala Forest Research Institute.

Sasidharan, N. 2013. *Flowering Plants of Kerala.* CD-ROM version 2.0. Thrissur: Kerala Forest Research Institute.

Sasidharan, N., and A. Jomy. 1999. A new species of *Syzygium* Gaertn. (Myrtaceae) from southern Western Ghats, India. *Rheedea* 9: 155–158.

Shareef, S. M., and A. R. Beegam. 2015. Lectotypification and status of *Syzygium myhendrae* (Bedd. ex Brandis) Gamble (Myrtaceae)—An endemic myrtle of southern Western Ghats, India. *Taiwania* 60: 59–62.

Shareef, S. M., M. P. Geetha Kumary, E. S. S. Kumar, and T. Shaju. 2010. *Syzygium claviflorum* (Myrtaceae)—A new record for south India. *Rheedea* 20: 53–55.

Shareef, S. M., E. S. S. Kumar, and P. E. Roy. 2012a. *Syzygium fergusonii* (Trimen) Gamble (Myrtaceae)—A new record for Kerala. *J. Econ. Taxon. Bot.* 36: 379–380.

Shareef, S. M., E. S. S. Kumar, and P. E. Roy. 2013a. A new species of *Syzygium* (Myrtaceae) from Kerala, India. *Phytotaxa* 129: 34–38.

Shareef, S. M., E. S. S. Kumar, and T. Shaju, 2012b. A new species of *Syzygium* (Myrtaceae) from the southern Western Ghats of Kerala, India. *Phytotaxa* 71: 28–33.

Shareef, S. M., P. E. Roy, and M. V. Krishnaraj. 2013b. *Syzygium munnarensis* sp. Nov. (Myrtaceae): An overlooked endemic species from southern Western Ghats of Kerala, India. *Webbia* 69: 1, 53–57.

Sheeba, I. J., D. Narasimhan, and R. Ganeshan. 2003. Status of *Syzygium gambleanum* Rathakr. & Chithra (Myrtaceae) from southern Western Ghats. *Bull. Bot. Surv. India* 45: 111–120.

Shenoy, H. S., G. Krishnakumar, and M. Ramakrishna. 2015. Rediscovery of *Syzygium kanarense* (Talbot) Raizada (Myrtaceae)—An endemic species of the Western Ghats, India. *J. Threatened Taxa* 7: 6833–6835.

Sujanapal, P., A. J. Robi, and N. Sasidharan. 2014. *Syzygium sahyadricum* (Myrtaceae), a new tree species from India, and notes on the distribution of *S. spathulatum* Thwaites. *Phytotaxa* 174: 285–290.

Sujanapal, P., A. J. Robi, P. S. Udayan, and K. J. Dantus. 2013. *Syzygium sasidharanii* sp. nov. (Myrtaceae)—A new species with edible fruits from Agasthyamala Hills of Western Ghats, India. *Int. J. Adv. Res.* 1: 44–48.

Sujanapal, P., and N. Sasidharan. 2002. Relocation of *Syzygium palghatense* Gamble (Myrtaceae) and description of its hither to unknown fruits. *Rheedea* 12(2): 189–191.

Thomas, B., A. Rajendran, R. P. Mathews, and K. M. Prabhu Kumar. 2012. Wild edible fruit of *Syzygium calophyllifolium* Walp. (Myrtaceae): A gift of shola forest of Kerala, India. *Int. J. Biol. Technol.* 3: 22–27.

Thomas, B., and V. S. Ramachandran. 2014. Additions to the flora of Tamil Nadu, southern India. *J. Biodiversity Photon* 113: 355–359.

Tina, P., D. Padmavathi, R. J. Sajini, and A. Sarala. 2011. *Syzygium samarangense*: A review on morphology, phytochemistry and pharmacological aspects. *Asian J. Biochem. Pharm. Res.* 4: 155–163.

Vignesh, R., P. Puhazhselvan, M. Sangeethkumar, J. Saranya, P. Eganathan, and P. Sujanapal. 2013. GC-MS analysis, antimicrobial, scavenging ability and cytotoxic activity of leaves of *Syzygium calophyllifolium* Walp. *J. Biol. Active Prod. Nat.* 3: 121–129.

Vinod Kumar, T. G. 2003. Systematic studies of family Myrtaceae in the Kerala state, 87–160. PhD thesis, Mahatma Gandhi University, Kerala.

Viswanathan, M. B., and U. Manikandan. 2008. A new species of *Syzygium* (Myrtaceae) from the Kalakkad-Mundanthurai Tiger Reserve in peninsular India. *Adansonia* 30: 113–118.

WGEEP (Western Ghats Ecology Expert Panel). 2011. Report of the Western Ghats Ecology Expert Panel Part I. Submitted to the Ministry of Environment and Forests, Government of India.

Wight, R. 1842. *Icones Plantarum Indiae Orientalis*. Vol. 2. Madras: J. P. Pharoah.

Wight, R. 1846. *Icones Plantarum Indiae Orientalis*. Vol. 3(4). Madras: J. P. Pharoah.

Wong, K. C., and F. Y. Lai. 1996. Volatile constituents from the fruits of four *Syzygium* species grown in Malaysia. *Flavour Fragr. J.* 11: 61–66.

4 Diversity, Distribution, and Life History of *Syzygium cumini*

Arun Kumar Kushwaha and Lal Babu Chaudhary

CONTENTS

INTRODUCTION

Syzygium cumini (L.) Skeels is an evergreen, medium to large-sized, fast-growing tropical tree, which is well known for its various uses as medicine, wood, fuel, food, fruit, pharmaceuticals, ornamental, and so forth. Almost all parts of the plant are considered highly important. The ripe fruits are edible and also used for preparing health drinks, squashes, preservatives, jellies, wine, etc. (Warrier et al. 1996). The seeds are used as a very effective medicine in various diseases. The extracts of *S. cumini* contain many pharmacological properties, such as antibacterial, antiulcerogenic, antiallergic, antiviral, antidiarrheal, antifungal, hepatoprotective, and cardioprotective (Sagrawat et al. 2006). Timber is used for building houses, and making agricultural tools, utensils, and furniture. Since *S. cumini* occurs almost throughout the world either in the wild or in cultivation, it has been given several vernacular names in different languages (Morton 1987; Binggeli 2006).

MORPHOLOGICAL DIVERSITY

Syzygium cumini exhibits a considerable amount of variation, especially in its habit, leaves, and fruits as it grows in a variety of soils in different climatic conditions. The tree attains its full size in nearly 40 years. The height of the fully matured trees varies from 15 to 40 m with the trunk girth of 0.5–2.5 m diameter. The main trunk usually forks into multiple trunks after a short distance from the ground and then profusely branches out, upward. The whole tree is generally covered with dense foliage and flowers (Figure 4.1a and b). The bark on the lower portion of the stem is rough, cracked, flaking, and discolored, while it is light gray to gray or grayish-brown and smooth toward the upper portion

FIGURE 4.1 *Syzygium cumini* (L.) Skeels: (a) Habit. (b) Close-up of habit laden with flowers. (c) Bark. (d) Blaze. (e) Young leaves. (f, g) Leafy twigs showing variations in leaves. (h) Seedlings in natural condition. (Photo Credits: Authors.)

(Figure 4.1c). Usually, the color of blaze is light pinkish-white (Figure 4.1d), but dark red–colored blaze is also seen in some plants. The wood is whitish, hard, and durable.

The leaves show extreme variations in their shape and size on different plants. The leathery, glabrous, and glossy leaves vary from 6 to 12 cm in length and 3.5 to 7 cm in width. The shape of the leaf lamina varies from oblong-ovate, ovate-elliptic, or narrowly elliptic to lanceolate, with an acute to acuminate or rarely obtuse apex. The distinct variations have also been observed in the leaf base, which may be cuneate or obtuse to subtruncate. The young leaves are generally reddish or pinkish, but turn green at maturity (Figure 4.1e and f).

The flowers do not exhibit much variation except in color. They may be white to creamy white or occasionally slightly green (Figure 4.2a and b). The flowers are

FIGURE 4.2 *Syzygium cumini* (L.) Skeels: (a) Close-up of inflorescence. (b) Flower. (c) Close-up of habit laden with fruits. (d) Fruits showing developmental stages. (e, f) Fully matured fruits. (Photo Credits: Authors.)

borne in clusters of a few to 40 in number on terminal or axillary panicled racemes (Figure 4.2a). The fruit is a berry, and remarkable variations have been observed in the shape, size, color, and thickness of the fruit and in the color and taste of the pulp. The berries initially appear green, and gradually turn to light violet-red or purplish-red, and finally dark purple to black at full maturity (Figure 4.2c–f). In India, the fruits are generally categorized in two forms based on size, taste, and color. In one form, the fruits are large, oblong, dark purple to black, with pink or violet and sweet pulp and small seed. In the other form, the fruits are small, rounded, green to light violet-red or black, with white, thin, and mildly sour pulp. The latter form is considered a wild type. A form with white fruits has been reported from Indonesia. Sometimes, fruits without seed have also been noticed.

PHENOLOGY AND LIFE HISTORY

S. cumini is an evergreen, medium to large-sized, mass-flowering tree species of the tropical forests. It keeps its full foliage during the summer (April to August). However, the seasonal reduction in foliage occurs during December to January, and the lowest amount of foliage (i.e., approximately 35%) is observed in the months of January and February. The trees never become completely leafless in moist localities, but in dry localities they may become leafless for a very short period of time. Leaf flushing starts at the end of February, continues up to March, and reaches maturity in April to May. Shedding of old leaves starts from January to February in dry localities and February to March in moist localities, followed by leaf flush. The time gap between completion of leaf fall and initiation of production of new leaves is very short; hence, the trees always remain almost evergreen.

The flowering starts in March and continues until May. Full blooming occurs in April to May. In most of the habitats, the fruiting starts in May and bunches of young fruits appear after 15–25 days from fruit initiation. Unripe mature fruits appear about 30–40 days after flowering. The fruits are green at first, turn from green to light violet-red or purplish-red, and then dark purple or nearly black as they ripen (Figure 4.2d). Fruits take about two weeks to change their color from green to complete black. Ripening of fruits occurs in June to July and may extend up to August. Shedding of fruits starts when they become fully mature. The fruiting timing varies from place to place. In Marquesas (French Polynesia), fruiting starts in April, while in Java, July to August is the flowering time and September to October is the fruiting time. In Sri Lanka, the blooming begins from May to August and the fruits ripen during November to December (Morton 1987).

The fruits and seeds fall in large quantities near the parent trees, out of which some are destroyed and some seeds germinate on moist soil in the rainy season in the same year, as the seeds lose viability quickly. The fruits are eaten by some birds, monkeys, squirrels, and human beings; therefore, they are widely dispersed. The natural regeneration is profuse around the mother trees as seeds fall in large quantities (Figure 4.1h). Sometimes, the seedlings of the previous years are also seen near the trees. The seedlings start growing slowly in the first year but grow rapidly in subsequent years and may reach up to about 4 m in two years and start flowering in 8–10 years.

DISTRIBUTION PATTERN AND ECOLOGY

It is considered that *S. cumini* originated in the Indian subcontinent (India, Bangladesh, Myanmar, Nepal, Pakistan, and Sri Lanka) and its other adjoining regions, such as Indonesia (Ayyanar and Subash-Babu 2012). In South Asia, the plants are also widely cultivated along roadsides as avenue trees and near temples, as they are considered sacred by Hindus and Buddhists. In the past, the species was introduced to many places where it has been utilized as a fruit producer, as an ornamental, and also for its timber, and today it is growing all over the Asian subcontinent, eastern Africa, South America, and Madagascar, and has been naturalized to many places in the Philippines, Malaysia, and Florida and Hawaii in the United States (Warrier et al. 1996; Li et al. 2009).

S. cumini is one of the most commonly distributed trees of India, a major component in all forest types, except in very arid regions. It thrives well in both moist and dry situations from sea level to hills up to 1200 m in the Himalayas and 1800 m in the Nilgiris. It grows in a variety of soils and geological formations, such as alluvial, lateritic, sandy alluvia, marl, and oolitic limestone, which has resulted in many variants within the species. It can also tolerate saline soil. Its fast-spreading nature has allowed the species to become invasive in Hawaii, the Cook Islands, French Polynesia, and other places where it prevents the reestablishment of native lowland forests. The species is commonly found on deep, rich, well-drained soils in areas witnessing a mean annual rainfall between 900 and 1000 mm. It grows abundantly in areas of heavy rainfall. It grows well on riverbanks and can withstand prolonged flooding. It is fast growing and has the ability to form a dense cover. In India, the trees are grown in the coffee fields to provide shade. They also act as wind resistance and sometimes are closely planted in rows as windbreak.

ACKNOWLEDGMENTS

The authors are thankful to the director, CSIR-National Botanical Research Institute, Lucknow, India, for facilities. Uttar Pradesh State Biodiversity Board, Lucknow, is also duly acknowledged for providing financial assistance for working on tree flora of Uttar Pradesh.

REFERENCES

Ayyanar, M., and P. Subash-Babu. 2012. *Syzygium cumini* (L.) Skeels: A review of its phytochemical constituents and traditional uses. *Asia Pac. J. Trop. Biomed.* 2: 240–246.

Binggeli, P. 2006. *Syzygium cumini* (tree). Global Invasive Species Database. www.issg.org /database/ (accessed December 22, 2014).

Li, L., L. S. Adams, S. Chen et al. 2009. *Eugenia jambolana* Lam. berry extract inhibits growth and induces apoptosis of human breast cancer but not non-tumorigenic breast cells. *J. Agric. Food Chem.* 57: 826–831.

Morton, J. 1987. Jambolan. In *Fruits of Warm Climates*, 375–378. Miami: J. F. Morton.

Sagrawat, H., A. S. Mann, and M. D. Kharya. 2006. Pharmacological potential of *Eugenia jambolana*: A review. *Pharmacog. Mag.* 2: 96–105.

Warrier, P. K., V. P. K. Nambiar, and C. Ramankutty. 1996. *Indian Medicinal Plants*, 225–228. Vol. 5. Hyderabad: Orient Longman Ltd.

5 Syzygium cumini in Ayurveda and Other Traditional Medicare Systems in India

S. Rajasekharan and Vinodkumar T. G. Nair

CONTENTS

INTRODUCTION

India is considered the homeland of the most diverse and rich traditional knowledge systems with unique cultural expressions that are directly or indirectly connected with ecology, the environment, and biodiversity, ranging from coastal, desert, and plains to mountainous regions. This varied and diverse topography marked by a typical climate, edaphic factors, and physiographic conditions resulted in a rich variety and variability of flora and fauna, especially of plants used for food and medicine. The astoundingly rich floral and faunal diversity played a major role in the evolution of a unique classical health tradition (CHT) and local health tradition (LHT) in India. The CHT consists of Ayurveda, Siddha, Unani, Yoga, and Naturopathy and is the recognized system of health care by the Government of India. It is highly organized, classified, and codified with a strong conceptual and theoretical foundation, having a philosophical and scientific background. LHT is very rich and diverse but not organized or codified fully. In short, LHT is the distilled knowledge of experience of people belonging to individuals, families, and communities, including the tribal healers.

Syzygium cumini is an important medicinal plant used in both CHT and LHT of India. According to Ayurveda, this medicinal tree is known as jambavam in Sanskrit, the official language of India in earlier times. The tree was widely distributed in ancient India and was in those days popularly known as jamboo, which means a tree where flowers and fruits are available irrespective of seasons. History reveals that people first located this tree species in the south of Mahameru and described the indigenous taxonomical features of the tree. In Purana, *S. cumini* is considered one of the divine trees (*mahavriksha*) on earth. Other mahavrikshas are kadamba (*Neolamarckia cadamba*), peral (*Ficus benghalensis*), and makanda (*Mangifera indica*).

HISTORICAL PERSPECTIVE

S. cumini is widely distributed in India and is popularly known as Indian black plum, Java plum, Indian blackberry, and jamun. In Sanskrit, it is known as jambavam, meaning that it was widely distributed in ancient India. According to the legend, there was a river known as Jamboo where Goddess Jamboodini (the divine mother) stayed for a long period. She was fond of eating the fruits of *S. cumini* growing on the banks of the river. According to the Indian mythology, all *Devas*, *Asuras*, and *Nagas* worshipped Jamboodini with devotion to attain salvation. Abundant distribution of *S. cumini* in a particular locality is considered an indicator of the existence of gold mines in the area by ancient mining experts. By applying this indicator, once they identified a gold mine located on the bank of the river Jamboo and struck high-quality gold, especially recommended for making beautiful ornaments (Rajasekharan et al. 2005).

Tree worshiping was the practice prevalent in ancient Indian culture. This practice is still alive in the southern part of India, and certain temples declare their trees, popularly known as sthalavriksha (tree of sacred place), to be associated with the deity of the temple. Some of the examples are Jambu (*S. cumini*) of Jambukeswara temple in Thiruchirapalli (Tamil Nadu), Tirunelli (*Phyllanthus emblica*) of Tirunelli temple in Wayanad (Kerala), and Mula (*Bambusa bambos*) of Aaranmula Parthasarathi temple in Pathanamthitta (Kerala).

VERNACULAR NAMES

The 16 vernacular names of *S. cumini* are given below (AYUSH 2008):

Assamese:	Jam
Bengali:	Jaam
Bodo:	Jam
Gujarati:	Jambu
Hindi:	Jamun
Kannada:	Neralamara
Konkani:	Jambul

Malayalam:	Njaaval
Manipuri:	Jam
Marathi:	Jambool
Oriya:	Jamukoli
Punjabi:	Jamu
Sanskrit:	Jambavam
Tamil:	Njaaval
Telugu:	Nesedu
Urudu:	Jamun

INDIGENOUS TAXONOMY

According to Ayurvedic literature, the plant kingdom has been broadly classified into four parts:

- Vanaspathi: Trees, shrubs, herbs that fruit from modified inflorescence, for example, *Artocarpus heterophyllus* Lam. (Moraceae)
- Veeruth: Weak stemmed plants or climbers, shrubs, and herbs, for example, *Evolvulus alsinoides* L. (Convolvulaceae)
- Vanaspathyam: True flowering and fruiting plants, for example, *S. cumini* (L.) Skeels (Myrtaceae)
- Oushadhi: Annual herbs, for example, *Leucas aspera* (Willd.) Spreng. (Lamiaceae)

Caraka, an eminent Ayurvedic scholar and physician, states that simply confirming the name of the plant does not help in any way for treatment, and that one should be well versed with its morphological characters. It denotes that an intensive field survey is highly essential. To acquire knowledge of plants, especially through their name, morphological characters, and uses, one should interassociate or interact with the hill tribes and cowherds, sages, and local healers living in and around the forest. This is the ethnomedical concept described in Ayurveda.

Identification and naming of plants in Ayurveda are based on 21 criteria. Some of the criteria adopted are the peculiar morphological characters, *Hygrophila schulli* (Ham.) M. R. Almeida & S. M. Almeida (Kokilaksha); the distribution of plants confined to a certain region or area (geographical indications), *Elettaria cardamomum* Maton (Dravidi) and *S. cumini* (Vetasi); peculiar signs visible on the plants, *Azimia tetracantha* Lam. (Kundali); the resemblance of plants with animals and birds, *Actinopteris radiata* Bedd. (Mayurasika); the number of leaves or typical characters of leaves, *Alstonia scholaris* (L.) R. Br. (Saptaparna); the floral appearance, *Gloriosa superba* L. (Langali); the size and shape of the stem of a plant, *Cissus quadrangularis* L. (Asthisringala); various active principles (rasa), *Ocimum tenuiflorum* L. (Surasa); various properties (guna) and potencies (veerya), *Baliosperum montanum* Mull-Arg. (Danti); and the habitat, *Trichopus zeylanicus* ssp. *travancoricus* Burkill. ex Narayanan (Varahi).

Jambava is the name of the fruit jambu (*S. cumini*). It is a common item in the prescription of tribal vaidyas of Singhbhumi in Bihar (Singh and Chunekar 1972).

Jambu-dvayam (twin plant), according to the physicians of Kerala, constitutes *S. cumini* and *Syzygium caryophyllatum* (L.) Alson, which are respectively called njaval and njara in the regional language (Malayalam—official language of Kerala state, India) (Mooss 1980).

According to indigenous taxonomy or Ayurveda, *S. cumini* belongs to the category vanaspathyam (true flowering and fruiting plants). The various Sanskrit names attributed to *S. cumini* in Ayurveda and the criteria adopted for each of these names are given in Table 5.1.

TABLE 5.1
Criteria for Naming of *S. cumini* in Ayurveda

Sanskrit Names	Criteria Adopted for Naming the Plants
Jambu	Flowers and fruits are available irrespective of seasons
Surabhipatra	Sweet-smelling leaves (based on aromatic character)
Mahaskanda	Long (stout) stem and branches (based on the structure and quality of wood)
Vetasi	Growing along riverbanks (based on the distribution as riverine plant)
Jalajambuka	Growing near water (based on the distribution as riverine plant)
Meghamodini	Luxuriating or thriving well during the rainy season (based on its growth rate
Ghanapriya	during the rainy season)
Bhramaresta	
Bhrgavallabha	Attracting bees and flies (based on the presence of bees)
Bhringesta	
Bhramesta	Attracting cuckoos and crows (based on the bird–tree interaction)
Dhavaniksa	
Pikabhaksya	Fruits eaten by cuckoos
Kakaphala	May refer to either the dark or blackish color of the fruits or that they are eaten by crows
Krishnaphala	
Kasta jambu	
Kakanila	Indicates the dark or blue-black color of the fruits (based on the color of fruits, which resembles the color of Lord Krishna)
Neelaphala	
Neelavarui	
Mahaphala	
Brhatphala	A comparatively large size of the fruits (based on size and taste of the fruit)
Sitavalibha	Cooling agent (based on therapeutic action)
Rajajambu	
Phalendra	Signifies that its fruits are the best and are tasty (based on taste)
Kaka jambu	A wild variety (vana jambu); its fruits are *naadeyi*, enriches musical skill; it is the favorite of crows (*kakavallabha*) and liked by large black bees (*bhringeshta*)
Ghanaparni	Broad and thick leaves and the plants grow closely (*vaidsasea*)

Source: Iyer, K. N., and Kolammal, M., *Syzygium jambolanum* DC., *Syzygium caryophyllum* Garten, *Pharmacognosy of Ayurvedic Drugs*, Series I, No. 5, Pharmacognosy Unit Ayurveda College, Thiruvananthapuram, 1993, 55–62.

ETHNO-MEDICO-BOTANICAL INFORMATION

All the parts of *S. cumini* are used for medicinal and edible purposes. These include the root bark, stem bark, heart wood, leaves, tender leaves, flower, fruit, and seed kernel. The ripe fruits are used for the preparation of health drinks, squashes, jellies, vines, and so forth. The tree is planted mainly to provide shade. The wood is hard and durable and is used in fuel wood and for making agriculture implements. The ethnomedical uses of *S. cumini* are given in Table 5.2.

MEDICINAL PROPERTIES IN TRADITIONAL MEDICINE SYSTEMS

Ayurveda

Ayurveda is a "sastra" different from modern science, dealing with the physical, mental, and spiritual world of the life system on our planet as a whole, supported by the nonliving components, such as air, soil, and water. It is not merely a system of medicine; in a broader sense, it is the "science of life" of the universe. Hence, it is universally applicable. Ayurveda teaches the science of life from the micro- to the macrolevel. Based on the Sankhya and Vaiseshika philosophies, Ayurveda, conceptualized with concrete fundamental theories, begins with the theory of evolution of the universe (brahmanda), with the entire life-forms (pindanda) prevailing in it—mainly human beings, plants, animals, and microbes. Ayurveda identifies man as an integral part of nature and stresses the necessity of maintaining complete harmony with all living and nonliving components of the surroundings (environment). Panchabhoutika characters and qualities (based on the five elementary principles) of our planet are deteriorating day by day. Changes in the environment and health take place in millions of life-forms existing in our planet, including humans. Ayurveda envisaged such calamities about 7000 years ago and recommended suitable revival or remedial measures in a holistic manner. Ayurveda envisaged a "global health concept."

S. cumini finds a prime place in several of the ancient Ayurvedic treatises, such as Caraka Samhita and Kashypa Samhita, and is a major ingredient in several modern Ayurvedic formulations. In Caraka Samhita, *S. cumini* is described as having an astringent and sweet taste, with cold characteristic properties, and it induces heaviness and fullness in the stomach. It also acts as a pacifying agent to control the aggravated functioning of *kapha* and *pitta* and vitiates functioning of *vatadosha* (Caraka Samhitha Sutra Sthanam 25.39; Pandeya 1997). Bark is used in a polyherbal formulation for treating diarrhea due to aggravation of *kaphadosha*. A cold infusion prepared from tender leaves of *S. cumini* and *Mangifera indica*, along with *Terminalia chebula* with honey, is administered orally to cure vomiting due to the aggravation of *pittadosha*. In combination, bark powder is administered orally to check vomiting. Bark is used in a polyherbal formulation in the form of an external application to treat erysipelas. Bark is also used in a polyherbal formulation administered orally and as an external application over the heart and forehead region to treat vertigo, polydipsia, and syncope (Caraka Samhitha Chikitsa Sthanam 19.116, 19.117, 20.30, 20.38, 21.85, 22.34; Pandeya 1997).

TABLE 5.2
Ethnomedical Uses of *S. cumini*

Serial No.	Indications and Uses	Reference
1	The juice of the leaves and bark is given for dysentery and chronic diarrhea.	Dastur (1962)
2	The decoction of the bark is used as mouthwash and to treat spongy gum.	
3	A paste prepared from the ash of the bark with some blend of oil heals burns, boils, and wounds.	
4	The juice extracted from the seed is taken to reduce sugar in urine.	
5	The leaves assimilate more carbon dioxide from the atmosphere and liberate oxygen, thus balancing oxygen in the atmosphere, and the tree is pollution tolerant.	Singh and Rao (1983)
6	The leaves are used as fodder and on distillation yield bright green oil.	Randhava (1983)
7	The juice of fresh jambu fruits and mangoes in equal parts relieves thirst very effectively in diabetes. Powdered seeds are used as a remedy for diabetes.	Anonymous (1988)
8	The bark, leaves, and fruits are astringent, and the leaves emetic. Young leaves are chewed to check vomiting. Fresh juice of the bark is given with goat's milk in diarrhea, especially in children.	Saxena and Tripathi (1989)
9	The tender leaves act as an antidote and the fruits cure bilious disorders.	Shanmugam (1989)
10	Leaf juice is given internally for stomach pain.	Kapur et al. (1992)
11	The ripe fruit is edible. The young leaves and fruit, if chewed and taken, enhance lactation in goats and cattle.	Misra (1992)
12	The leaves of *S. cumini* and *Gymnema sylvestre* (10 g each) are boiled in 0.5 L of water until it is reduced to about 60 ml. The filtered extract mixed with 1 teaspoon of honey is given twice daily for two months to cure diabetes.	Satapathy (1992)
13	Leaf paste is applied on the forehead to cure redness of the eyes.	Aminuddin (1993)
14	Fruit pulp is useful in treating diabetes and is also used in controlling bleeding, stimulating digestion, control enlargement of the spleen, and activating urination, and as a tonic to improve appetite.	Sudharshan et al. (1993)
15	Leaf juice and stem bark juice are given in the treatment of diarrhea and diabetes.	Singh and Prakash (1994)
16	Stem bark prepared in the form of a decoction is administered orally to check diarrhea.	Saradamma et al. (1995)
17	The expressed juice of fresh bark is recommended to treat hoof disease (foot and mouth disease) in cattle by the Kurichiyan tribe of Kerala.	
18	Ripened fruit is eaten to reduce fatigue by the Irular tribe of Kerala.	

Some of the polyherbal formulations containing *S. cumini* as an ingredient and their indication or usage as prescribed in classic Ayurvedic texts are shown in Table 5.3.

The Kashyapa Samhita prescribes the following mode of uses of different parts of *S. cumini* for various ailments: Bark is used in a polyherbal formulation in the form of a powder, administered orally to purify breast milk. Bark, prepared in the form of medicated milk, is administered orally to treat the side effects of *vasthi*, one of the therapeutic procedures known as panchakarma in Ayurveda. The bark and tender leaves of *S. cumini*, along with bark of *Punica granatum* and *Spondias pinnata*, are prepared in the form of a decoction. The watery portion obtained from

---

TABLE 5.3

Ayurverdic Formulations with *Syzygium cumini* as an Ingredient and Their Indications and Usages as Prescribed in Classical Ayurveda Texts

Serial No.	Formulation: Indications and Usages	Reference
1	**Kudajaphanitham:** Bark of *S. cumini* is one of the ingredients; prescribed for piles, diarrhea, and dysentery	Latlithamma and Shyleswariyamma (1996)
2	**Useeraasavam:** Consisting of 25 ingredients; prescribed to treat skin diseases, diabetes, and anemia	
3	**Yavagu:** A polyherbal formulation with seed kernels of *S. cumini* as one of the ingredients; applied as medicated gruels; prescribed for diarrhea and dysentery	Caraka Samhitha (Sutra Sthanam 2.27); Pandeya (1997)
4	**Mahakashaya:** A medicated decoction with leaves of *S. cumini* along with 9 other ingredients; administered orally to check vomiting	Caraka Samhitha (Sutra Sthanam 4.28); Pandeya (1997)
5	**Purisha virechania mahakashaya:** A medicated decoction containing bark of *S. cumini* along with 9 other ingredients; administered orally for removing the facial pigments	Caraka Samhitha (Sutra Sthanam 4.32); Pandeya (1997)
6	**Mootra samgrhaniya:** A polyherbal formation with bark of *S. cumini* as an ingredient; antilithic	Caraka Samhitha (Sutra Sthanam 4.33); Pandeya (1997)
7	**Vachadyadi choornam:** A polyherbal formulation with fruit kernels of *S. cumini* as one of the ingredients; prescribed specifically for chronic dysentery and celiac diseases	Caraka Samhitha (Chikitsa Sthanam 15.136); Pandeya (1997)
8	**Pushyanuga choornam:** A polyherbal formulation with *S. cumini* seed kernel as one of the ingredients; specifically prescribed for vaginal and menstrual disorders	Caraka Samhitha (Chikitsa Sthanam 30.77, 30.107); Pandeya (1997)
9	**Sarvaathisara nasaka gritham:** A polyherbal formulation with *S. cumini* bark; used in the form of medicated clarified butter for treating diarrhea and dysentery	Caraka Samhitha (Siddhi Sthanam 8.35); Pandeya (1997)

the curd is administered in the form of *yoosham* (medicated soup) to cure diarrhea and dysentery. Tender leaves of *S. cumini* and *Mangifera indica* are prepared in the form of a decoction administered orally, along with sugar or honey, to check nausea and vomiting during pregnancy. The flower is used in a polyherbal formulation prepared in the form of *lehya* (jelly-type semisolid preparation) and administered orally to cure heart pain due to vitiation of *kapha dosha* during pregnancy. The flower is also used as mouthwash for persons suffering from typhoid. Bark is used as one of the ingredients in a polyherbal formulation prepared in the form of paste and mixed with medicated butter obtained from the *Nalpamara* (bark of four species of *Ficus*) applied externally to cure erysipelas due to aggravated *pitha dosha* (Vaidyar 1995).

According to *Bhavaprakasa*, an Ayurvedic text written during the sixteenth century, the jambu fruit (*S. cumini*) is considered to be sweet, an anticarminative, and an appetizer, and it has the potential to overcome *kapha*, *raktapitta* (epistaxis), and burning sensation. The medicinal uses of jambu have also been described in several lexicographic works on medicinal plants. For example, *Dhanvantari Nighantu* describes the expectorant property of jambu, while *Rajanighantu* prescribes *S. cumini* as an effective remedy to treat worm infestation, asthma, diarrhea, cough, diseases of the throat, and fatigue. *Nighanturatnakaram* identifies *S. cumini* as one of the effective remedies for diabetes (Sharma 1990).

Sharma (1990) described the following distinct properties of *S. cumini* and the consequent actions on the human body system: Due to its *rooksha* (dry) property and *kashayarasa* (astringent taste), it pacifies *kapha*. Because of *seetaveerya* (potentiated with cold in action) and its astringent taste, it pacifies *pitta*. It also aggravates *vatadosha* owing to *rooksha* (properties), *seeta* (action), and *kashaya*. Fruit enhances digestion and appetite and stimulates the functions of the liver. The leaves mainly act as an antiemetic, and bark acts as a *sthambanam*. In addition to its cooling, drying, and astringent properties, *S. cumini* helps easy digestion and causes stasis and condensation in the tissue fluid.

The medicinal properties and mode of uses of different parts of *S. cumini* in the Ayurvedic medicine systems are summarized in Table 5.4.

VRIKSHAYURVEDA

Medicinal plants constitute one of the important components effectively used by the physicians of Ayurveda, not only for treating various disorders but also for preventive, promotional, and corrective purposes. Animal care (Mrigayurveda) and plant care (Vrikshayurveda) uses were developed in later years. There are about 70 authentic treatises and books with more than 8000 recipes either in a single form or as combinations available in Ayurveda.

An indigenous seed germination technique of *S. cumini* has been mentioned in Vrikshayurvedic literature, along with other medicinal plant species, like *Artocarpus heterophyllus*, *Mangifera indica*, *Flacourtia montana*, and *Artocarpus hirsutus*. The technique involves the following steps. Seeds are soaked in cow's milk and dried properly. This is to be smeared with the paste prepared from cow dung, powder of *Embelia ribes*, and clarified butter. Subsequently, seeds are sown in the soil for producing saplings. *S. cumini* is considered one of the auspicious trees described in

TABLE 5.4

Medicinal Properties and Uses of Different Plant Parts of *Syzygium cumini* in Modern Ayurveda

Serial No.	Plant Part Used	Indications and Usages	Reference
I	Leaves	1. Tender leaves are one of the ingredients in the formulation known as *jambuamrapallavadi kashyam*, administered orally along with honey to check vomiting with blood stain.	Kurup (1968)
		2. Leaves mainly act as antiemetic and bark acts as *sthambanam*. (In addition to its cooling, drying, and astringent properties of *S. cumini*, it aids in easy digestion and causes stasis and condensation in the tissue fluid.)	Sharma (1990)
		3. Juice obtained from the slightly heated fresh leaves of *S. cumini* is one of the ingredients in a polyherbal formulation administered orally to check bleeding due to fever (virus and bacterial infection).	
		4. Tender leaves are used in a polyherbal formulation in the form of a decoction administered orally to check vomiting and also to treat fever and polydipsia.	
		5. Tender leaves prepared in the form of paste, along with coconut kernel juice, are administered orally to treat fever and diarrhea.	
		6. Juice obtained from the tender leaves of *S. cumini* and fresh fruit of *Emblica officianalis*, along with goat's milk and honey, is administered orally to check bleeding due to dysentery.	Menon (1993)
		7. Tender leaves are used in a polyherbal formulation in the form of a decoction administered orally, along with powder of *Cuminum cyminum* and rock salt, to cure piles in children.	
		8. Tender leaves are used in a polyherbal formulation known as *madhukadhaya avaleham*, consisting of 26 ingredients for treating uterine bleeding, bleeding due to dysentery and piles, epistaxis, urinary disorders, and nausea.	

(Continued)

TABLE 5.4 (CONTINUED)
Medicinal Properties and Uses of Different Plant Parts of *Syzygium cumini* in Modern Ayurveda

Serial No.	Plant Part Used	Indications and Usages	Reference
		9. Tender leaves of *S. cumini* and *Mangifera indica* prepared in the form of a decoction and administered orally to check vomiting and hyperacidity.	
		10. The tender leaves, along with *Gymnema silvestre* and *Aloe vera*, are prepared in the form of pills and administered orally to control diabetes.	
		11. Leaves in combination with fresh turmeric, *Lawsonia inermis*, and *Sesamum orientale* are applied externally in the form of paste to remove tumor and also to treat abscess.	
		12. The decoction prepared from the tender leaves of *S. cumini* is used for purposes of washing the eye to check excess secretion from the lachrymal glands in children.	
		13. Tender leaves are one of the ingredients in an Ayurvedic formulation known as *amritha brihatyadi gritham*, prescribed for treating erysipelas, skin diseases, leprosy, and scabies.	Variyar (2011)
II	Bark	1. Bark of *S. cumini* is one of the ingredients in a polyherbal formulation known as *thengumpookulathi kashayam*, which is exclusively prescribed for uterine bleeding.	Kurup (1968)
		2. Bark is used in a polyherbal formulation known as *lajavilwadi kashayam*, which is prescribed to check vomiting.	
		3. Bark is one of the ingredients in a polyherbal formulation known as *sreephaladi sheetha kashyam* (cold infusion), administered orally to check vomiting.	
		4. Bark is one of the ingredients in a polyherbal formulation known as *kanchadathi kashyam*, administered orally to check vomiting with diarrhea.	
		5. Bark is one of the ingredients in *varahyadi grithim*, administered orally to cure vaginal diseases.	

(Continued)

TABLE 5.4 (CONTINUED)
Medicinal Properties and Uses of Different Plant Parts of *Syzygium cumini* in Modern Ayurveda

Serial No.	Plant Part Used	Indications and Usages	Reference
		6. Bark is used to detoxify the ingredients (iron fillings and sulfur) of the formulation known as *swayam bhasmam*, which is prescribed for treating jaundice and anemia.	
		7. Bark is used as one of the ingredients in *useerasavam*, administered orally to treat epistaxis, anemia, skin diseases, diabetes, piles, and intestinal parasites, and it acts as anti-inflammatory.	
		8. Bark is used in a polyherbal formulation prepared in the form of a decoction and administered orally to treat chronic dysentery.	Menon (1993)
		9. Bark is used in a polyherbal formulation prepared in the form of pills administered orally and applied externally to check vomiting and diarrhea.	
		10. Juice is obtained from the fresh bark administered orally, along with cow's milk, to check vomiting and also to induce sound sleep.	
		11. Bark is prepared in the form of paste, along with milk, administered orally to check bleeding.	
		12. Bark powder is used, along with honey, to check vomiting due to the aggravation of *kapha*.	
		13. Bark in combination is administered orally in the form of a decoction to treat internal abscesses.	
		14. Bark is prepared in the form of paste by adding cow's milk and is administered orally to treat the complications of chicken pox.	
		15. Bark in combination is administered orally to treat chronic ulcerative wounds.	
		16. Bark in combination is prepared in the form of a decoction used for gargling and to check bleeding gums.	

(Continued)

TABLE 5.4 (CONTINUED)
Medicinal Properties and Uses of Different Plant Parts of *Syzygium cumini* in Modern Ayurveda

Serial No.	Plant Part Used	Indications and Usages	Reference
III	Flower	1. Flowers and leaves are the ingredients in a polyherbal formulation known as *dashamoolarishtam*, prescribed for the treatment of an enlarged spleen, hyperacidity, and fever, and it acts as anti-inflammatory and improves the functioning of peristaltic movements.	Kurup (1968)
		2. Flowers are one of the ingredients in a polyherbal formulation prepared in the form of medicated clarified butter, along with honey, and applied externally to improve the complexion of skin (especially on the face).	
		3. Flowers and leaves are among the ingredients in a polyherbal formulation, prepared in the form of a paste and applied along with honey to treat throat cancer.	
		4. Flower is one of the ingredients in the polyherbal formulations—*mahalodhradhi choornam*, and *lodhradhi choornam*—sprinkled over the cut wounds to arrest bleeding.	Variyar (2011)
		5. Flower is one of the ingredients in a polyherbal formulation known as *lodhradi choornam*, in the form of a powder that is sprinkled over the cut wounds to arrest bleeding.	
IV	Fruit	1. Fruit juice is administered on alternate days to the person suffering from jaundice and other liver diseases.	Menon (1993)
		2. Fruit juice, along with a little salt, a half teaspoon thrice daily, is prescribed for liver disorders. Avoid curd and butter milk during the administration.	
		3. Fruit kernel is prepared in the form of powder administered orally to treat dysuria.	

(*Continued*)

TABLE 5.4 (CONTINUED)
Medicinal Properties and Uses of Different Plant Parts of *Syzygium cumini* in Modern Ayurveda

Serial No.	Plant Part Used	Indications and Usages	Reference
		4. The asava (self-generated alcoholic preparation) from the fruit, prepared along with honey, is used to control diabetes.	
		5. Drinks prepared from the fruit of *S. cumini*, along with sugar and water, perfumed with pepper and the leaves of bringaraja (*Eclipta prostrata*), act as an appetizer.	Sarma (2009)
		6. The juice obtained from the fruit alleviates *kapha*, slightly aggravates *vata*, and is astringent, sweet and sour, appetizing, and *grahi* (dries up substances due to its hot nature).	
		7. The jamun fruit is astringent and sweet, and used in fatigue and the vitiation of pitta, burning sensation, throat pain, worm infestation, dyspnea, diarrhea, and cough.	Suri (2012)
		8. Consumption of fruit causes constipation. It also imparts taste and favors digestion.	
V	Seed	1. Seed kernel is one of the ingredients in a polyherbal formulation known as *patajambuathi choornam* (powder), administered orally to treat menstrual disorders.	Kurup (1968)
		2. Seed kernel is one of the ingredients in *pushyanuga choornam* (powder), mentioned in the Ayurvedic textbook known as *Bhaishajya Retnavali*. This drug is administered orally to treat various types of vaginal disorders and the female reproductive system.	
		3. Seed kernel is one of the ingredients of a polyherbal formulation known as *chandanadi choornam*, indicated for epistaxis, to control vomiting and to check diarrhea, uterine bleeding, and piles.	Menon (1993)
		4. Seed kernel is prepared in the form of powder and sprinkled over the ulcer as a healing agent.	

Vrikshayurveda. Preference has to be given to plant such trees. Big trees should not be planted very close to houses. It is also included in a group of five trees known as panchapallavam, which possess high therapeutic value. Other trees are *Mangifera indica*, *Flacourtia montana*, *Citrus limon*, and *Aegle marmelos*. A liquid biomanure is prepared from the mixture of curd, the watery portion of curd, and one-day-old rice-washed water. An alcoholic beverage is obtained from rice powder, *Ziziphus mauritiana*, *Sesamum orientale*, and *Trigonella foenum-graecum*. Kunapajalam, a liquid biofertilizer prepared from herbs and the fat of animals, fishes, and so forth; an alcoholic beverage prepared from *Saccharum officinarum* and milk is considered one of the most effective liquid biomanures for enhancing the aroma of the flowers of *S. cumini*, *Neolamarckia cadamba*, and *Mesua ferrea*. A polyherbal mixture consisting of bark of *S. cumini* and six other ingredients, along with clarified butter, is poured to the bottom of the *Magnolia champaca* and *Couroupita guianensis* for healthy growth. It is believed that planting *S. cumini* within the vicinity of residences will be auspicious and bring prosperity to the family (Sastri 1945).

S. *CUMINI* IN OTHER CLASSICAL TRADITIONAL INDIAN SYSTEMS OF MEDICINE

Unani Medicine

Unani is the Arabic name for Greece, which denotes the origin of the system. The Unani system considers the human body to be composed of matter and spirit, and gives importance to the proper balance between the bodily and spiritual functions to maintain a harmonious life. The system was further enriched by the Arabs and Persians (Brinda 1997).

The Unani-Tibb (Greek medicine) or Greco-Arab system of medicine originated in Greece with the god Apollo, the inventor of healing art, "who chases away all ills." Then via the centaur Chiron and his pupil Aesculapius came Pythagoras, Alemneon, and Hippocrates (Said 1983; Alok 1988b; Siddhiqui 1995). This medicine system was introduced to India about a thousand years ago (1000 CE) by Muslim rulers and became indigenous to the country. It is now practiced in the Indo-Pakistan subcontinent (Alok 1988b). The Unani physicians, who settled in India, were not content with the ancient knowledge, and subjected new drugs to clinical tests and experimentation, thus adding new drugs to the system. Therefore, the Unani system practiced in India is different from the original Greek form.

As per the Unani system of medicine, *S. cumini* acts as a liver tonic, enriches blood, strengthens teeth and gums and cures ringworm infection of the head (tinea capitis). Various extracts of fruit and seeds of *S. cumini* were reported to have antidiabetic, anti-inflammatory, hepatoprotective, antihyperlipidemic, diuretic, and antibacterial activities (Ayyanar and Subash Babu 2012).

Siddha Medicine

There are many theories that postulate the origin of the Siddha system of medicine. The Siddha system owes its origin to the Dravidian culture, which is of the prevedic period. An examination of the ancient literature reveals that the vedic Aryans owed allegiance to the cult of Shiva and the worship of the phallus (linga), which was later absorbed by and incorporated into the vedic culture. Siddha system has its literature written on palm leaves in the Tamil language. It is believed that this system flourished in the Indian subcontinent during the first Tamil Sangam period (700–600 BCE), and is exclusively linked with the Tamil culture and civilization. The term *Siddha* means achievement, and the siddhars were saintly figures who achieved the results in medicine through the practice of Yoga. Eighteen siddhars seemed to have contributed to the development of this medicine system. The system is also called the Agasthyar system, in the name of its famous exponent sage Agasthya, who is known as the first physician of Siddha medical science. He is believed to have conceived the system incorporating many elements of Saiva tantra to the medical practices of the time (Kurup 1983; Alok 1988a).

The chronological age of this system is prehistoric and prevedic, that is, 10,000–4000 BCE. The findings of historians and the Tamil literary works such as Tholkappiam and Thirumandhiram reveal many facts about the growth of this Siddha medicine. Although this system declined in later years by the changing modes of life and by the interventions of Western colonial influences, it has still continued to exist in many parts of south India (Arunachalam 2011). According to Siddha-bheshajamanimala, the juice of jambu fruit or its vinegar was given in sluggish digestion (Khare 2004).

Other Uses and Legends

According to Hindu mythology *S. cumini* is sacred to Lords Shiva, Krishna, and Ganapati. This tree is also associated with Venus, the goddess of love. The tree is held in veneration by the Buddhists. Buddha Kapilavastu was closely associated with *S. cumini*. The thirteenth Jain trithankara, Vimalanatha, obtained his divine knowledge under this tree (Bhattacharya 1974). The god Megh is said to have been transformed into the jambu tree. The color of fruit is like that of Krishna, hence, this plant is worshiped, and *brahmins* are fed under it. The leaves are used as platters for pouring ritualistic liquid materials such as water and milk during the sacrifice to attain liberation (Lisboa 1886).

The jamun tree is described in the Mahabaratha as a cosmic tree, standing to the south of Mount Meru, the axis of the universe. When the ripe fruit of this gigantic tree bursts, the juice falls as a waterfall, forming rivers that become the boundaries of a land known as Jambudvipa, or the continent of the jambu trees, populated by epic heroes who gained immortality by drinking the juice. The juice of ripe fruits that fell flowed to form a stream known as the Jambu Nathi, and the people of Ilavarta fed on it. Although the jamun tree receives no ostensible worship, it is still invoked in Indian rituals, which always commences with the words "In the continent of the jamun trees, where the land of India lies" (Patnaik 1993). The use of *S. cumini* (fruit) acts as an antidote for the side effect due to the large-quantity consumption of mango (Verma 1980).

The seeds received considerable attention in folk medicine, the Ayurveda and Unani traditional systems of medicine, as they are antidiabetic (Satyavati and Gupta 1973; Zafar 1994; Shukla et al. 2000). Several studies using modern techniques have authenticated their use in diabetes and shown promising results (Brahmachari and Augusti 1961; Kar et al. 2003; Ravi et al. 2004; Pepato et al. 2005).

REFERENCES

Alok, S. K. 1988a. Siddha system of medicine. In *Indian System of Medicine and Homeopathy, National and State Profiles*. New Delhi: Ministry of Health and Family Welfare, India, 48–52.

Alok, S. K. 1988b. Unani system of medicine. In *Indian System of Medicine and Homeopathy, National and State Profiles*. New Delhi: Ministry of Health and Family Welfare, India, 53–70.

Aminuddin, G. R. D. 1993. Observations on ethnobotany of the Bhunjia—A tribe of Sonabera plateau. *Ethnobotany* 5: 83–86.

Anonymous. 1988. *Sacred Plants*. Bangalore: Karnataka Forest Department, 32.

Arunachalam, A. V. 2011. Traditional and folk practices—A contemporary relevance and future prospects, ed. S. Rajasekharan and P. G. Latha. Thiruvananthapuram: KSCSTE and JNTBGRI, 108–118.

AYUSH. 2008. *The Ayurvedic Pharmacopoeia of India*. Part I, Vol. II. New Delhi: Department of AYUSH, Ministry of Health and Family Welfare, India, 56–57.

Ayyanar, M., and P. Subash Babu. 2012. Syzygium cumini (L.) Skeels: A review of its phytochemical constituents and traditional uses. *Asian Pac. J. Trop. Biomed.* 2: 240–246.

Bhattacharya, B. C. 1974. *The Jain Iconography*. New Delhi: Motilal Banarasi Das.

Brahmachari, H. D., and K. T. Augusti. 1961. Hypoglycemic agents from Indian indigenous plants. *J. Pharm. Pharmocol.* 13: 381–382.

Brinda, S. 1997. Angiotensin converting enzyme inhibitors from Indian medicinal plants. PhD dissertation. Copenhagen: Department of Medicinal Chemistry, Royal Danish School of Pharmacy, 30.

Dastur, J. F. 1962. *Medicinal Plants of India and Pakistan*. Bombay: D. B. Taraporevala Sons & Co., 116–117.

Iyer, K. N., and M. Kolammal. 1993. *Syzygium jambolanum* DC., *Syzygium caryophyllum* Garten*, Pharmacognosy of Ayurvedic Drugs*. Series I, No. 5. Thiruvananthapuram: Pharmacognosy Unit Ayurveda College, 55–62.

Kapur, S. K., S. Nanon, and Y. K. Sarin. 1992. Ethnobotanical uses of RRL Herbarium 1. *J. Ecol. Taxon. Bot. Addl. Ser.* 10: 479–493.

Kar, A., B. K. Choudhary, and N. G. Bandyopadhyay. 2003. Comparative evaluation of hypoglycaemic activity of some Indian medicinal plants in alloxon diabetic rats. *J. Ethnopharmacol.* 84: 105–108.

Khare, C. P., ed. 2004. *Indian Herbal Remedies. Rational Western Therapy, Ayurvedic and Other Traditional Usage, Botany*. Berlin: Springer.

Kurup, K. V. 1968. *Sahasrayogam* [Textbook of 1000 Herbal Formulations]. Malayalam translation with Vaidya Priya commentary. Kollam: Sree Rama Vilasam Press & Book Depo.

Kurup, P. N. V. 1983. Ayurveda. In *Traditional Medicine and Health Care Coverage*, ed. R. H. Bannerman, J. Burton, and C. Wen-Cheih. Geneva: World Health Organization, 50–60.

Latlithamma, K., and B. Shyleswariyamma. 1996. *Pharmacopeia* [Bhishajya Ratnavali, in Malayalam]. Thiruvananthapuram: Government Ayurveda College, 31, 54.

Lisboa, T. C. 1886. Useful plants of Bombay presidency. *Gazetteer Bombay Pres.* 25: 284.

Menon, K. V. M. 1993. *Chikitsa Kauthukam*. Textbook written in Malayalam through compilation of selected single, simple, large combinations prescribed for various diseases from the original text references from the classical books. Coimbatore: Loka Swasthya Parambara Samvardhan Samithi, 65, 77, 81, 105, 107, 173, 207, 222, 227, 229–230, 235–236, 338, 340–341, 581, 592, 594–595, 600, 609, 618, 620, 665, 689, 694, 703, 709, 1061, 1010.

Misra, R. C. 1992. Medicinal plants among the tribals of upper Bonda region. Koraput (Orissa). *J. Econ. Taxon. Bot.* 10: 275–279.

Mooss, V. N. S. 1980. *Ganas of Vahata (Astanga Hridaya Samhitha Sutra Sthana)*, ed. and trans. V. N. S. Mooss. Kottayam, Kerala: Vidyasarathy Press, chap. XV.

Pandeya, G. S., ed. 1997. Caraka Samhita. Agni Vesa. Revised by Caraka and Drdhabala with Ayurveda-Dipika commentary of Cakrapani Datta and with Vidyotini Hindi commentary by Kasi Natha Sasthri. Varanasi: Chaukhambha Sanskrit Sansthan. Sutra Sthanam 2.27, p. 42–48; 4.28, p. 65; 4.32, p. 65; 4.33, p. 66; 25.39. Chikitsa Sthanam 8.12, 8.129–130, p. 243; 15.136, p. 339; 19.109–111, p. 498; 19.117, 20.30, p. 507; 20.38, p. 509; 21.85, p. 511–535; 22.34, p. 543; 30.77, p. 763; 30.107, p. 766. Siddhi Sthanam 8.35, p. 970.

Patnaik, N. 1993. In the continent of jamun trees, where the land of India lies. In *The Garden of Life—An Introduction to the Healing Plants of India*. New Delhi: New Delhi Aquarian, 116–117.

Pepato, M. T., D. M. Mori, A. M. Baviera, J. B. Harami, R. C. Vendramini, and L. L. Brunetti. 2005. Fruit of the jambolan tree (*Eugenia jambolana* Lam.) and experimental diabetes. *J. Ethanopharmacol*. 96: 43–48.

Rajasekharan, S., N. Mohanan, P. G. Latha, K. G. Mohanlal, and G. M. Nair. 2005. *Star and Trees—Trees of Nakshatravanam*. Thiruvananthapuram: Thenmala Ecotourism Society and Tropical Botanic Garden and Research Institute, 24–26.

Randhava, M. S. 1983. *Flowering Trees*. New Delhi: National Book Trust of India.

Ravi, K., D. S. Sekar, and S. Subramanian. 2004. Hypoglycemic activity of inorganic constituents in *Eugenia jambolana* seed on streptozotocin-induced diabetes in rats. *Biol. Trace Elements Res*. 99: 145–155.

Said, H. M. 1983. The Unani system of health and medicare. In *Traditional Medicine and Health Care Coverage*, ed. R. H. Bannerman, J. Burton, and C. Wen-Cheih. Geneva: World Health Organization, 61–67.

Saradamma, L., R. Nair, A. V. Bhat, and S. Rajasekharan. 1995. Final technical report of All India coordinated research project on ethnobiology. New Delhi: Ministry of Environment and Forests, India, 127, 162.

Sarma, K. 2009. *Ksema Kutuhalam* [Work on dietetics and well-being]. Bangalore: Institute of Ayurveda & Integrative Medicine, 362.

Sastri, K. P. P. 1945. Sarangadharapathadhi-Upavanavinoda [Malayalam translation with commentary by Mahopadhyaya K. Padmanabha Pillai Sastri]. Thiruvananthapuram: Vidvan K. Vasu Panickar, 15, 24, 25, 30–32, 64.

Satapathy, K. B. 1992. Medicinal uses of some plants among the tribals of Sundargarh District, Orissa. *J. Econ. Taxon. Bot. Addl. Ser*. 10: 241–249.

Satyavati, G. V., and Gupta A. K. 1973. *Medicinal Plants of India*. New Delhi: Indian Council of Medical Research, 696.

Saxena, S. K., and J. P. Tripathi. 1989. Ethnobotany of Bundelkhand. I. Medicinal uses of wild trees by tribals inhabitants of Bundelkhand region. *J. Econ. Taxon. Bot*. 13: 2: 381–388.

Shanmugam, N. K. 1989. *Moolgai Kalai Kulanjiyam*. Madras: Nakkeeran Publishers.

Sharma, P. V. 1990. *Dravyaguna-Vijnana* [Vegetable Drugs]. Vol. 2. Varanasi: Chaukhambha Bharati Academy, 659–660.

Shukla, R., S. B. Sharma, D. Pari, K. M. Prabhu, and P. S. Murthy. 2000. Medicinal plants for treatment of diabetes mellitus. *Ind. J. Clin. Biochem*. 15: 169.

Siddhiqui, M. K. 1995. State of Unani Tibb [Unani medicine] in India. In *Glimpses of Indian Ethnopharmacology*, ed. P. Pushpangadan, U. Nyman, and V. George. Thiruvananthapuram: Tropical Botanic Garden and Research Institute, 85–98.

Singh, B., and K. C. Chunekar. 1972. *Glossary of Vegetable Drugs in Brihat Trayi*. Varanasi: Chowkhamba Sanskrit Series Office, 164–165.

Singh, K. K., and A. Prakash. 1994. Tree wealth in the life and economy of the tribals of Uttar Pradesh, India. *Ind. J. Forestry* 17: 154–160.

Singh, S. K., and Rao, D. N. 1983. Evaluation of the plants for their tolerance to air pollution. In *Proceedings of the Symposium on Air Pollution Control*. Delhi: IIT, 218–224.

Sudharshan, S. R., A. N. Y. Reddy, and B. Gowda 1993. *Oushadhi Kosha* [Encyclopedia of Medicinal Plants]. Vol. 1. Bangalore: Kalpatharu Research Academy.

Suri, R. 2012. *Bhojana Kutuhalam—An Encyclopedic Work on Various Aspects on Food from the Perspective of Ayurveda.* Bangalore: Institute of Ayurveda & Integrative Medicine, 155.

Vaidyar, M. N. 1995. *Kashyapa Samhita* [in Malayalam]. Kannur, Kerala: Dhanwantari Printers, 13, 259, 440, 527–528, 558, 583.

Variyar, K. R. 2011. *Aarogyakalpadrumam* [Textbook of pediatrics in Ayurveda based on original text written in Sanskrit by Vak Dasa-Malyalam translation], ed. B. R. Syamla. Thrisivaperoor: Sambrat Publications, 192, 202–203, 308, 342, 357, 383, 436.

Verma, G. S. 1980. *Miracles of Fruits.* New Delhi: Rasayan Pharmacy, 178.

Zafar, R. 1994. *Medicinal Plants of India.* New Delhi: CBS Publications and Distributors, 105.

6 Chemistry of Syzygium cumini

Sudhir Kumar and Rakesh Maurya

CONTENTS

INTRODUCTION

Syzygium cumini (Jambul) is one of several species of the Myrtaceae family that are cultivated for their edible fruits. They are medicinally useful and are largely used by the local population to meet their health promotion and protection needs. In association with its dietary and commercial uses, the entire plant has been attributed in various traditional systems of medicine in India to possess several medicinal properties. In Ayurveda, the bark is described as acrid, sweet, antihelminthic, digestive, and astringent to the bowels (Ravichandiran and Ilango 2012). The tree has long been considered to have medicinal properties against dysentery, inflammation, diabetes mellitus, constipation, leucorrhea, stomachalgia, fever, gastropathy, strangury, and dermopathy, and to inhibit blood discharges in the feces (Shafi et al. 2002; Srivastava and Chandra 2013). Modern experimental studies have shown that polyherbal preparations made using *S. cumini* and other plants used in Ayurveda are effective in preventing both hyperglycemia and its complications (Bopp et al. 2009; Baliga et al. 2013). However, some studies in experimental and clinical models have also shown the absence of an antihyperglycemic effect of jambolan (Teixeira et al. 1997, 2000, 2004).

The phytochemistry of *S. cumini* is well described in the literature, mainly focusing on composition of essential oil (EO) and phenolic compounds. The objective of this chapter is to review existing information on the phytochemical composition of different plant parts of *S. cumini* concerning the isolation, structural elucidation, and biological activity evaluation.

ESSENTIAL OILS

EOs (or volatile oils or ethereal oils) are aromatic oily liquids obtained mainly from plant material by steam distillation, expression, fermentation, enfleurage, or extraction by means of solvents, but steam distillation is the most common method for producing EOs on a commercial basis. A more recent method of extraction by means of liquid carbon dioxide at low temperature and high pressure produces a more natural profile of EO content (Guan et al. 2007). Chemically, EOs are a complex mixture of volatile constituents and have a characteristic essence of the plant material. They are extracted from almost all parts of the plant, that is, flowers, buds, seeds, leaves, twigs, bark, wood, fruits, and roots. Their presence and function are still a question; probably, they play an important role in direct and indirect plant defense against herbivores and pathogens. Apart from their importance in plant defense, EOs have a vital role in the attraction of pollinators and seed disseminators, thereby playing an important role in the reproduction and survival of species (Langenheim 1994). EOs are well recognized for their cosmetic, pharmaceutical, agricultural, and industrial applications. They have a beneficial impact on humans as health-promoting compounds. EOs or their components have exhibited antiviral, antimycotic, antitoxigenic, antiparasitic, and insecticidal properties. Volatile oils are also used as natural flavor and aroma compounds (Burt 2004).

EOs are dominated by volatile terpenoids, along with some aromatic compounds and nonterpenoidal hydrocarbons. Nonterpenoidal hydrocarbon derivatives include short-chain alcohols and aldehydes, formed by the metabolic conversion or degradation of fatty acids and phospholipids (Hüsnü et al. 2007). Volatile terpenoids are represented by mainly isoprenes (C5), monoterpenes (C10), and sesquiterpenes (C15).

CHEMICAL COMPOSITION OF ESSENTIAL OILS

Numerous publications have presented data on the composition of the EOs from different plant parts of *S. cumini*. The EO is generally obtained by hydrodistillation and analyzed by gas chromatography–mass spectrometry (GC-MS), and the constituents are identified by comparison of their retention indexes (RIs) and mass spectra with published data. The amount of each component is given as a percentage of the total of oil, and in general, most of the oils are identified in this way.

Leaves have highest content of EO due to the presence of oil glands in leaves, a documented feature of the genus *Syzygium*. EOs from all aboveground parts of the plant, for example, fruit, flower, stem, and leaves, have been extracted and analyzed. Most of the reports are on the volatile oils obtained from the leaves, and only a few reports appeared on the remaining parts. The composition of the EO is characterized by significant variation depending on the habitat and ecology of the plant. Notably, major components of the EO differ significantly depending on the origin of the material (Table 6.1). Variability is also observed due to the method of extraction, as reflected by the variation of the activity profile of EOs. The difference in the composition of oils obtained by solvent extraction, as opposed to distillation, may also influence biological properties. This would appear to be confirmed by the fact that EOs extracted from plants by hexane have been shown to exhibit greater antimicrobial activity than the corresponding steam-distilled volatile oil (Packiyasothy and Kyle 2002). Moreover, EOs are volatile and compositional change may also be due to loss of more volatile components or decomposition and oxidation. EOs therefore need to be stored in airtight containers in the dark, in order to prevent compositional changes. However, many compounds are found to present only in small amounts, and the number of compounds with a higher percentage are relatively small. The major components of a number of EOs from different plant parts from different origins are presented in Table 6.1.

Chemical Composition of Essential Oils from Leaves

The major component of leaf EO is usually α-pinene, but the proportions vary widely. Studies carried out on different populations of *S. cumini* commonly showed variations. Leaves collected from Egypt yielded 0.125% (v/w) EOs. The main constituents of oil were α-pinene (17.53%), α-terpineol (16.67%), alloocimene (13.55%), α-bornyl acetate (6.37%), 2-β-pinene (5.34%), and caryophyllene (5.41%) (Elansary et al. 2012). A similar composition of components found in the Egyptian *S. cumini* was also reported earlier by Power and Callan (1912). Additionally, a recent finding expressed that α-pinene (32.32%), β-pinene (12.44%), trans-caryophyllene (11.19%), 1,3,6-octatriene (8.41%), Δ-3-carene (5.55%), α-caryophyllene (4.36%), and α-limonene (3.42%) are present in EO (Mohamed et al. 2013). Mahmoud et al. (2002) identified α-terpineol (26.79%), caryophyllene (21.98%), α-humulene (11.58%), and bornyl acetate (6.3%) out of a total of 27 components from fresh leaves by GC-MS, showing significant contribution of the sesquiterpenoids in EOs.

The analysis with GC-MS of EO from leaves of *S. cumini* growing in northeastern Brazil showed high levels of monoterpenes. It was found that 85% of leaf EO was composed of α-pinene (30%), β-pinene (20%), cis-ocimene (9%), trans-ocimene (9.5%), and α-humulene (2.8%) (Craveiro et al. 1983). Moreover, Dias et al. (2013) identified α-pinene (31.85%), (Z)-β-ocimene (28.98%), and (E)-β-ocimene (11.71%) as the major compounds present in the EO of leaves growing in northeastern Brazil.

TABLE 6.1
Major Components of Essential Oils

Serial No.	Plant Part	Major Components (Percentage of Oil)	Origin	References
1	Leaves	α-Pinene (30.10%), β-pinene (20.5%), limonene (8.50%), cis-ocimene (9.00%), and trans-ocimene (9.50%)	Brazil	Craveiro et al. (1983)
2	Leaves	α-Pinene (31.85%), (Z)-β-ocimene (28.98%), and (E)-β-ocimene (11.71%)	Brazil	Dias et al. (2013)
3	Leaves	α-Pinene (17.53%), 2-β-pinene (5.3%), α-terpineol (16.67%), alloocimene (13.55%), and caryophyllene (5.4%)	Egypt	Elansary et al. (2012)
4	Leaves	Pinocarveol (15.1%), α-terpineol (8.9%), myrtenol (8.3%), eucarvone (6.6%), muurolol (6.4%), myrtenal (5.8%), geranyl acetone (5.6%), α-cadinol, and pinocarvone (4.4%)	India (south region)	Jirovetz et al. (1999)
5	Leaves	α-Terpineol (5.71%), β-caryophyllene (6.34%), α-humulene (12.30%), β-selinene (4.19%), calacorene (4.43%), α-muurolol (4.28%), α-santalol (3.55%), and cis-farnesol (5.04%)	India (north region)	Kumar et al. (2004)
6	Leaves	α-Caryophyllene (25.24%), β-caryophyllene (16.00%), α-terpeneol (9.08%), and epiglobulol (5.23%)	Brazil	Machado et al. (2013)
7	Leaves	α-Terpineol (26.79%), caryophyllene (21.98%), α-humulene (11.58%), and bornyl acetate (6.3%)	Egypt	Mahmoud et al. (2002)
8	Leaves	α-Pinene (32.32%), β-pinene (12.44%), trans-caryophyllene (11.19%), 1,3,6-octatriene (8.41%), Δ^3-carene (5.55%), α-caryophyllene (4.36%), and α-limonene (3.42%)	Egypt	Mohamed et al. (2013)
9	Leaves	α-Pinene (17.53%), α-terpineol (16.67%), and alloocimene (13.55%)	Egypt	Power and Callan (1912)
10	Leaves	α-Pinene (15.80%), limonene (6.90%), cadinene (15.10%), Me salicylate (1.15%), 1,8-cineole (3.78%), α-terpineol (1.61%), perillaldehyde (21.76%), and geraniol (4.2%)	India (central region)	Ramaiah and Nigam (1969)
11	Fruits	cis-Ocimene (30%), trans-ocimene (23%), β-myrcene (7%), and α-terpineol (6.5%)	India (south region)	Vijayanand et al. (2001)
12	Fruits	Me eugenol (22.5%), limonene (14.43%), and α-terpineol (12.04%)	Egypt	Abdelhady (2012)

(Continued)

TABLE 6.1 (CONTINUED)
Major Components of Essential Oils

Serial No.	Plant Part	Major Components (Percentage of Oil)	Origin	References
13	Fruits	α-Pinene (30.89%), β-pinene (10.81%), cis-ocimene (18.5%), and trans-ocimene (12.10%)	Brazil	Craveiro et al. (1983)
14	Stem	α-Pinene (18.56%), β-pinene (12.61%), limonene (6.48%), cis-ocimene (14.83%), trans-ocimene (12.24%), and α-humulene (6.51%)	Brazil	Craveiro et al. (1983)
15	–	α-Pinene (17.26%)	Egyptian plants	Badawy and Abdelgaleil (2014)

S. cumini growing in the southwestern region of Brazil produces an oil rich in sesquiterpenes. Machado et al. (2013) reported the dominance of α-caryophyllene (25.24%), β-caryophyllene (16%), α-terpineol (9.08%), and epiglobulol (5%) in leaf EO. On the other hand, Shafi et al. (2002) identified dissimilar chemical constituents in leaf EO, with the predominant presence of pinocarveol (15.1%), α-terpineol (8.9%), myrtenol (8.3%), eucarvone (6.6%), muurolol (6.4%), myrtenal (5.8%), geranylacetone (5.6%), α-cadinol (4.6%), and pinocarvone (4.4%) in Indian *S. cumini*. Ramaiah and Nigam (1969) reported isolation of 0.01% EO from leaves by steam distillation, which contained perillaldehyde (21.76%), α-pinene (15.80%), cadinene (15.10%), limonene (6.90%), geraniol (4.2%), 1,8-cineole (3.78%), α-terpineol (1.61%), and methyl salicylate (1.15%). Moreover, Kumar et al. (2004) observed the presence of sesquiterpenoids like α-terpineol (5.71%), β-caryophyllene (6.34%), α-humulene (12.30%), β-selinene (4.19%), calacorene (4.43%), α-muurolol (4.28%), α-santalol (3.55%), and cis-farnesol (5.04%) as major constituents among 42 constituents of the EO, obtained by hydrodistillation of leaves growing in the Lucknow region of north India.

Chemical Composition of Essential Oils from Fruits

The first description of the EO from fruits appeared in the early 1980s. Craveiro et al. (1983) found a high proportion of α-pinene (30.89%) and β-pinene (10.81%) in EOs of fruits. Some other monoterpenes, like myrcene and (E)-ocimene, including α-pinene and β-pinene, were recognized by nuclear magnetic resonance (NMR) as major constituents (Lustosa et al. 1999). Moreover, extraction of volatile components from the fruits by simultaneous distillation and the solvent extraction method using modified Likens and Nickerson apparatus showed the presence of 30 compounds. In another study, the volatile components trans-ocimene, cis-ocimene, β-myrcene, and α-terpineol were found to be the major compounds, including three esters, dihydrocarvyl acetate, geranyl butyrate, and terpinyl valerate, which may be responsible for the characteristic flavor of the jamun fruit (Vijayanand et al. 2001). However, the most recent study by Abdelhady (2012) showed 24 components and afforded 1.2% of EOs from fruits. The major components of the volatile oil of fruits were methyl eugenol (22.5%), limonene (14.43%), and α-terpineol (12.04%).

Chemical Composition of Essential Oils from Stem

The scientific literature contains only a few research articles on the composition of EO from the stem. α-Pinene (12.61%), β-pinene (18.56%), and bornyl acetate are determined in the largest amounts (Craveiro et al. 1983).

BIOLOGICAL ACTIVITY OF ESSENTIAL OILS

EOs and their constituents are reported to possess a wide range of interesting pharmacological activities. The antibacterial activity of EOs has been demonstrated against both Gram-positive and Gram-negative bacteria, including *Escherichia coli*, *Staphylococcus aureus*, and *Pseudomonas aeruginosa* (Lustosa et al. 1999; Abdelhady 2012). The leaf EOs showed moderate zones of inhibition for *E. coli*, *P. aeruginosa*, *Neisseria gonorrhoeae*, *Bacillus subtilis*, *S. aureus*, and *Enterococcus faecalis* (12, 13, 13, 14, 12, and 13 mm, respectively, at a concentration of 10.0 μL of the EO) (Mohamed et al. 2013). In an another study, zones of inhibition were 16, 14, 12, 14, 17, 12, and 10 mm against *B. subtilis*, *S. aureus*, *E. coli*, *P. aeruginosa*, *Sternbergia lutea*, *Agrobacterium tumefaciens*, and *Pectobacterium carotovorum*, respectively (Elansary et al. 2012). Leaf EO showed antibacterial activity against *Salmonella typhimurium* by the disk diffusion method, showing a zone of inhibition of 20 mm at 5 μL of the EO (Shafi et al. 2002). The mechanism of antimicrobial action of the EO is possibly related to the fact that EOs may disrupt the permeability of cell membranes (Ultee et al. 2002). Antimycobacterial activity of the EO of leaves against *Mycobacterium bovis* was detected *in vitro* (Machado et al. 2013).

EOs as liposome in the zymosan, tested in lipopolysaccharide (LPS)-induced inflammatory models was found effective in inhibiting total leukocyte (up to 56% at 200 mg/kg treatment as liposome) and eosinophil migration in LPS-induced pleurisy (up to 74% at 100 mg/kg) (Ramos et al. 2006). Furthermore, Machado et al. (2013) found that the treatment of 300 μL of EO per kilogram of body weight every day with gavage applications of 0.1 mL of sterile phosphate-buffered saline (PBS) solution (concentration of 66 mL of oil per 1 L of solution) in BALB/c mice, intravenously infected with *Mycobacterium bovis* Bacillus Calmette–Guérin (BCG), showed reduction in the area of granulomas formed in the hepatic parenchyma. This suggests the oil treatment decreases the development of granulomatous inflammation caused by BCG (Machado et al. 2013).

The EO of leaf showed moderate molluscicidal activity with a median lethal concentration observed after 48 h of 90 mg/L against *Biomphalaria glabrata* snails. It has leishmanicidal rather than leishmanistatic activity against *Leishmania amazonensis* promastigotes with a minimal inhibitory concentration (IC_{50}) of 36 mg/L observed at 24 h of exposure (Dias et al. 2013). The oils also demonstrated a cytoprotective effect against mercury chloride (Sobral-Souza et al. 2014).

CHEMICAL CONSTITUENTS

The widespread therapeutic use of *S. cumini* in the traditional system of medicine evinced interest in phytochemical investigation of the plant to identify the constituents responsible for the observed activity in corresponding plant parts. Several scientific reports have been published to push the traditional observation-based crude drug into an evidence-based modern drug. In this context, several classes of phytoconstituents have been reported from different parts of the plant. Phytochemicals can be classified in a number of ways, on the basis of their chemical structure, botanical origin, biosynthesis, or biological properties. The characteristic chemical skeletons are considered for classifying them throughout the text. The main classes of compounds that have been reported from different parts of *S. cumini* include oxygenated and nonoxygenated monoterpenoids and sesquiterpenoids, triterpenoids, steroids, carotenoids, phenolic acids, phenyl-propanoids, anthocyanins, tannins, flavonols, flavonol glycoside, lignans, fatty acids, and alkaloids. Additionally, vitamins and tanning substances have also been reported. Chemical constituents of *S. cumini* are outlined in Figures 6.1 through 6.16 and classed in main groups, such as monoterpenoids, sesquiterpenoids, triterpenoids and steroids, tetraterpenoids, lipids and hydrocarbons, polyphenols, and miscellaneous compounds.

MONOTERPENOIDS

Most of the results of the identification process described the monoterpenoids as dominating constituents of EOs. The monoterpenoids may be subdivided into acyclic, monocyclic, and bicyclic. Monoterpenoids that have been reported from the plant can be further subdivided into oxygenated and nonoxygenated. Reported monoterpenoids are shown in Tables 6.2 through 6.4 with their names, the corresponding plant sources, and references.

Acyclic Monoterpenoids

Acyclic monoterpenoids are derivatives of 2,6-dimethyloctane, a head-to-tail combination of two isoprenes. β-Ocimene (**1**) and its geometrical isomer (E)-β-ocimene (**2**), (Z)-β-ocimene (**3**), β-myrcene (**4**), and alloocemene (**5**) are the most common nonoxygenated acyclic monoterpenoids that are formed by elimination of the pyrophosphate group from geranyl pyrophosphate. Geranyl acetate (**6**), geranyl butyrate (**7**), and geranyl isobutyrate (**8**) are esters of geraniol reported in EOs. Additional rearrangements and oxidation of geranyl pyrophosphate provide compounds such as citronellal (**9**), citronellyl acetate (**10**), linalool (**11**), geraniol (**12**), linalyl acetate (**13**), and linalyl propionate (**14**), along with neryl acetate (**15**). A complete list of acyclic monoterpenoids spotted in EOs from different morphological parts is given in Table 6.2 and Figure 6.1.

Monocyclic Monoterpenoids

In addition to linear attachments, the core skeleton of 10 carbons can form rings. The most common ring size in monoterpenoids is a six-membered ring. α-Limonene (**16**),

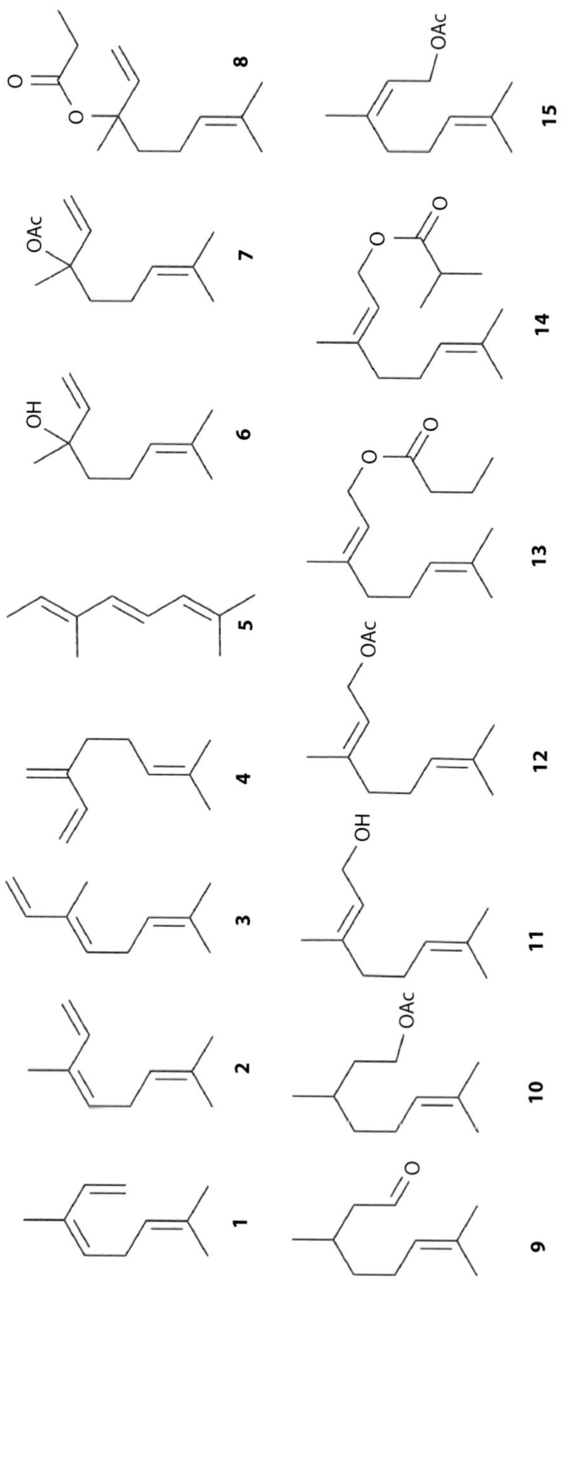

FIGURE 6.1 Chemical structure of acyclic monoterpenoids.

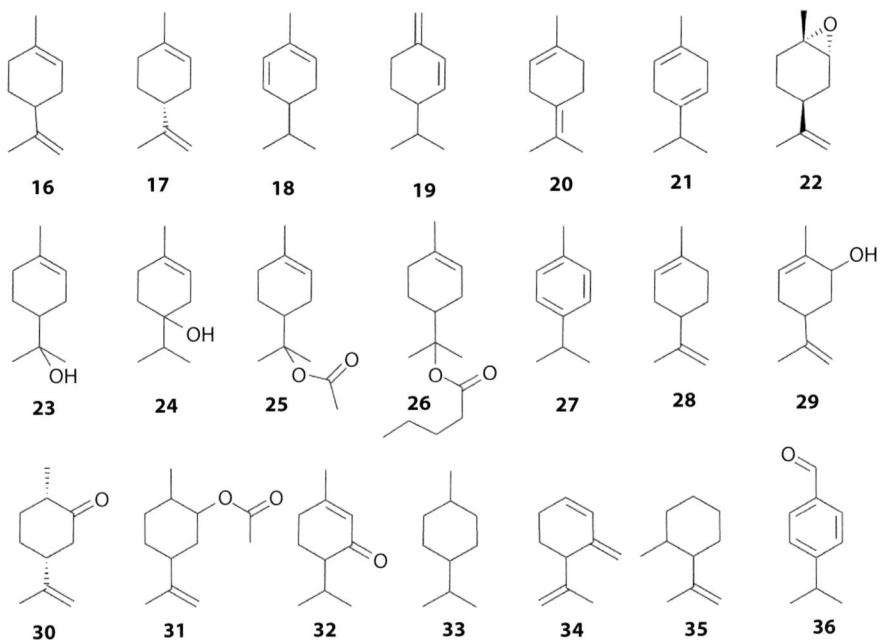

FIGURE 6.2   Chemical structure of monocyclic monoterpenoids.

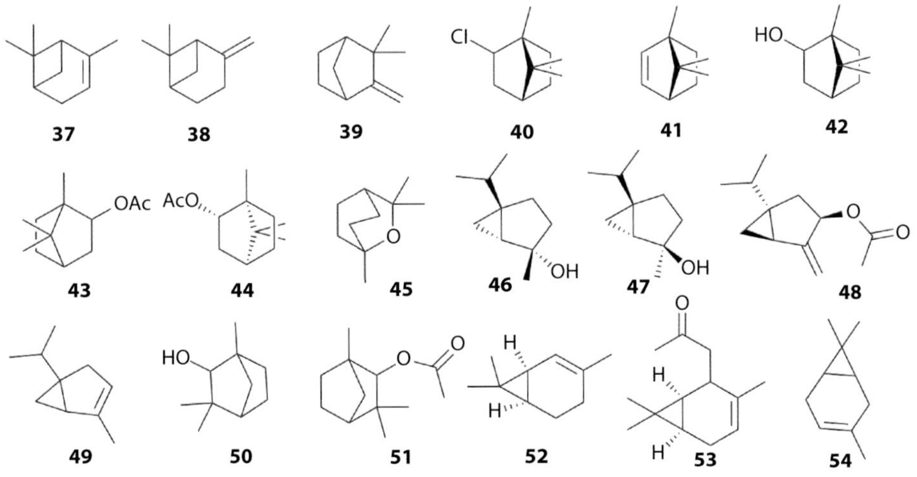

FIGURE 6.3   Chemical structure of bicyclic monoterpenoids.

L-limonene (**17**), α-phellandrene (**18**), β-phellandrene (**19**), α-terpinolene (**20**), and γ-terpinene (**21**) are among the most typical monocyclic monoterpenes, along with oxygenated monoterpenoids cis-limonene oxide (**22**), α-terpineol (**23**), 4-terpineol (**24**), terpineol acetate (**25**), and terpinyl valerate (**26**), which are formed biosynthetically by cyclization of geranyl pyrophosphate. Compounds **17** and **23** are

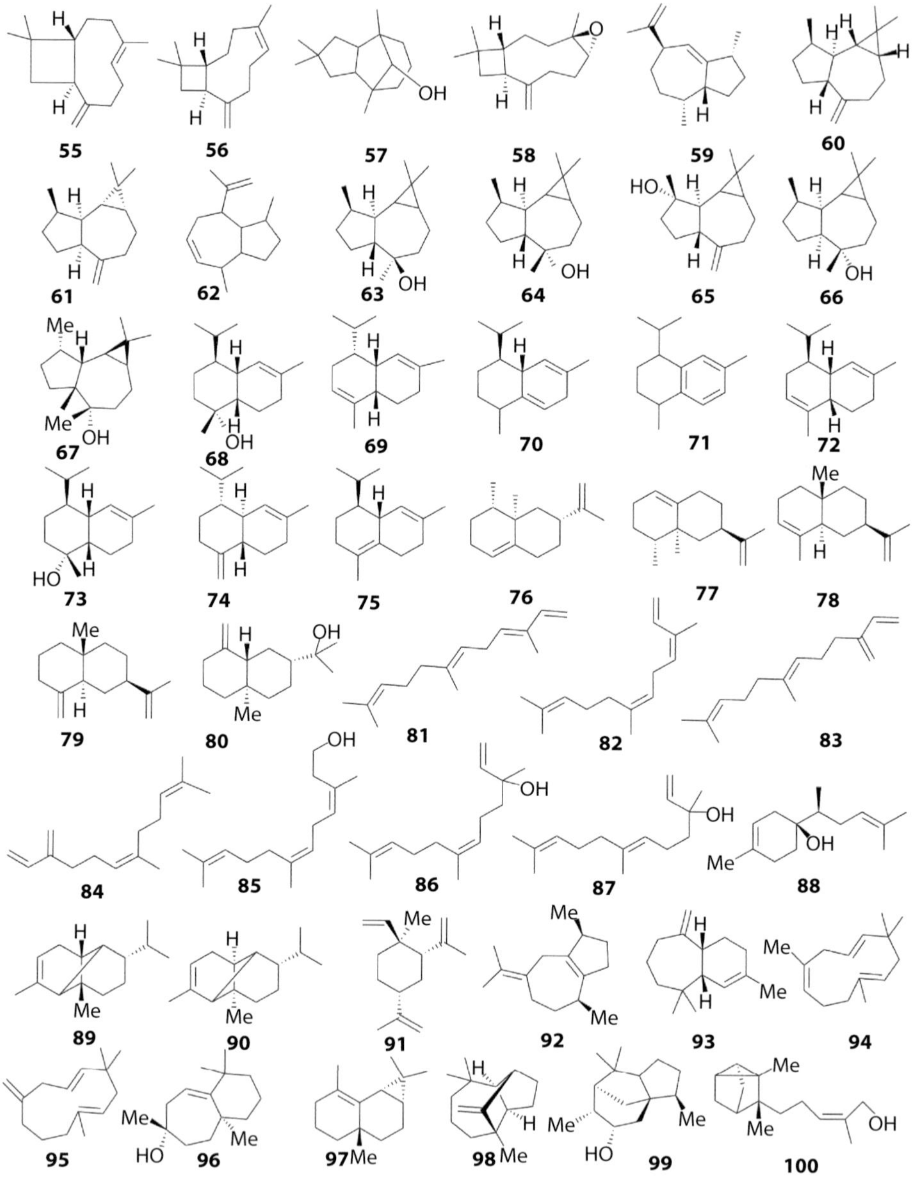

FIGURE 6.4 Structure of different groups of sesquiterpenoidal compounds.

among the major components in leaf EOs. Furthermore, hydroxylation of any of these compounds followed by dehydration can lead to the aromatic *p*-cymene (**27**). Other important monocyclic terpenoids are cinene or *p*-cymol (**28**), carveol (**29**), cis-dihydrocarvone (**30**), dihydrocarvyl acetate (**31**), piperitone (**32**), *p*-menthane (**33**), 1(7),5,8-*O*-menthatriene (**34**), *O*-menth-8-ene (**35**), and cuminaldehyde (**36**).

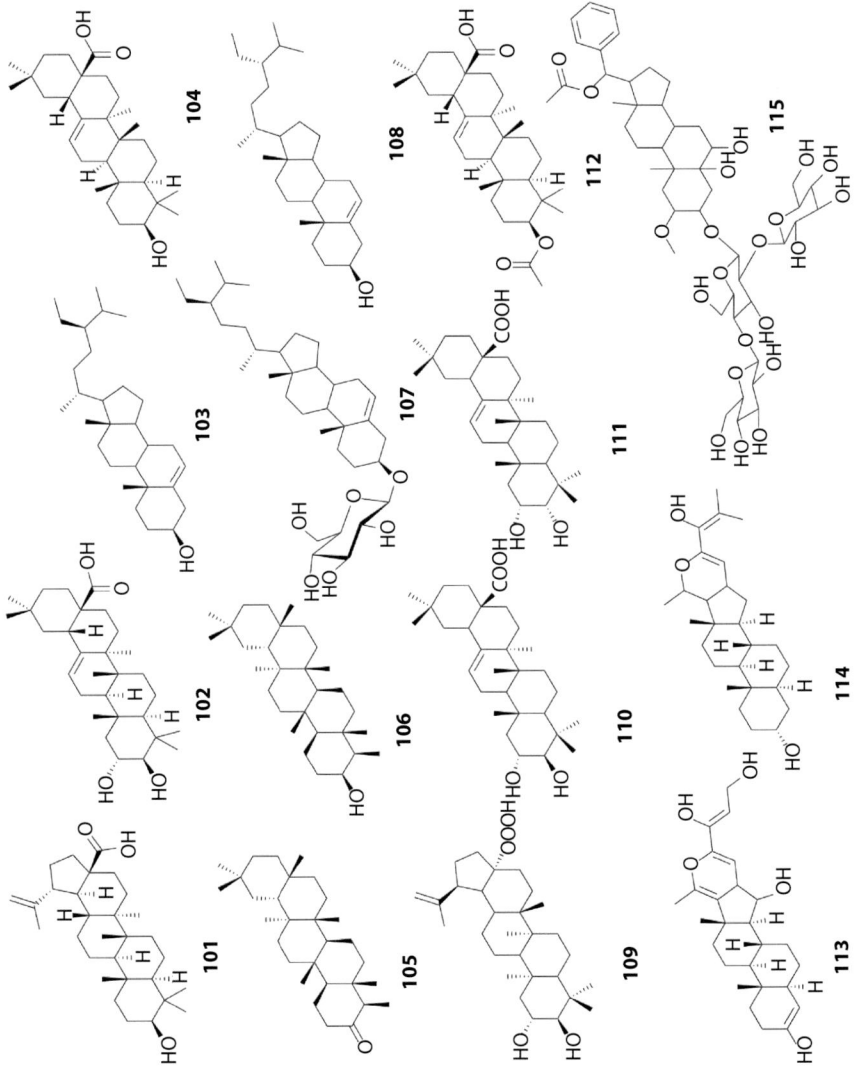

FIGURE 6.5 Structure of triterpenoids and steroids.

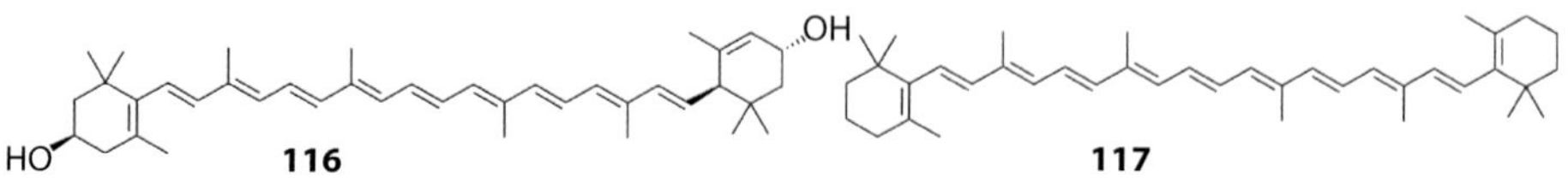

FIGURE 6.6 Tetraterpenoids detected in fruits.

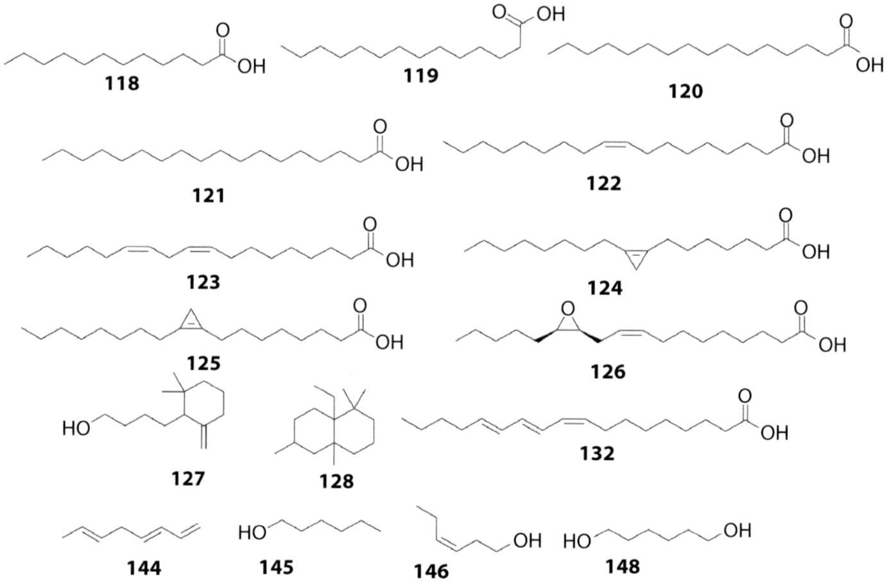

FIGURE 6.7 Structure of some lipids and hydrocarbon compounds.

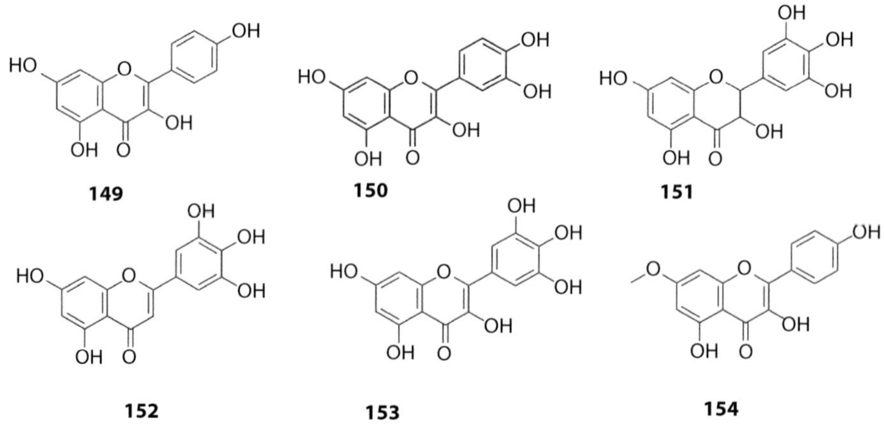

FIGURE 6.8 Flavones and flavonols.

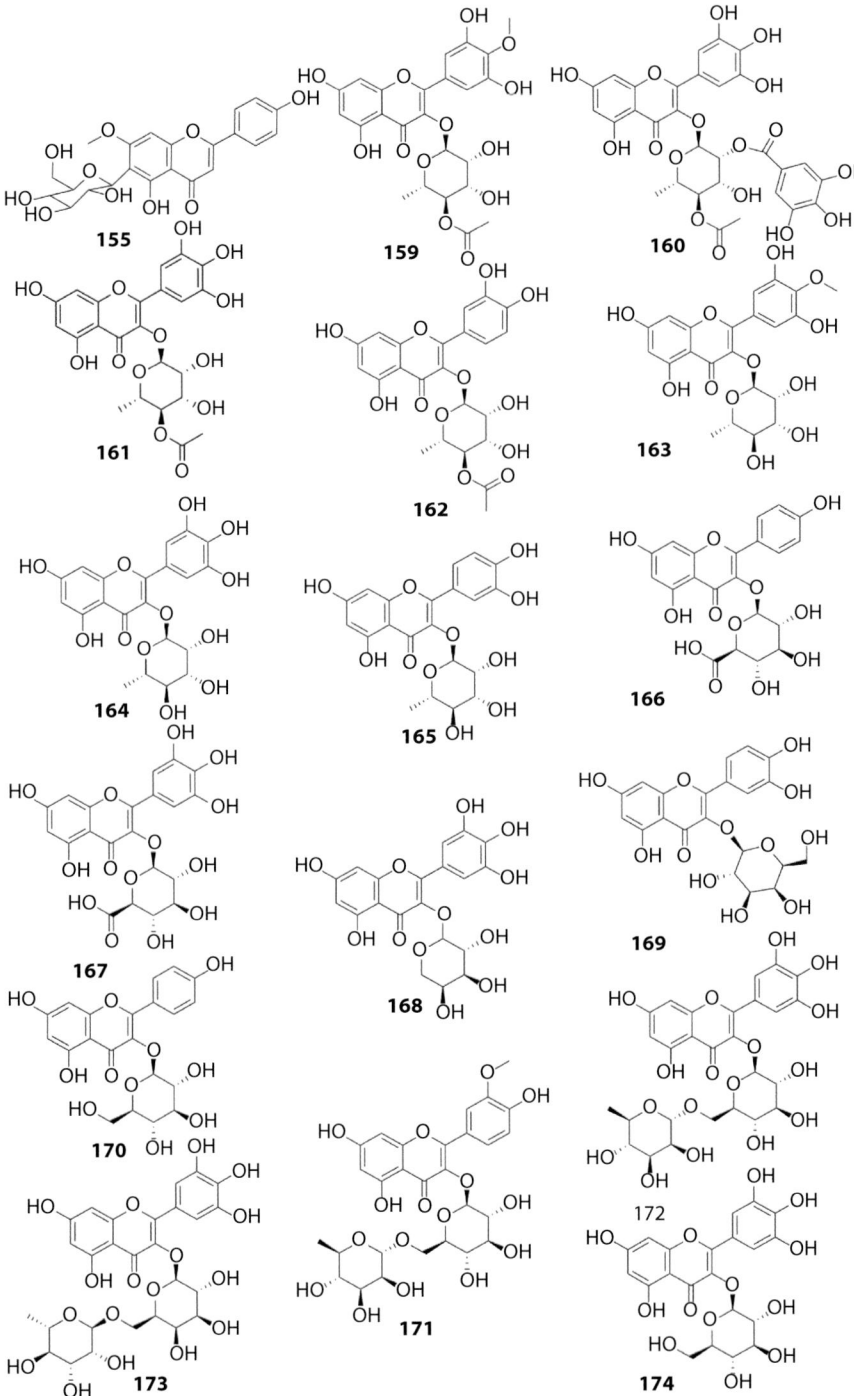

FIGURE 6.9   Structures of flavonoid glycosides.

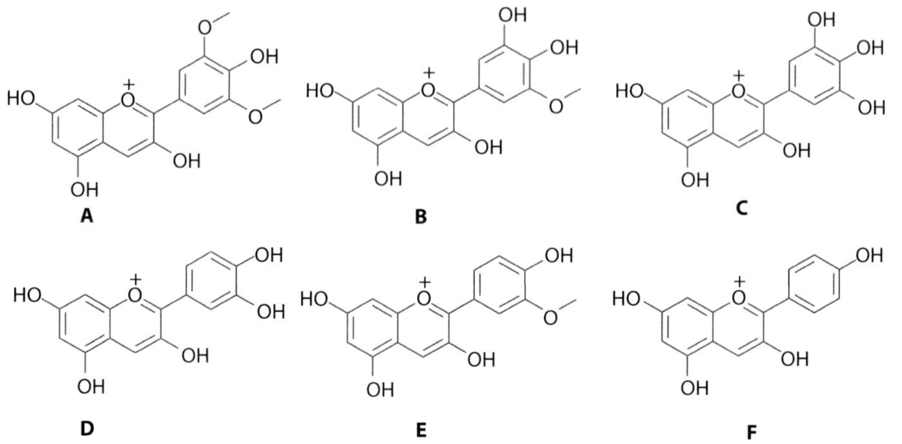

FIGURE 6.10   Anthocyanidins.

Bicyclic Monoterpenoids

Bicyclic monoterpenoids are formed by two sequential cyclization reactions of geranyl pyrophosphate. α-Pinene (**37**), β-pinene (**38**), camphene (**39**), 2-chlorocamphane (**40**), bornylene (**41**), borneol (**42**), α-bornyl acetate (**43**), isobornyl acetate (**44**), 1,8-cineole (**45**), cis-sabinene hydrate (**46**), trans-sabinene hydrate (**47**), sabinyl acetate (**48**), α-thujene (**49**), fenchol (**50**), fenchyl acetate (**51**), fenchyl alcohol, β-fenchyl alcohol, (+)-2-carene (**52**), 2-acetylmethyl-(+)-3-carene (**53**), and δ-3-carene (**54**) are examples of bicyclic monoterpenoids containing ketone, alcohol, ether, and other functional groups.

SESQUITERPENOIDS

Sesquiterpenes are a broad group having a core skeleton of 15 carbons derived from the coupling of three isoprene subunits. Most sesquiterpenes are formed from farnesyl diphosphate (FPP), a C_{15} diphosphorylated intermediate of the mevalonate biosynthetic pathway. A number of natural sesquiterpenoids from *S. cumini* have been reported (Table 6.5). According to their structures, this group is further classified into 13 subgroups: caryophyllanes, aromadendranes, cadinanes, eudesmanes, farnesanes, bisabolanes, copaanes, elemanes, guaianes, himachalanes, humulanes, widdrane, and others.

The caryophyllane type of sesquiterpenoids, such as β-caryophyllene (**55**), isocaryophyllene (**56**), caryophyllene alcohol (**57**), and β-caryophyllene epoxide (**58**), has been detected in EOs. Additionally, nine aromadendranes, γ-gurjunene (**59**), (+)-aromadendrene (**60**), alloaromadendrene (**61**), isoaromadendrene V (**62**), globulol (**63**), epiglobulol (**64**), spathulenol (**65**), viridiflorol (**66**), and ledol (**67**), have been detected in EOs of various morphological parts, as shown in Table 6.5. Other sesquiterpenes of the cadinane group, namely, torreyol (**68**), α-amorphene (**69**), cadina-1,4-diene (**70**), calacorene (**71**), α-muurolene (**72**), α-muurolol (**73**),

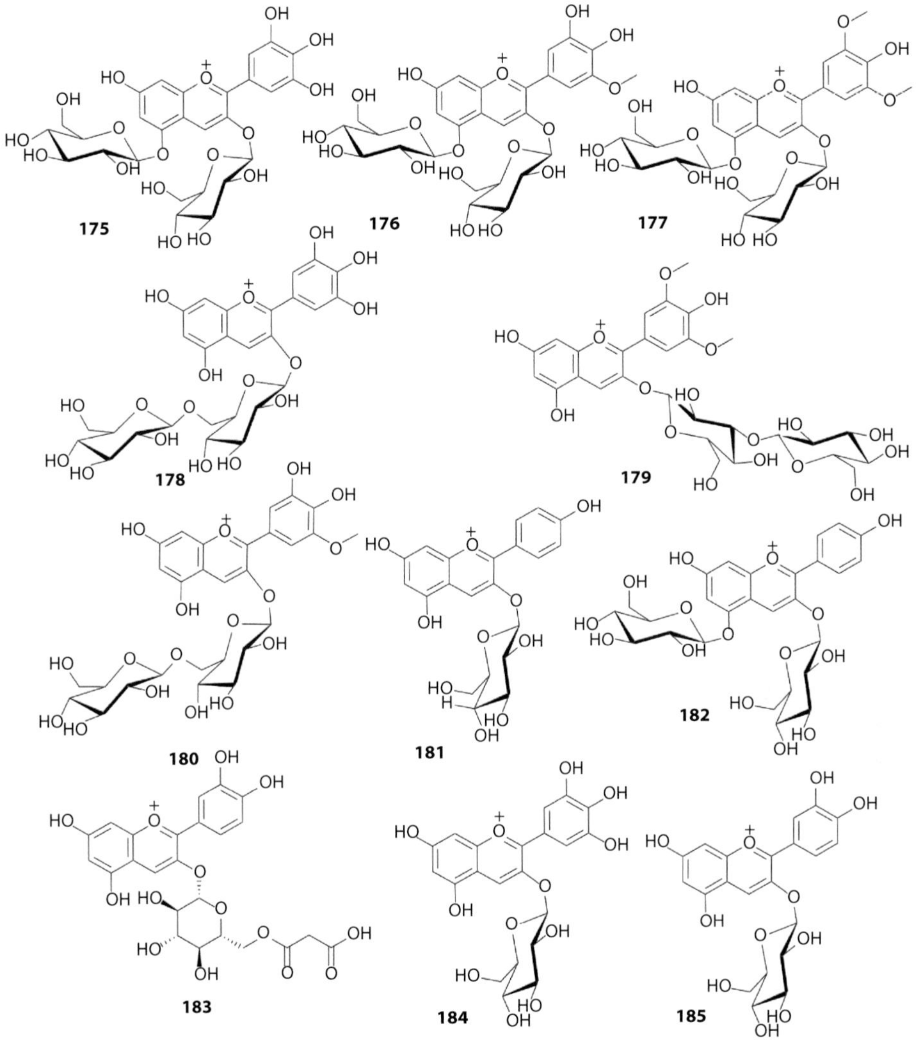

FIGURE 6.11 Structure of anthocyanins from fruits.

γ-cadinene (**74**), and δ-cadinene (**75**), have been reported. Eudesmanes such as eremophilene (**76**), valencene (**77**), α-selinene (**78**), β-selinene (**79**), and β-eudesmol (**80**) have been reported. In addition, seven more sesquiterpenes of the farnesene group, α-farnesene (**81**), cis-α-farnesene (**82**), β-farnesene (**83**), cis-β-farnesene (**84**), cis-farnesol (**85**), cis-nerolidol (**86**), and trans-nerolidol (**87**), have been detected. A bisabolane, β-bisabolol (**88**), has been detected from fruits. Two copaane sesquiterpenoids, namely, α-ylangene (**89**) and α-copaene (**90**), have been found in EOs of leaves and aerial parts, respectively. The leaf EOs have shown the presence of β-elemene (**91**), β-guaiene (**92**), and α-himachalene (**93**), which belong to the elemene, guaiene, and himachalane types of sesquiterpenoids, respectively. Widdrol (**96**) is a widdrane

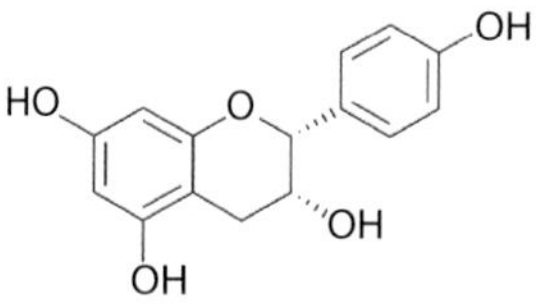

187 A 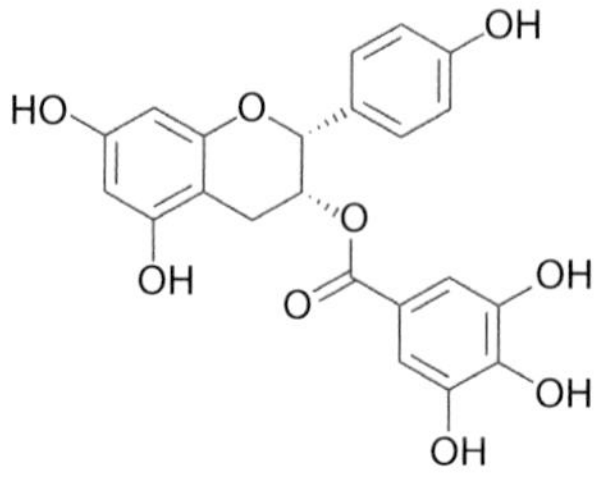   **187 B** 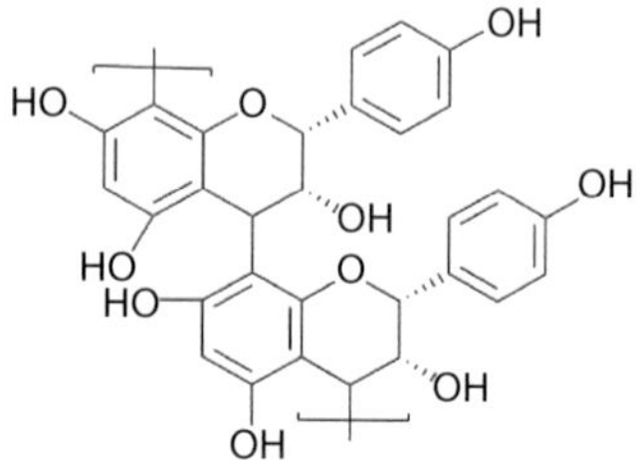   **187**

FIGURE 6.12   Propelargonidin and its constitutive monomer.

type of sesquiterpenoids found in fruits. Other sesquiterpenoids, like β-maaliene (**97**), junipene (**98**), neocedranol (**99**), and α-santalol (**100**), have been reported from different parts of the plant, as shown in Table 6.5.

TRITERPENOIDS AND STEROIDS

Triterpenoids, steroids, and consequently sterols represent a large diverse group of natural products and structurally represent cyclization products of squalene (Xu et al. 2004). The steroids are derived from tetracyclic triterpenes and possess a cyclo-pentaperhydrophenanthrene backbone. A core of steroids is composed of four fused rings of 17 carbon atoms bonded together to form three cyclohexane rings (rings A, B, and C) and one cyclopentane ring (D ring). Sterols are hydroxylated forms of steroids, containing a hydroxyl group at position 3.

Ursane, oleanane, and friedlane types of triterpenoids, along with steroids, are reported from many plant parts of jamun. Its leaves yielded pentacyclic triterpenoids, betulinic acid (**101**), and crotegolic (maslinic) acid (**102**) (Gupta and Sharma 1974). Phytochemical investigation of seeds yielded β-sitosterol (**103**), along with oleanolic acid (**104**), from the pet-ether and carbon tetrachloride soluble fractions of a methanol extract, and structures were elucidated through spectroscopic studies (Al et al. 2012). Further, from the pet-ether extract of the stem bark, two friedlane type of triterpenoids, friedelin (**105**) and epifriedelanol (**106**), along with β-sitosterol-D-glucoside (**107**), **101**, and **103**, were isolated by column chromatography (Sengupta and Das 1965; Bhargava et al. 1974). S. Li et al. (2009) isolated 24(S)-stigmast-5-en-3-β-ol (**108**), 2α-hydroxybetulinic acid (**109**), arjunolic acid (**110**), 2α,3α,24-trihydroxy-olean-12-en-28-oic acid (**111**), **105**, and **106** from roots. The glucose consumption in L6 muscle cells was increased 17.35% by compound **105** at a concentration of 10 µg/mL without insulin. Compound **110** showed multifunctional therapeutic applications like wound healing, antimutagenic, and antimicrobial activity (Hemalatha et al. 2010). Two oleanane-type triterpenoids, acetyloleanolic acid (**112**) (Nair and Subramanian 1962) and oleanolic acid (**104**), have been isolated from flowers. Administration of **104** to male rats has decreased spermatogenesis without causing any abnormality to spermatogenic cells, Leydig interstitial cells, or Sertoli cells (Rajasekaran et al. 1988).

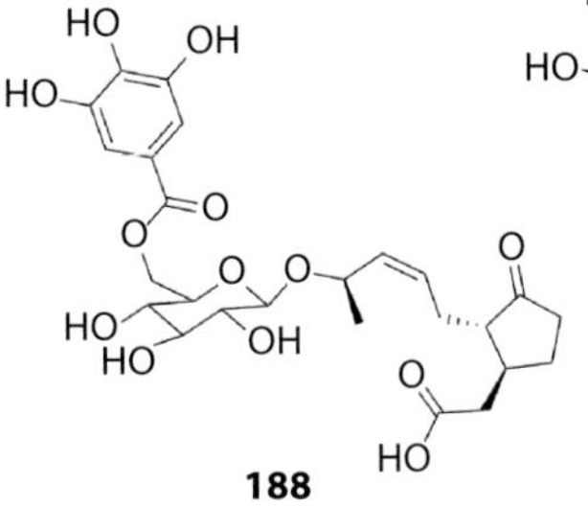

FIGURE 6.13 Structure of hydrolyzable tannins.

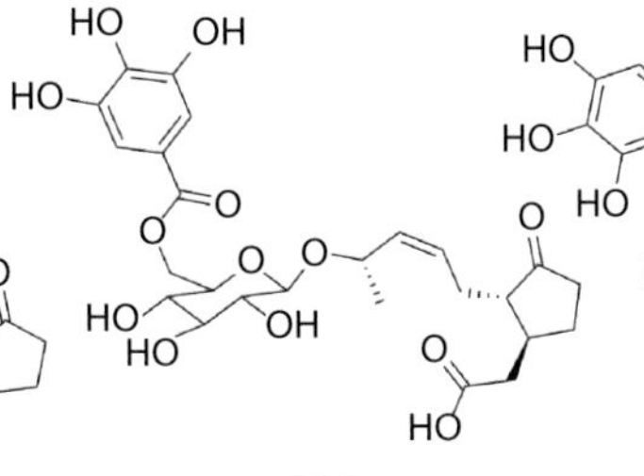

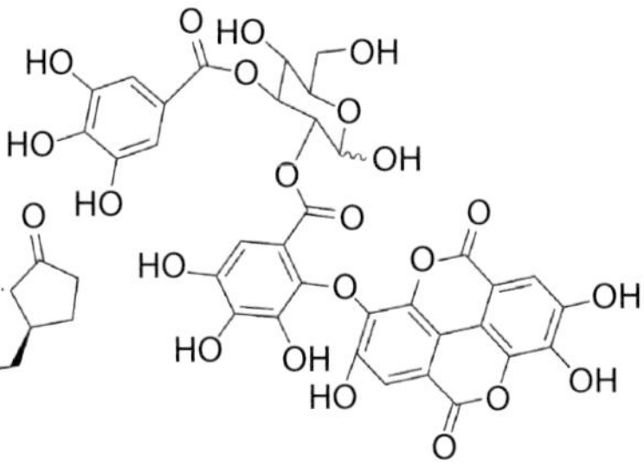

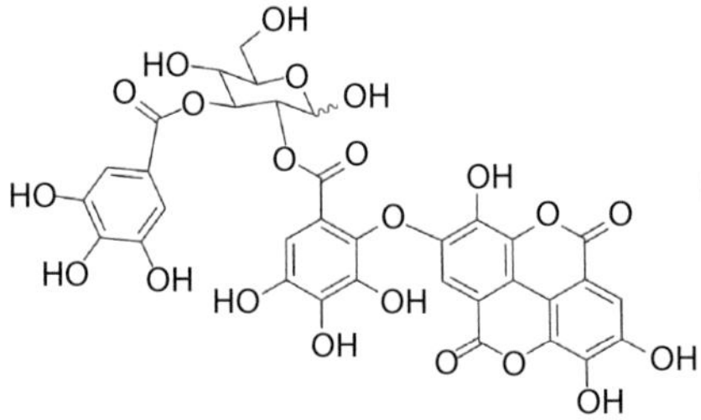

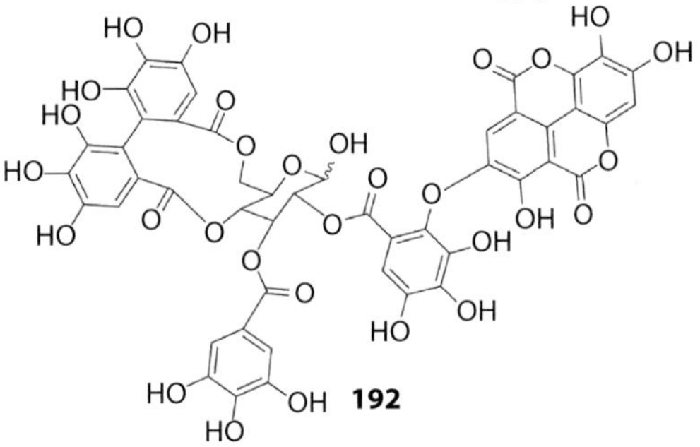

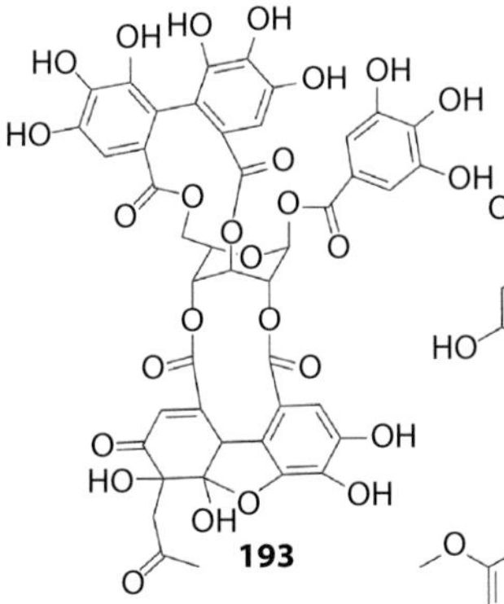

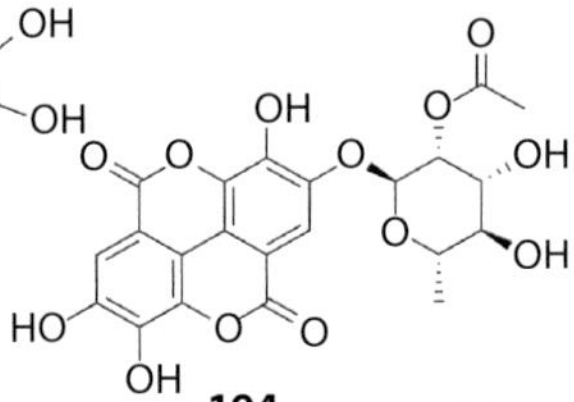

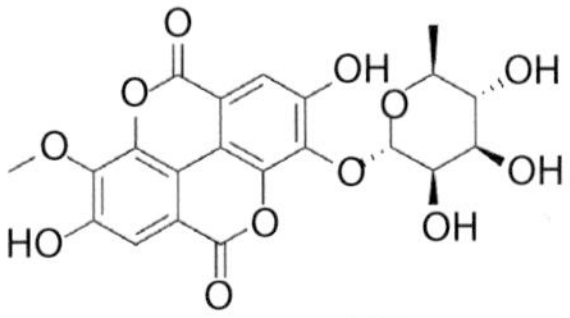

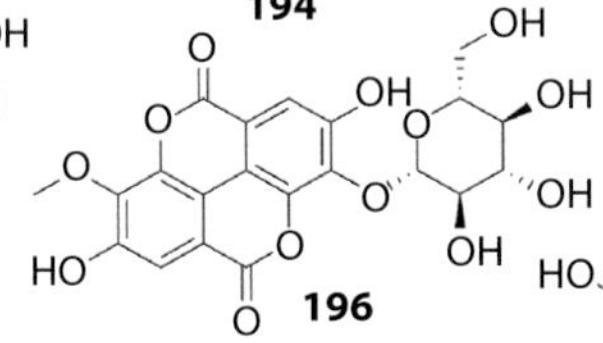

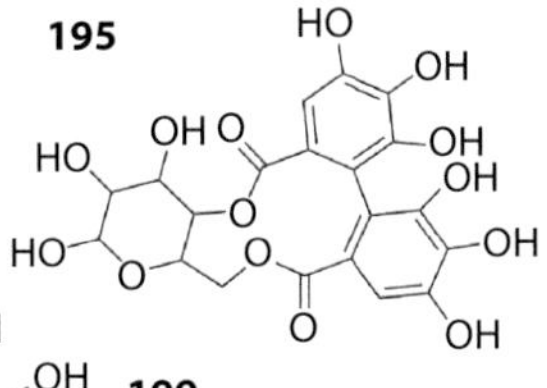

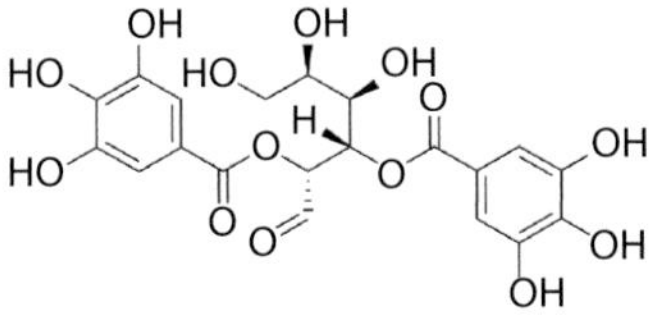

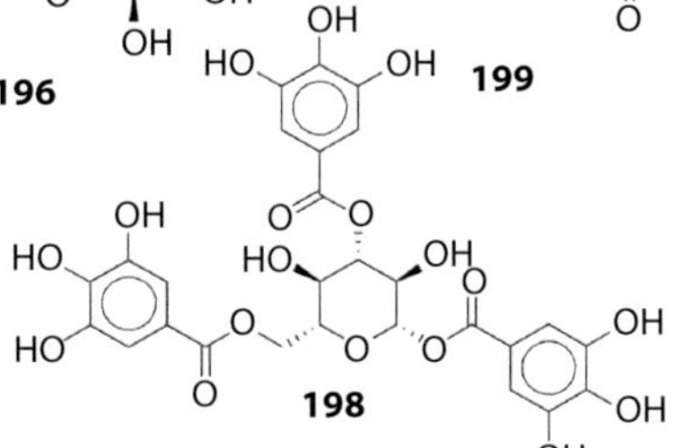

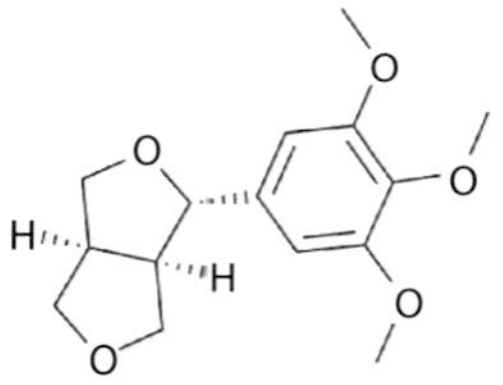

FIGURE 6.14 Chemical structure of lignans isolated from stem bark.

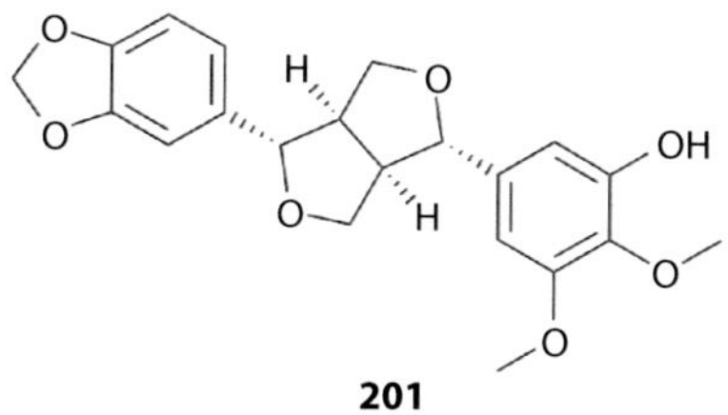

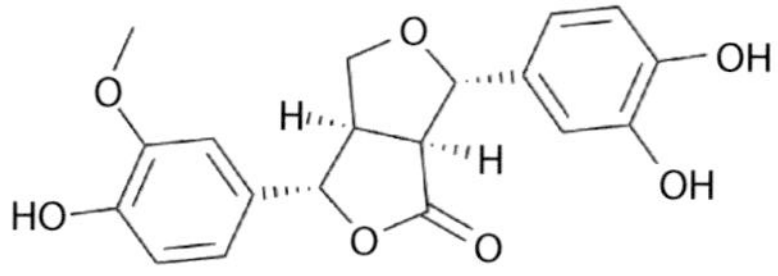

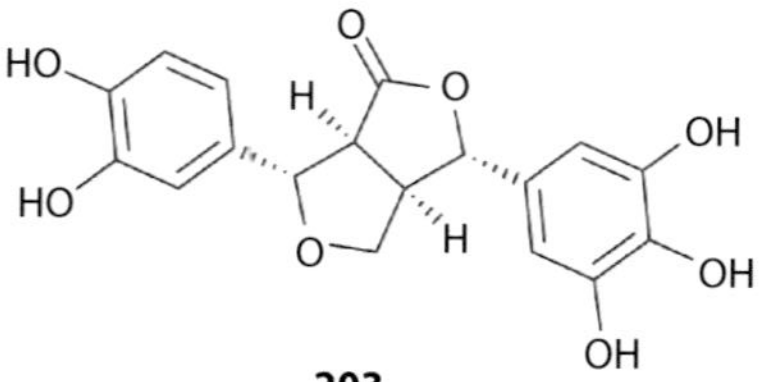

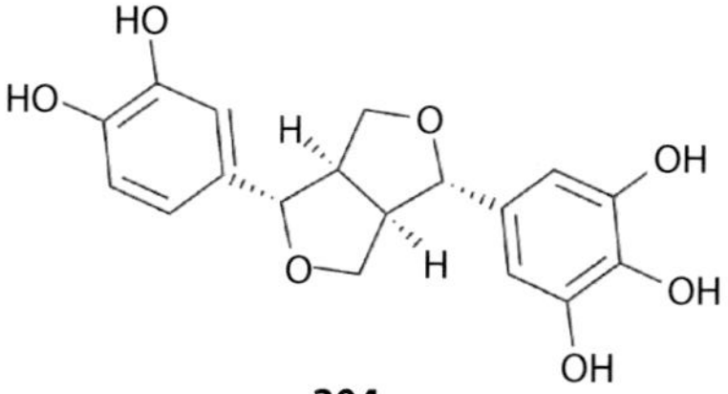

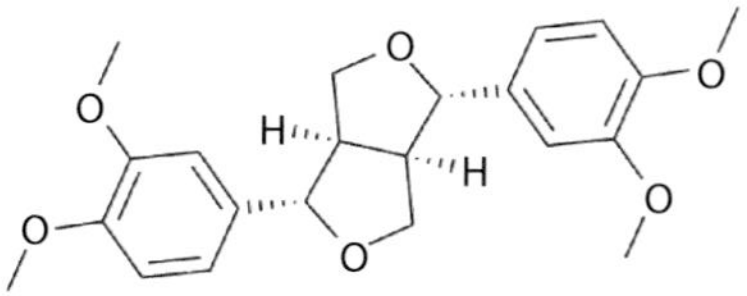

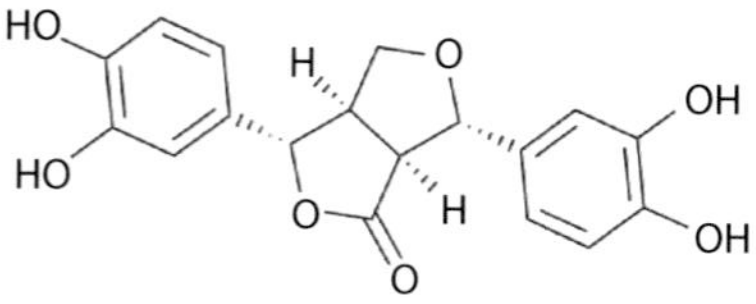

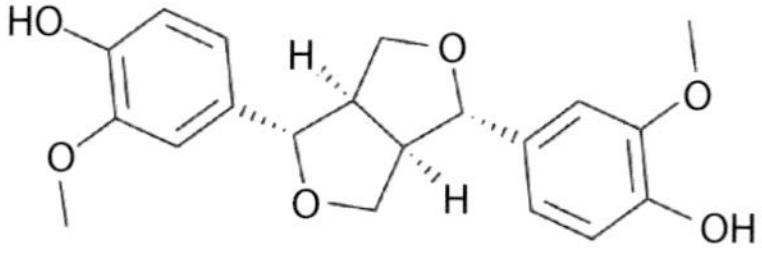

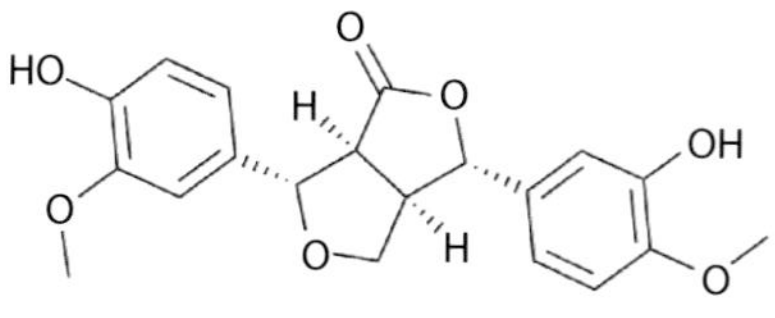

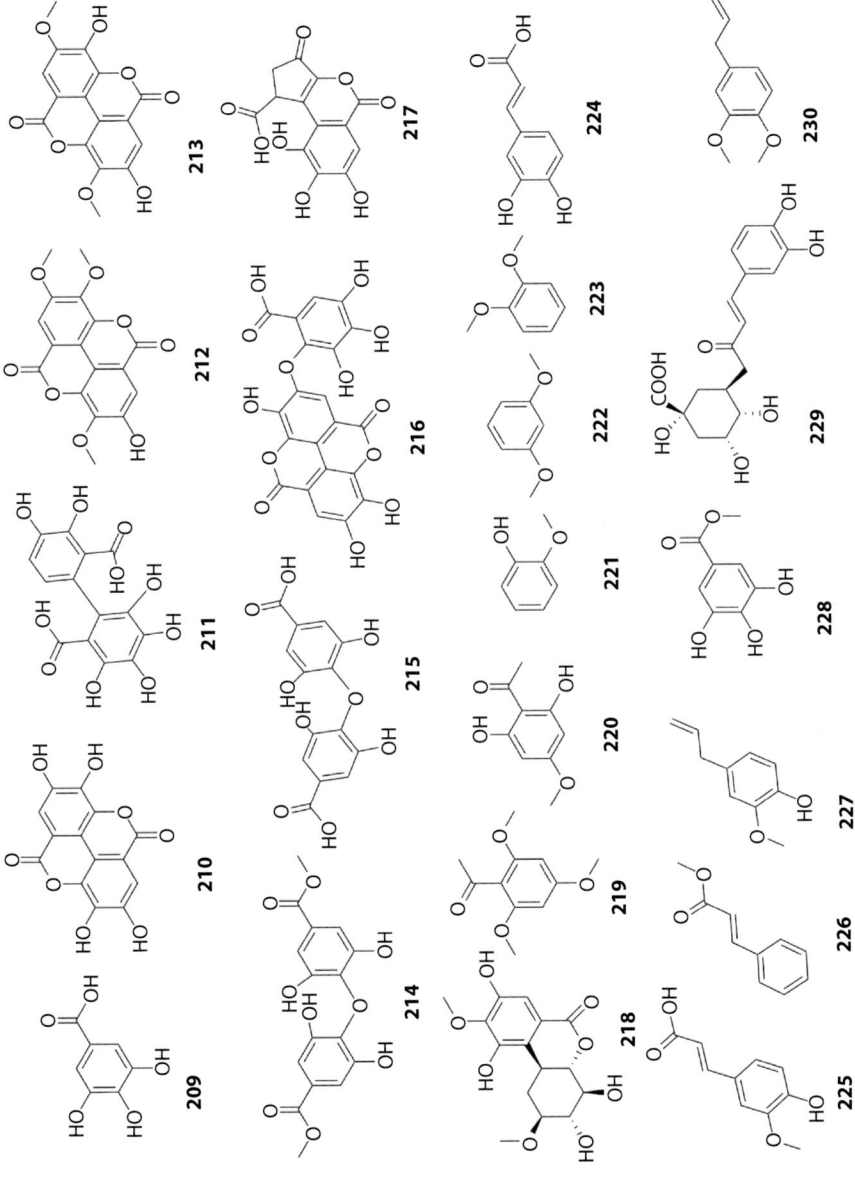

FIGURE 6.15 Chemical structure of other phenolic compounds.

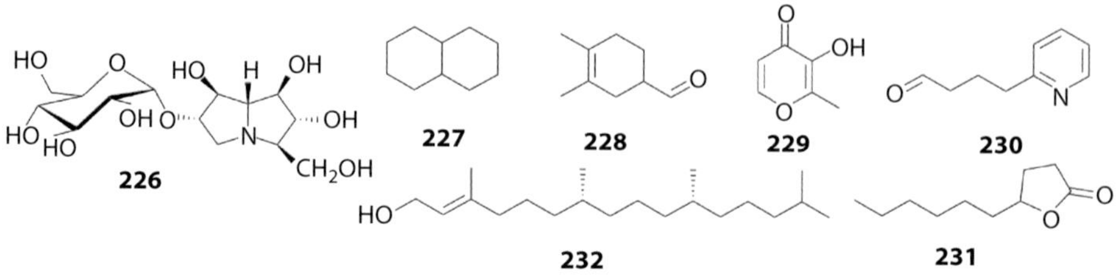

FIGURE 6.16   Chemical structure of miscellaneous compounds.

TABLE 6.2
Acyclic Monoterpenoids and Their Corresponding Source

Compound No.	Compound Name	Plant Part	References
1	(E)-β-Ocimene	Leaves	Kumar et al. (2004), Dias et al. (2013), Machado et al. (2013), Mohamed et al. (2013), Sobral-Souza et al. (2014)
		Fruits	Craveiro et al. (1983), Vijayanand et al. (2001)
		Stem	Craveiro et al. (1983)
		Leaves	Craveiro et al. (1983)
2	(Z)-β-Ocimene	Leaves	Craveiro et al. (1983), Elansary et al. (2012), Dias et al. (2013)
		Fruits	Craveiro et al. (1983), Vijayanand et al. (2001)
		Stem	Craveiro et al. (1983)
3	β-Ocimene	Leaves	Mohamed et al. (2013), Sobral-Souza et al. (2014)
4	β-Myrcene	Leaves	Craveiro et al. (1983), Khanna (1991), Kumar et al. (2004), Dias et al. (2013), Mohamed et al. (2013), Sobral-Souza et al. (2014)
		Fruits	Craveiro et al. (1983), Vijayanand et al. (2001)
		Stem	Craveiro et al. (1983)
5	Alloocimene	Leaves	Elansary et al. (2012), Mohamed et al. (2013)
6	Linalool	Fruits	Vijayanand et al. (2001), Abdelhady (2012)
		Leaves	Khanna (1991), Kumar et al. (2004)
7	Linalyl acetate	Leaves	Khanna (1991)
8	Linalyl propionate	Fruits	Abdelhady (2012)
9	Citronellal	Fruits	Abdelhady (2012)
10	Citronellyl acetate	Fruits	Abdelhady (2012)
11	Geraniol	Fruits	Abdelhady (2012)
12	Geranyl acetate	Fruits	Abdelhady (2012)
13	Geranyl butyrate	Fruits	Vijayanand et al. (2001)
14	Geranyl isobutyrate	Fruits	Abdelhady (2012)

TABLE 6.3
Monocyclic Monoterpenoids from Essential Oils

Compound No.	Compound Name	Plant Part	References
15	Neryl acetate	Fruits	Abdelhady (2012)
16	α-Limonene	Fruits	Craveiro et al. (1983), Abdelhady (2012)
		Stem	Craveiro et al. (1983)
		Leaves	Craveiro et al. (1983), Khanna (1991), Kumar et al. (2004), Dias et al. (2013), Mohamed et al. (2013)
17	L-Limonene	Leaves	Elansary et al. (2012), Sobral-Souza et al. (2014)
18	α-Phellandrene	Leaves	Mohamed et al. (2013)
		Fruits	Abdelhady (2012)
19	β-Phellandrene	Leaves	Khanna (1991)
20	Terpinolene	Fruits	Craveiro et al. (1983), Vijayanand et al. (2001), Abdelhady (2012)
		Leaves	Craveiro et al. (1983), Khanna (1991), Kumar et al. (2004), Elansary et al. (2012), Mohamed et al. (2013), Sobral-Souza et al. (2014)
		Stem	Craveiro et al. (1983)
21	γ-Terpinene	Leaves	Craveiro et al. (1983), Khanna (1991), Kumar et al. (2004)
		Fruits	Craveiro et al. (1983)
		Stem	Craveiro et al. (1983)
22	cis-Limonene oxide	Leaves	Kumar et al. (2004)
23	α-Terpineol	Leaves	Khanna (1991), Kumar et al. (2004), Elansary et al. (2012), Dias et al. (2013), Machado et al. (2013), Mohamed et al. (2013)
		Fruits	Vijayanand et al. (2001), Abdelhady (2012)
24	4-Terpineol	Leaves	Mohamed et al. (2013)
		Fruits	Vijayanand et al. (2001), Abdelhady (2012)
25	Terpineol acetate	Leaves	Khanna (1991), Kumar et al. (2004)
26	Terpinyl valerate	Fruits	Vijayanand et al. (2001)
27	p-Cymene	Leaves	Khanna (1991), Kumar et al. (2004), Elansary et al. (2012), Sobral-Souza et al. (2014)
28	Cinene or p-cymol	Leaves	Mohamed et al. (2013)
29	Carveol	Leaves	Machado et al. (2013)
30	cis-Dihydrocarvone	Fruits	Vijayanand et al. (2001)
31	Dihydrocarvyl acetate	Fruits	Vijayanand et al. (2001)
32	Piperitone	Leaves	Khanna (1991)
33	p-Menthane	Leaves	Khanna (1991), Kumar et al. (2004)
34	1(7),5,8-o-Menthatriene	Leaves	Elansary et al. (2012)
35	O-Menth-8-ene	Leaves	Elansary et al. (2012)
36	Cuminaldehyde	Leaves	Khanna (1991)

TABLE 6.4

Bicyclic Monoterpenoids from Essential Oils of Various Plant Parts

Compound No.	Compound Name	Plant Part	References
37	α-Pinene	Fruits	Craveiro et al. (1983), Vijayanand et al. (2001), Abdelhady (2012)
		Leaves	Craveiro et al. (1983), Khanna (1991), Kumar et al. (2004), Elansary et al. (2012), Dias et al. (2013), Mohamed et al. (2013), Sobral-Souza et al. (2014)
		Stem	Craveiro et al. (1983)
38	β-Pinene	Leaves	Craveiro et al. (1983), Khanna (1991), Elansary et al. (2012), Dias et al. (2013), Mohamed et al. (2013), Sobral-Souza et al. (2014)
		Fruits	Craveiro et al. (1983), Vijayanand et al. (2001)
		Stem	Craveiro et al. (1983)
39	Camphene	Stem	Craveiro et al. (1983)
		Fruits	Craveiro et al. (1983)
		Leaves	Craveiro et al. (1983), Dias et al. (2013), Mohamed et al. (2013), Sobral-Souza et al. (2014)
40	2-Chlorocamphane	Leaves	Mohamed et al. (2013)
41	Bornylene	Leaves	Khanna (1991)
42	Borneol	Leaves	Khanna (1991), Kumar et al. (2004), Mohamed et al. (2013), Elansary et al. (2012)
43	Bornyl acetate	Leaves	Craveiro et al. (1983), Kumar et al. (2004), Elansary et al. (2012), Machado et al. (2013), Mohamed et al. (2013), Sobral-Souza et al. (2014)
		Fruits	Craveiro et al. (1983)
		Stem	Craveiro et al. (1983)
44	Isobornyl acetate	Leaves	Dias et al. (2013)
45	1,8-Cineole	Fruits	Abdelhady (2012)
		Leaves	Khanna (1991)
46	cis-Sabinene hydrate	Leaves	Kumar et al. (2004)
47	trans-Sabinene hydrate	Leaves	Kumar et al. (2004)
48	Sabinyl acetate	Leaves	Kumar et al. (2004)
49	α-Thujene	Leaves	Khanna (1991)
50	Fenchol	Leaves	Kumar et al. (2004), Mohamed et al. (2013), Sobral-Souza et al. (2014)
51	Fenchyl acetate	Leaves	Kumar et al. (2004), Elansary et al. (2012)
52	(+)-2-Carene	Leaves	Mohamed et al. (2013)
53	2-Acetylmethyl-(+)-3-carene	Leaves	Mohamed et al. (2013)
54	Δ^3-Carene	Leaves	Mohamed et al. (2013)

TABLE 6.5
Sesquiterpenoidal Compounds Detected from Essential Oils of Different Morphological Parts

Compound No.	Compound Name	Plant Part	References
55	β-Caryophyllene	Fruits	Craveiro et al. (1983), Vijayanand et al. (2001)
		Stem	Craveiro et al. (1983)
		Leaves	Craveiro et al. (1983), Khanna (1991), Kumar et al. (2004), Elansary et al. (2012), Dias et al. (2013), Machado et al. (2013), Mohamed et al. (2013), Sobral-Souza et al. (2014)
56	Isocaryophyllene	Leaves	Mohamed et al. (2013)
57	Caryophyllene alcohol	Leaves	Elansary et al. (2012), Machado et al. (2013)
58	Caryophyllene oxide	Leaves	Kumar et al. (2004), Elansary et al. (2012), Machado et al. (2013), Mohamed et al. (2013)
59	γ-Gurjunene	Leaves	Kumar et al. (2004), Mohamed et al. (2013)
60	Aromadendrene	Leaves	Kumar et al. (2004), Elansary et al. (2012), Mohamed et al. (2013)
		Fruits	Abdelhady (2012)
61	Alloaromadendrene	Leaves	Kumar et al. (2004)
62	Isoaromadendrene V	Leaves	Mohamed et al. (2013)
63	Globulol	Fruits	Vijayanand et al. (2001), Abdelhady (2012)
		Leaves	Elansary et al. (2012), Mohamed et al. (2013)
64	Epiglobulol	Fruits	Abdelhady (2012)
		Leaves	Elansary et al. (2012), Machado et al. (2013)
65	Spathulenol	Fruits	Abdelhady (2012)
66	Viridiflorol	Fruits	Abdelhady (2012)
		Leaves	Mohamed et al. (2013)
67	Ledol	Fruits	Vijayanand et al. (2001)
68	Torreyol	Fruits	Vijayanand et al. (2001)
69	α-Amorphene	Leaves	Mohamed et al. (2013)
70	Cadina-1,4-diene	Leaves	Mohamed et al. (2013)
71	Calacorene	Leaves	Kumar et al. (2004)
72	α-Muurolene	Leaves	Kumar et al. (2004), Machado et al. (2013), Mohamed et al. (2013)
73	α-Muurolol	Leaves	Kumar et al. (2004)
74	γ-Cadinene	Fruits	Craveiro et al. (1983)
		Stem	Craveiro et al. (1983)
		Leaves	Craveiro et al. (1983), Mohamed et al. (2013)
75	δ-Cadinene	Fruits	Craveiro et al. (1983)
		Stem	Craveiro et al. (1983)
		Leaves	Craveiro et al. (1983), Kumar et al. (2004), Mohamed et al. (2013)
76	Eremophilene	Leaves	Mohamed et al. (2013)

(Continued)

TABLE 6.5 (CONTINUED)

Sesquiterpenoidal Compounds Detected from Essential Oils of Different Morphological Parts

Compound No.	Compound Name	Plant Part	References
77	Valencene	Leaves	Mohamed et al. (2013)
78	α-Selinene	Leaves	Mohamed et al. (2013)
79	β-Selinene	Leaves	Kumar et al. (2004), Mohamed et al. (2013)
80	β-Eudesmol	Fruits	Vijayanand et al. (2001)
81	trans-α-Farnesene	Fruits	Vijayanand et al. (2001), Abdelhady (2012)
82	cis-α-Farnesene	Fruits	Vijayanand et al. (2001)
83	β-Farnesene	Leaves	Kumar et al. (2004)
84	cis-β-Farnesene	Fruits	Vijayanand et al. (2001)
85	cis-Farnesol	Leaves	Kumar et al. (2004)
86	cis-Nerolidol	Fruits	Vijayanand et al. (2001)
87	trans-Nerolidol	Leaves	Kumar et al. (2004)
88	β-Bisabolol	Fruits	Vijayanand et al. (2001)
89	α-Ylangene	Leaves	Mohamed et al. (2013)
90	α-Copaene	Fruits	Craveiro et al. (1983)
		Stem	Craveiro et al. (1983)
		Leaves	Craveiro et al. (1983), Kumar et al. (2004), Mohamed et al. (2013)
91	β-Elemene	Leaves	Kumar et al. (2004)
92	β-Guaiene	Leaves	Mohamed et al. (2013)
93	α-Himachalene	Leaves	Kumar et al. (2004)
94	α-Caryophyllene or α-humulene	Leaves	Elansary et al. (2012), Machado et al. (2013), Craveiro et al. (1983), Kumar et al. (2004), Elansary et al. (2012), Dias et al. (2013), Mohamed et al. (2013), Sobral-Souza et al. (2014)
		Fruits	Craveiro et al. (1983), Vijayanand et al. (2001)
		Stem	Craveiro et al. (1983)
95	β-Humulene	Leaves	Machado et al. (2013)
96	Widdrol	Fruits	Vijayanand et al. (2001)
97	β-Maaliene	Leaves	Kumar et al. (2004)
98	Junipene	Leaves	Elansary et al. (2012)
99	Neocedranol	Fruits	Vijayanand et al. (2001)
100	α-Santalol	Leaves	Kumar et al. (2004)

Active marker compounds from ethanolic extract of seeds of *S. cumini* were isolated as androstane derivatives 3,15-dihydroxy-Δ^3androstene [16,17-C](6′-methyl, 2′-1,3-dihydroxy-1-propene)-4H-pyran (**113**) and 3-hydroxy-androstane [16,17-C] (6′-methyl, 2′-1-hydroxy-isopropene-1-yl)-4,5,6H-pyran (**114**). Compounds **113** and **114** were isolated by normal-phase column chromatography using toluene–ethyl acetate (80:20), and reduced elevated blood glucose levels in alloxan-induced diabetes

TABLE 6.6
Triterpenoids and Steroids from Different Plant Parts of *S. cumini*

Compound No.	Compound Name	Plant Part	References
101	Betulinic acid	Leaves	Gupta and Sharma (1974)
		Stem bark	Sengupta and Das (1965), Bhargava et al. (1974)
102	Crotegolic (maslinic) acid	Leaves	Gupta and Sharma (1974)
103	β-Sitosterol	Stem bark	Sengupta and Das (1965), Bhargava et al. (1974)
		Seeds	Al et al. (2012)
104	Oleanolic acid	Seeds	Al et al. (2012)
		Fruit pulp	Omar et al. (2012)
		Flowers	Rajasekaran et al. (1988)
105	Friedelin	Stem bark	Sengupta and Das (1965)
		Roots	S. Li et al. (2009)
106	Epifriedelanol	Roots	S. Li et al. (2009)
		Stem bark	Sengupta and Das (1965), Bhargava et al. (1974)
107	β-Sitosterol-D-glucoside	Leaves	Gupta and Sharma (1974)
108	24 (S)-Stigmast-5-en-3-β-ol	Roots	S. Li et al. (2009)
109	2α-Hydroxybetulinic acid	Roots	S. Li et al. (2009)
110	Arjunolic acid	Roots	S. Li et al. (2009)
111	2α,3α,24-Trihydroxy-olean-12-en-28-oic acid	Roots	S. Li et al. (2009)
112	Acetyloleanolic acid	Flowers	Nair and Subramanian (1962)
113	3,15-Dihydroxy Δ^3 androstene [16,17-C](6′-methyl, 2′-1,3-dihydroxy-1-propene)-4H-pyran	Seeds	Shankar et al. (2007)
114	3-Hydroxy androstane [16,17-C] (6′-methyl, 2′-1-hydroxy-isopropene-1-yl)-4,5,6H-pyran	Seeds	Sapana et al. (2009)
115	5,6-Dihydroxy-3-[(4-hydroxy-6-(hydroxymethyl)-3,5-di[3,4,5-trihydroxy-6-(hydroxymethyl) tetrahydro-2H-2-pyranyl]oxytetrahydro-2H-2-pyranyl)oxy]-2-methoxy-10,13-dimethylperhydrocyclopenta[a] phenanthren-17-yl(phenyl)methyl acetate	Seeds	Daisy et al. (2007)

(Shankar et al. 2007). Compound **114** was quantified to 7.38% in seed powder (Sapana et al. 2009). 5,6-Dihydroxy-3-[(4-hydroxy-6-(hydroxymethyl)-3,5-di[3,4,5-trihydroxy-6-(hydroxymethyl)tetrahydro-2H-2-pyranyl]oxytetrahydro-2H-2-pyranyl)oxy]-2-methoxy-10,13-dimethylperhydrocyclopenta[a]phenanthren-17-yl(phenyl) methyl acetate (**115**) was isolated from seeds of *S. cumini* based on bioassay-guided fractionation (Daisy et al. 2007). Structural details of all the triterpenoidal and steroidal compounds are given in Figure 6.5 and Table 6.6.

Tetraterpenoids

Tetraterpenoids of the carotenoid category are reported from fruits of *S. cumini*. The carotenoids are red or yellow-colored natural pigments that are synthesized by plants and are responsible for the bright colors of various fruits and vegetables. Faria et al. (2011) determined the presence of a xanthophyll, all-trans-lutein (**116**), and a carotene, all-trans-β-carotene (**117**), in the fruits by high-performance liquid chromatography (HPLC)–diode array detector (DAD)–MS/MS from fruits (Figure 6.6).

Lipids and Hydrocarbons

Saturated, unsaturated, epoxy, and cyclopropenoid fatty acids have been detected in seed oil of *S. cumini*. The usual fatty acids from seed oil of *S. cumini* include lauric acid (**118**), myristic acid (**119**), palmitic acid (**120**), stearic acid (**121**), oleic acid (**122**), and linoleic acid (**123**), along with cyclopropenoid fatty acid malvalic acid (**124**), sterculic acid (**125**), and an epoxy fatty acid, vernolic acid (**126**), having abundances of 2.8%, 31.7%, 4.7%, 6.5%, 32.2%, 16.1%, 1.2%, 1.8%, and 3.0%, respectively, in seed oil (Saeed et al. 1987; Daulatabad et al. 1988). Conversely, more recent GC analysis of seed extract revealed that the percentage of **122** is the highest, followed by **123** (Kumar et al. 2007). In the seed extract, 4-(2-2-dimethyl-6-6-methylenecyclohexyl) butanol (**127**), decahydro-8a-ethyl-1,1,4a,6-tetramethylnapthalene (**128**), octadecane (**129**), 1-chlorooctadecane (**130**), and tetratetracontane (**131**) were identified by gas chromatography (Kumar et al. 2009), and the presence of eleostearic acid (**132**) was detected by spectroscopic methods from seed oil (Das and Banerjee 1995).

Although major components in the leaves are monoterpenoids and sesqueterpenoids, some fatty acids and paraffins like heptacosane (**133**), nonacosane (**134**), triacontane (**135**), and hentriacontane (**136**), along with aliphatic alcohols octacosanol (**137**), triacosanol (**138**), and dotriacosanol (**139**) (Gupta and Sharma 1974); octacosane (**140**); tricontane (**141**) (Kumar et al. 2009); eicosane (**142**); and hexadecane (**143**) (Rao et al. 2012), were detected from leaves. Additionally, some simple aliphatic hydrocarbons, 1,3,6-octatriene (**144**), and alcohols, 1-hexanol (**145**), cis-3-hexenol (**146**), nonyl alcohol (**147**), and hexylene glycol (**148**), have been detected in EOs.

Polyphenols

Flavonoids

The flavonoids, ever present in plants, are the largest class of polyphenols, with a diphenylpropane (C6-C3-C6) carbon skeleton consisting of two aromatic rings linked through three carbons. The major subclasses of flavonoids from *S. cumini* are flavones, flavonols, dihydroflavonols, and anthocyanidins. A comprehensive study by HPLC-DAD-electrospray ionization (ESI)-MS/MS revealed the occurrence of around 74 individual phenolic compounds in the edible parts of jambolan, including 9 anthocyanins (mainly based on delphinidin, petunidin, and malvidin), 9 flavonols (myricetin, laricitrin, and syringetin glycosides), 19 flavanonols (dihexosides of dihydromyricetin and its methylated derivatives), 8 flavan-3-ol monomers (mainly gallocatechin), 13 gallotanins, and 13 ellagitanins, together with some proanthocyanidins (highly galloylated prodelphinidins) and free gallic and ellagic acids (de Carvalho Tavares et al. 2016).

Flavonols and Flavones

A few numbers of flavonoids free of a sugar unit are reported from jamun. Flavonoid glycosides are relatively large in number and are discussed under Flavonoid Glycosides. Four flavonols, one dihyhydroflavonol, and one flavone have been reported from *S. cumini*. Kaempferol (**149**) and its 3′ hydroxylated derivative, quercetin (**150**), are isolated from petroleum ether extract by column chromatography from stem bark (Bhargava et al. 1974). S. Li et al. (2009) isolated a dihydroxyflavonoid, dihydromyricetin (**151**), and a flavone, 5,7,3′,4′,5′-pentahydroxyflavone (**152**), from roots. Compound **152** increased glucose uptake in insulin-resistant L6 muscle cells by 51.11% at a concentration of 0.1 μg/mL with insulin (S. Li et al. 2009). The presence of dihydromyricetin (**151**) was demonstrated in alcoholic extract from fresh flowers by preparative partition chromatography (Subramanian and Nair 1972). Moreover, the leaves also account for quercetin (**150**), myricetin (**153**) (Timbola et al. 2002), and kaempferol (**149**); all three compounds are members of a close group of well-known flavonols (Mahmoud et al. 2001). Furthermore, kaempferol-7-O-methylether (**154**) has been identified by comparison of its mass spectra from ethanolic extracts of fruits (Afify et al. 2011).

Flavonoid Glycosides

This section covers O-glycosylated and C-glycosylated flavonoids. Most of the sugar attachment is found at 3C-OH of the flavone backbone in O-glycosides. C-glycosides are not common; however, swertisin (**155**) is the only reported flavone C-glycoside. Compound **155** has shown better α-glycosidase inhibitory activity than acarbose (IC$_{50}$ 146.5 μM compared with IC$_{50}$ 208.6 μM for acarbose) (Omar et al. 2012). 3,5-Diglucosides of dihydromyricetin (**156**), methyl-dihydromyricetin (**157**), and dimethyl-dihydromyricetin (**158**) are detected by HPLC-DAD-MS/MS from jambul fruits (Faria et al. 2011). Acylation of sugars has also been observed. Three new acylated flavonol glycosides, mearnsetin 3-O-(4″-O-acetyl)-α-L-rhamnopyranoside (**159**), myricetin 3-O-(4″-O-acetyl-2″-O-galloyl)-α-L-rhamnopyranoside (**160**), and myricetin 3-O-(4″-acetyl)-α-L-rhamnopyranoside (**161**) (i.e., myricetrin 4″-O-acetate), along with known acylated flavonol glycoside quercetrin 4″-O-acetate (**162**), have been isolated and identified from the leaves. Other glycosides from leaves include myricetin 4′-methyl ether 3-O-α-L-rhamnopyranoside (**163**), myricetrin (**164**), quercetin 3-O-α-L-rhamnopyranoside (**165**) (quercet), kaempferol 3-O-β-D-glucuronopyranoside (**166**), and myricetin 3-O-β-D-glcuronopyranoside (**167**)

(Mahmoud et al. 2001; Timbola et al. 2002). The presence of myricetin-3-L-arabinoside (**168**) and quercetin 3-D-galactoside (**169**) was established in alcoholic extract of fresh flowers by preparative partition chromatography (Subramanian and Nair 1972). Kaempferol-3-O-glucoside (**170**) was isolated from petroleum ether extract by column chromatography from stem bark (Bhargava et al. 1974).

Isorhamnetin 3-O-rutinoside (**171**) was isolated from the ethyl acetate soluble portion of ethanol extract of air-dried roots (Vaishnav and Jain 2004), along with myricetin-3-O-rutinoside (**172**), myrcetin-3-O-robinoside (**173**) (Vaishnav and Sahu 2005), and myricetin-3-O-glucoside (**174**) from the root of jamun (Vaishnava and Gupta 1990; Vaishnava et al. 1992).

Anthocyanins

Anthocyanins are plant pigments that are potential candidates for use as natural food colorant. The chemoprotective action of colored fruits, particularly berries, can be attributed to anthocyanins, which are glycosides of anthocyanidins. Additionally, antioxidant, antiproliferative, and anti-inflammatory activities are associated with anthocyanins (Banerjee et al. 2004; Veigas et al. 2008; Devkar et al. 2012). *S. cumini* fruit, particularly fruit skin, has been used as an anthocyanin source. Usually, incompletely ripe fruits have more anthocyanidins than fully ripe fruits. The anthocyanidin content at the 8% ripe stage is 5.96%, which is 133.04% times that of fully ripen fruits (Zhang and Diao 2012a). The pulp powder contains 0.54% anthocyanins, but seeds contain no detectable amount of anthocyanins. The highest anthocyanin yield (763.80 mg per 100 mL of fruit peel) can be obtained when 20% ethanol is used in combination with 1% acetic acid (Chaudhary and Mukhopadhyay 2013).

All six major types of anthocyanidins are identified by HPLC (Figure 6.10). Acid-hydrolyzed pulp extract showed five anthocyanidins by HPLC: malvidin (**A**) (44.4%), petunidin (**B**) (24.2%), delphinidin (**C**) (20.3%), cyanidin (**D**) (6.6%), and peonidin (**E**) (2.2%) (Aqil et al. 2012), suggesting the presence of five types of anthocyanidins. However, de Sousa et al. (2007) detected pelargonidin (**F**) from fruit by HPLC-MS/MS.

The major anthocyanins identified in jamun fruits are the 3,5-diglucosides of delphinidin, malvidin, petunidin, cyanidin, and peonidin (Li et al. 2009a; Faria et al. 2011; de Carvalho Tavares et al. 2016). Additionally, delphinidin-3,5-diglucoside (**175**), petunidin-3,5-diglucoside (**176**), and malvidin-3,5-diglucoside (**177**) were detected and quantified by quantitative NMR (qNMR) (Chauthe et al. 2012). Further, 3,5-, diglucosides substitution on anthocyanidin have been reported whereas, 3-diglucosides substitution on anthocyanidin have not been reported in fruits (L. Li et al. 2009b). Several other sugar attachments to the six anthocyanidins produced a number of anthocyanins (Figure 6.11). From the fruits, delphinidin-3-gentiobioside (**178**), malvidin-3-laminaribioside (**179**), and petunidin-3-gentiobioside (**180**) have been isolated (Jain and Seshadri 1975). Pelargonidin-3-O-glucoside (**181**), pelargonidin-3,5-O-diglucoside (**182**), cyanidin-3-O-malonyl glucoside (**183**), and delphenidin-3-O-glucoside (**184**) are isolated from the acidic alcoholic extract of fruits growing in Egypt (Nazif 2007). Monoglycosides of anthocyanidins are relatively few. The monoglycoside of cyaniding, namely, cyanidin 3-glucoside (**185**), occurs in high amounts in Java plum (Bobbio and Scamparini 1982). Glacial AcOH extract of fruit led to the isolation of cyanidin rhamnoglucoside (**186**), and the structure was established by color reaction (Venkateswarlu 1952). Moreover, the glucoglucosides of delphinidin, petunidin, and malvidin were detected by HPLC-ESI-MS from fruit pulp (Veigas et al. 2007).

Tannins

Tannins (vegetable tannin, natural organic tannins, or sometimes tannoid) are a large group of polyphenolic compounds containing sufficient hydroxyl groups and other suitable groups to either bind or precipitate proteins (Lorenz et al. 2014). The tannins are astringent, bitter plant secondary metabolites that cause the dry and puckery feeling in the mouth after the consumption of an unripened fruits (Ashok and Upadhyaya 2012). Tannins are water-soluble phenolic compounds that give the usual phenolic reactions, emphasizing the multiplicity of phenolic groups; the term *polyphenolics* is used for tannins. Tannin can form complexes with a molecular weight up to 20,000, not only with proteins and alkaloids, but also with some polysaccharides. The tannin compounds are widely distributed in many parts of plants, and play a protective role from predators, and also in plant growth regulation. The destruction or modification of tannins with time plays an important role in the ripening of fruit (Li et al. 2006).

The tannin compounds are extensively distributed in several parts of jamun. It contains appreciable amounts of ellagitannins, with high antioxidant and antiproliferative activities (Aqil et al. 2012). Tannins have been reported in almost all parts of the plant, such as fruits (Joshi et al. 2012), having a content of tannins of 4.2% (Migliato et al. 2007); aqueous extracts of leaves (Owolabi et al. 2013); seeds (Omar et al. 2012; Sarita et al. 2012); stems (Gopinath et al. 2012); methanolic extracts of bark (Kannabiran et al. 2011); and pollen (Ghoshal and Saoji 2013). Condensed or nonhydrolyzable and hydrolyzable are the two major classes of tannins reported from higher plants.

Condensed Tannins

Condensed tannins are based on flavan structures, and they are also known as proanthocyanidins, which give rise to anthocyanidins on oxidative depolymerization. Particularly in the flavone-derived tannins, they must be heavily hydroxylated and polymerized in order to import protein precipitation character. They consist of 2–50 or more monomeric units of flavanoids that are coupled by carbon–carbon bonds, which are resistant to cleavage by hydrolysis. Proanthocyanidins from fruits have strong free radical scavenging and antioxidant activities *in vitro*. The IC_{50} values of proanthocyanidins obtained from fruits in 2,2-diphenyl-1-picrylhydrazyl (DPPH), superoxide anion, and hydroxyl radical assays were 4.448, 5.718, and 60.490 µg/mL, respectively. An IC_{50} of 117.091 µg/mL was obtained in an anti–linoleic acid peroxidation assay (Zhang and Diao 2012b).

Acetone–water (7:3 v/v) extract of mature fruits of jamun has shown the presence of hydrolyzable and condensed tannins. NMR, matrix-assisted laser desorption/ionization time-of-flight (MALDI-TOF) MS, and HPLC analyses identified condensed tannins as B-type oligomers of epiafzelechin (propelargonidin) with a degree of polymerization up to 11. NMR analysis of a condensed tannin fraction of acetone–water (7:3 v/v) extract of fruit skin demonstrated that condensed tannins predominantly consist of propelarogonidin (**187**), with (–)-epiafzelechin as the main constitutive monomer. Further, MALDI-TOF MS revealed the addition of one galloyl group at the heterocyclic ring. Thus, the pro-pelarogonidins and galloylated propelarogonidin, containing not more than one galloyl group, have been detected as condensed tannins (Zhang and Lin 2009). Moreover, condensed tannin comprising gallic acid and leucoanthocyanin was isolated by Mura et al. (2000).

Hydrolyzable Tannins

Hydrolyzable tannins, a classic example of secondary metabolites, are based on gal-lic or ellagic acid moieties. Hydrolyzable tannins are hydrolyzed by weak acids or weak bases to produce carbohydrate and phenolic acids. On heating with hydrochlo-ric or sulfuric acids, hydrolyzable tannin yields gallic and ellagic acids. A carbohy-drate (usually D-glucose) makes the central core of a hydrolyzable tannin molecule. The hydroxyl groups of the carbohydrate are partially or totally esterified with phe-nolic groups such as gallic acid (in gallotannins) or ellagic acid (in ellagitannins) (Pouysegu et al. 2011).

Hydrolyzable tannins were identified as ellagitannins, consisting of a glucose core surrounded by ellagic acid units. Two new hydrolyzable gallotannins, jamu-tannins A (**188**) and jamutannins B (**189**), and a new ellagitannin, iso-oenothein C (**190**), along with known ellagitannins, oenothein C (**191**), cornusiin B (**192**), and phyllanthusiin E (**193**), were isolated from the n-butanol fraction of the methanol extract of seeds. Compound **190** exists as an equilibrium mixture of α and β forms. Compounds **190**, **191**, and **192** were more active than acarbose (IC_{50} 208.6 μM) with IC_{50} values of 8.2, 75.1, and 12.2 μM, respectively (Omar et al. 2012). Ellagic acid 4-O-α-L-2″-acetylrhamnopyranoside (**194**), 3-O-methylellagic acid 3′-O-α-L-rhamnopyranoside (**195**), and the new derivative 3-O-methylellagic acid 3′-O-β-D-glucopyranoside (**196**) were isolated from stem bark (Simoes-Pires et al. 2009). Ellagitannin, nilocitin (**197**) (Mahmoud et al. 2001), corilagin (**198**), and hexahy-droxydibenzoyl glucose (**199**) (Bhatia et al. 1971) were identified from the leaves and seeds, respectively.

Although tannins are a taste-reducing substance found in fruits, they have medic-inal values as well (Severo et al. 2010), and utilization of the fruit is a significant source of natural antioxidants. Tannins extracted from fruit showed a very good DPPH radical scavenging activity and ferric reducing or antioxidant power (Zhang and Lin 2009). The vibriocidal activity of bark of *S. cumini* is due to tannin content, indicating that the plant is rich in gallic acid and tannin can be used as an alternative to search for new vibriocidal drugs (Sharma et al. 2009). Major antioxidants found in fruit wine are a complicated mixture of hydrolyzable tannins and the fruit acids. The active compounds were identified as hydrolyzable tannins and their derivatives, that is, caffeoylquinic acid, gallic acid, ellagic acid, and methoxymethylgallate in fruit wine (Nuengchamnong and Ingkaninan 2009). Recent data suggest that ellagitan-nins found in some fruits and nuts may have beneficial effects against colon cancer (Sharma et al. 2010).

Lignans

Lignans are one of the major classes of phytoestrogens, owing to their estrogenic mimetic activity. They are naturally occurring plant phenols that are derived biosynthetically from phenylpropanoids in which the phenylpropane units are bound by the C8 central carbons of the propyl side chains. The dibenzylbutane skeleton of lignans is derived from phenylalanine through dimerization of substituted cinnamic alcohols, known as monolignols (Nakatsubo et al. 2007). The majority of lignans occur without sugar attachment, but a small proportion exists as glycosides. They occupy quite a large area in the plant world, and are distributed in almost all parts of the plants, including the wood, roots, leaves, flowers, fruits, rhizomes, stem bark, and seeds. Various types of lignans have numerous pharmacological features, such as antitumor, hepatoprotective, platelet-activating factor (PAF), antagonistic, antifungal, antihypertensive, sedative, antioxidant, insecticidal, and estrogenic activities (Dar and Arumugam 2013; Zhang et al. 2014). Some of the lignans are considered to be lead structures for new drugs, and some of them have been developed into approved therapeutics.

Mir et al. (2009) isolated lignan derivatives of the furofuran group: (7α,8α, 2′)-3,4,5-trimethoxy-7,3′,1′,9′-diepoxylignan (cuminiresinol, **200**), (7α,7′α,8α,8′)-3,4-dioxymethylene-3′,4′-dimethoxy-7,9′,7′,9-diepoxylignan-5′-ol (5′-hydroxy-methyl-piperitol, **201**), 3-demethyl-9-oxo-pinoresinol (syzygiresinol A, **202**), and 3,3′-didemethyl-9-oxo-pinoresinol (syzygiresinol B, **203**), along with known compounds di-demethyl-5-hydroxypinoresinol (**204**), dimethylpinoresinol (**205**), didemethoxypinoresinol (**206**), pinoresinol (**207**), and 4′-methyl-5′-hydroxypinoresinol (**208**) from stem bark. The chemical structure of all isolated lignans is portrayed in Figure 6.14.

Other Phenolic Compounds

Gallic acid and ellagic acid are the building blocks of many tannin materials (gallotannins and elligatannins). An ethyl acetate solution of hydrolyzed alcoholic extract yielded gallic acid (**209**) and ellagic acid (**210**) from stem bark (Bhargava et al. 1974). Ellagic acid was also tentatively identified from flowers by color reactions and formation of the tetraacetate (Nair and Subramanian 1962). Many other derivatives of gallic and ellagic acid have been isolated, including hexahydroxydiphenic acid (**211**) (Bhatia et al. 1971), 3,3,4′-tri-O-methylellagic acid (**212**), 3,4′-di-O-methylellagic acid (**213**) (Bhatia and Bajaj 1975), dimethyl-4,4′-oxybis (3,5-dihydroxy benzoate) (**214**), and 4,4′-oxybis(3,5-dihydroxy benzoic acid) (**215**). Compounds **214** and **215** produced irreversible antihyperglycemic efficacy in prolonged therapeutic application and also restored normal body weight and the lipid profile in an animal model, and have been patented as safe and effective antidiabetic therapeutic agents (Ghosh et al. 2008). Moreover, seeds yielded valoneic acid dilactone (**216**), brevifolin carboxylic acid (**217**), ellagic acid (**210**), gallic acid (**209**) (Mahmoud et al. 2001; Omar et al. 2012), and bergenin (**218**) from bark (Kopanski and Schnelle 1988).

Linde (1983) isolated methylxanthoxyline (**219**) and 2,6-dihydroxy-4-methoxy-acetophenone (**220**) from the unsaponifiable portion of the petroleum ether extract of blossoms. Gallic acid, ellagic acid, guaiacol (**221**), resorcinol dimethyl ether (**222**), veratrole (**223**), caffeic acid (**224**), and ferulic acid (**225**) have been identified by co-chromatography with authentic samples on silica gel G plates in seeds. A pair of solvent systems toluene–ethyl formate–formic acid (5:4:1) and C_6H_6-MeOH-HOAc (45:8:4) has been found to be useful for separation of phenolic compounds by both one- and two-dimensional thin-layer chromatography on silica gel G plates (Bhatia and Bajaj 1975). Some phenylpropanoids, like methyl cinnamate (**226**), eugenol (**227**) (Khanna 1991), methylgallate (**228**), and chlorogenic acid (**229**) (Mahmoud et al. 2001), were detected from leaf EOs and methyl eugenol (**230**) and **227** from fruit EOs (Abdelhady 2012).

MISCELLANEOUS COMPOUNDS

Casuarine-6-α-D-glucopyranose (**226**) is a member of the growing group of polyhydroxylated 3-hydroxymethylpyrrolizidine natural products. Casuarine-6-O-α-D-glucoside (Wormald et al. 1996) is a natural pseudodisaccharide and powerful inhibitor of trehalases. Trehalose (α-D-glucopyranosyl(1→1)α-D-glucopyranoside) constitutes the major blood sugar in insects and is particularly important for insects, as it is hydrolyzed into glucose to provide energy during flight. Trehalase inhibitor activity is promising for the development of new insecticides, as α-trehalase is required in order to utilize trehalases (Cardona et al. 2010).

Besides the above-mentioned components, *S. cumini* also contains decalina (**227**), 3,4-dimethyl-3-cyclohexen-1-carboxaldehyde (**228**), maltol (**229**), 4-butyl-1-oxide pyridine (**230**), γ-decalactone (**231**), a diterpenoid, phytol (**232**), and phenol.

CONCLUSION

The main secondary metabolite constituents of *S. cumini* are terpenoids and flavonoids. Volatile terpenoids are almost embodied in the aerial parts, especially in the leaves of the plant. Fruits are also characterized by high levels of total phenolics, and hence are a significant source of phenolic antioxidants, which may have potential positive effects on health. Extensive ethanopharmacological studies confirm the pharmacological potential of *S. cumini*. Different types of biological activities are reported in this plant, like antibacterial, antifungal, antiviral, antigenotoxic, antiallergic, anti-inflammatory, antiulcerogenic, antidiarrheal, anticancer, antioxidant, hypoglycemic, and antidiabetic effects. Even though good pharmacological or therapeutic effects are observed, there is still a need for more phytochemical studies to explore the active principle and elucidate their possible mechanisms of action. Further studies are necessary on the pharmacology of isolated compounds. EOs from leaves are explored well, and a number of components have been detected, but there are few reports of EOs from stem, fruits, and flowers. Most pharmacological works on diabetes were carried out with stem and leaf, but the pharmacological potential of the other parts of the plant needs to be explored.

ACKNOWLEDGMENT

Sudhir Kumar is grateful to the University Grant Commission, government of India, for the grant of a senior research fellowship. CSIR-CDRI communication number for this manuscript is 9381.

REFERENCES

Abdelhady, M. I. S. 2012. Essential oil extracted from fruits of Egyptian *Eugenia jambolana* has antimicrobial activity. *Nat. Prod.* 8: 68–71.

Afify, A. E.-M. M. R., S. A. Fayed, E. A. Shalaby, and H. A. El-Shemy. 2011. *Syzygium cumini* (pomposia) active principles exhibit potent anticancer and antioxidant activities. *Afr. J. Pharm. Pharmacol.* 5: 948–956.

Al, A. S. M., M. A. Kaisar, M. S. Rahman, C. M. Hasan, A. J. Al-Rehaily, and M. A. Rashid. 2012. Secondary metabolites from seed extracts of *Syzygium cumini* (L.). *J. Phys. Sci.* 23 (1): 83–87.

Aqil, F., A. Gupta, R. Munagala, J. Jeyabalan, H. Kausar, R. J. Sharma, I. P. Singh, and R. C. Gupta. 2012. Antioxidant and antiproliferative activities of anthocyanin/ellagitannin-enriched extracts from *Syzygium cumini* L. (jamun, the Indian blackberry). *Nutr. Cancer* 64 (3): 428–438.

Ashok, P. K., and K. Upadhyaya. 2012. Tannins are astringent. *J. Pharmacogn. Phytochem.* 1 (3): 45–50.

Badawy, M. E. I. and S. A. M. Abdelgaleil. 2014. Composition and antimicrobial activity of essential oils isolated from Egyptian plants against plant pathogenic bacteria and fungi. *Ind. Crop. Prod.* 52: 776–782.

Baliga, M. S., S. Fernandes, K. R. Thilakchand, P. D'Souza, and S. Rao. 2013. Scientific validation of the antidiabetic effects of *Syzygium jambolanum* DC. (black plum), a traditional medicinal plant of India. *J. Altern. Complement. Med.* 19 (3): 191–197.

Banerjee, A., N. Dasgupta, and B. De. 2004. In vitro study of antioxidant activity of *Syzygium cumini* fruit. *Food Chem.* 90 (4): 727–733.

Bhargava, K. K., R. Dayal, and T. R. Seshadri. 1974. Chemical components of *Eugenia jambolana* stem bark. *Curr. Sci.* 43 (20): 645–646.

Bhatia, I. S., and K. L. Bajaj. 1975. Thin-layer chromatography of the phenolic constituents of *Eugenia jambolana* seeds. *J. Inst. Chem.* 47 (4): 127–130.

Bhatia, I. S., K. L. Bajaj, and G. S. Ghangas. 1971. Tannins in black plum seeds. *Phytochemistry* 10 (1): 219–220.

Bobbio, A. F. O., and R. P. Scamparini. 1982. Carbohydrates, organic acids and anthocyanin of *Eugenia jambolana* Lamarck. *Ind. Aliment.* 21 (4): 296–298.

Bopp, A., K. S. De Bona, L. P. Belle, R. N. Moresco, and M. B. Moretto. 2009. *Syzygium cumini* inhibits adenosine deaminase activity and reduces glucose levels in hyperglycemic patients. *Fundam. Clin. Pharm.* 23 (4): 501–507.

Burt, S. 2004. Essential oils: Their antibacterial properties and potential applications in foods—A review. *Int. J. Food Microbiol.* 94 (3): 223–253.

Cardona, F., A. Goti, C. Parmeggiani, P. Parenti, M. Forcella, P. Fusi, L. Cipolla, S. M. Roberts, G. J. Davies, and T. M. Gloster. 2010. Casuarine-6-O-α-D-glucoside and its analogues are tight binding inhibitors of insect and bacterial trehalases. *Chem. Commun.* 46 (15): 2629–2631.

Chaudhary, B., and K. Mukhopadhyay. 2013. Solvent optimization for anthocyanin extraction from *Syzygium cumini* L. Skeels using response surface methodology. *Int. J. Food Sci. Nutr.* 64 (3): 363–371.

Chauthe, S. K., R. J. Sharma, F. Aqil, R. C. Gupta, and I. P. Singh. 2012. Quantitative NMR: An applicable method for quantitative analysis of medicinal plant extracts and herbal products. *Phytochem. Anal.* 23 (6): 689–696.

Craveiro, A. A., C. H. S. Andrade, F. J. A. Matos, J. W. Alencar, and M. I. L. Machado. 1983. Essential oil of *Eugenia jambolana. J. Nat. Prod.* 46 (4): 591–592.

Daisy, P., S. Ignacimuthu, and R. S. Jasmine. 2007. A process for preparation of a novel compound 5,6-dihydroxy-3-[(4-hydroxy-6-(hydroxymethyl)-3,5-di[3,4,5-trihydroxy-6-(hydroxymethyl)tetrahydro-2H-2-pyranyl]oxytetrahydro-2H-2-pyranyl)oxy]-2-methoxy-10,13-dimethylperhydrocyclopenta[a]phenanthren-17-yl(phenyl) methyl acetate from *Syzygium cumini* (L) Skeels seeds with antibacterial and antidiabetic activity. Indian Patent Application 244666.

Dar, A. A., and N. Arumugam. 2013. Lignans of sesame: Purification methods, biological activities and biosynthesis—A review. *Bioorg. Chem.* 50: 1–10.

Das, S., and A. K. Banerjee. 1995. Studies on *Syzygium cumini* seed oil. *J. Oil Technol. Assoc. India* 27 (4): 243–244.

Daulatabad, C. M. J. D., A. M. Mirajkar, K. M. Hosamani, and G. M. M. Mulla. 1988. Epoxy and cyclopropenoid fatty acids in *Syzygium cumini* seed oil. *J. Sci. Food Agric.* 43 (1): 91–94.

de Carvalho Tavares, I. M., E. S. Lago-Vanzela, G. L. P. Rebello, A. M. Ramos, S. Gomez-Alonso, E. Garcia-Romero, R. Da-Silva, and I. Hermosin-Gutierrez. 2016. Comprehensive study of the phenolic composition of the edible parts of jambolan fruit (*Syzygium cumini* (L.) Skeels). *Food Res. Int.* 82: 1–13.

de Sousa, B. E., M. C. de Araújo, R. E. Alves, C. Carkeet, B. A. Clevidence, and J. A. Novotny. 2007. Anthocyanins present in selected tropical fruits: Acerola, jambolao, jussara, and guajiru. *J. Agric. Food Chem.* 55 (23): 9389–9394.

Devkar, R. V., A. V. Pandya, and N. H. Shah. 2012. Protective role of *Brassica olerecea* and *Eugenia jambolana* extracts against HO induced cytotoxicity in H9C2 cells. *Food Funct.* 3 (8): 837–843.

Dias, C. N., K. A. F. Rodrigues, F. A. A. Carvalho, S. M. P. Carneiro, J. G. S. Maia, E. H. A. Andrade, and D. F. C. Moraes. 2013. Molluscicidal and leishmanicidal activity of the leaf essential oil of *Syzygium cumini* (L.) Skeels from Brazil. *Chem. Biodiv.* 10 (6): 1133–1141.

Elansary, H. O., M. Z. M. Salem, N. A. Ashmawy, and M. M. Yacout. 2012. Chemical composition, antibacterial and antioxidant activities of leaves essential oils from *Syzygium cumini* L., *Cupressus sempervirens* L. and *Lantana camara* L. from Egypt. *J. Agric. Sci.* 4 (10): 144–152.

Faria, A. F., M. C. Marques, and A. Z. Mercadante. 2011. Identification of bioactive compounds from jambolao (*Syzygium cumini*) and antioxidant capacity evaluation in different pH conditions. *Food Chem.* 126 (4): 1571–1578.

Ghosh, D., D. K. Panda, and B. Bhat. 2008. Dimethyl 4,4′ oxybis (3,5 dihydroxy benzoate), abbreviated as P1, and 4,4′ oxybis (3,5 dihydroxy benzoic acid), abbreviated as P2, two novel diabetic therapeutic agents from the seeds of *Eugenia jambolana*. Indian Patent Application IN 2008KO00436 A 20080502.

Ghoshal, K. P., and A. A. Saoji. 2013. Phytochemical screening of the pollen of some selected plants with antidiabetic properties. *Aust. J. Basic Appl. Sci.* 7 (7): 105–109.

Gopinath, S. M., C. K. Rakesh, P. G. M. Ashwini, and K. S. Dayananda. 2012. Preliminary phytochemical evaluation of leaf extracts of *Euphorbia hirta*, *Syzygium cumini* of Siddarabetta, Tumkur district, Karnataka. *Int. J. Pharma Bio Sci.* 3 (2): 431–435.

Guan, W., S. Li, R. Yan, S. Tang, and C. Quan. 2007. Comparison of essential oils of clove buds extracted with supercritical carbon dioxide and other three traditional extraction methods. *Food Chem.* 101 (4): 1558–1564.

Gupta, G. S., and D. P. Sharma. 1974. Triterpenoid and other constituents of *Eugenia jambolana* leaves. *Phytochemistry* 13 (9): 2013–2014.

Hemalatha, T., S. Pulavendran, C. Balachandran, B. M. Manohar and R. Puvanakrishnan. 2010. Arjunolic acid: A novel phytomedicine with multifunctional therapeutic applications. *Indian J. Exp. Biol.* 48 (3): 238–247.

Hüsnü, K., C. Başer, and F. Demirci. 2007. Flavours and fragrances: Chemistry, bioprocessing and sustainability. In *Chemistry of Essential Oils*, ed. R. G. Berger, 43–86. Berlin: Springer.

Jain, M. C., and T. R. Seshadri. 1975. Anthocyanins of *Eugenia jambolana* fruits. *Indian J. Chem.* 13 (1): 20–23.

Jirovetz, L., G. Buchbauer, C. Puschmann, W. Fleischhacker, P. M. Shafi and M. K. Rosamma. 1999. Analysis of the essential oils of the fresh leaves of *Syzygium cumini* and *Syzygium travancoricum* from South-India. *J. Essent. Oil-Bear. Plants* 2 (2): 68–77.

Joshi, V. K., R. Sharma, A. Girdher, and G. S. Abrol. 2012. Effect of dilution and maturation on physico-chemical and sensory quality of jamun (black plum) wine. *Indian J. Nat. Prod. Resour.* 3 (2): 222–227.

Kannabiran, K., M. Mariappan, and J. Kuncham. 2011. Preliminary phytochemical screening, anthelmintic activity of methanolic and aqueous extract of *Syzygium cumini* Linn. bark (Myrtaceae). *J. Pharm. Sci. Res.* 3 (9): 1460–1465.

Khanna, R. K. 1991. Chemical examination of the essential oil from the leaves of *Syzygium cumini* Skeel. *Indian Perfum.* 35 (2): 112–115.

Kopanski, L., and G. Schnelle. 1988. Isolation of bergenin from barks of *Syzygium cumini*. *Planta Med.* 54 (6): 572.

Kumar, A., T. Jayachandran, P. Aravindhan, D. Deecaraman, R. Ilavarasan, and N. Padmanabhan. 2009. Neutral components in the leaves and seeds of *Syzygium cumini*. *Afr. J. Pharm. Pharmacol.* 3 (11): 560–561.

Kumar, A., A. A. Naqvi, A. P. Kahol, and S. Tandon. 2004. Composition of leaf oil of *Syzygium cumini* L. from north India. *Indian Perfumer* 48 (4): 439–441.

Kumar, A., N. Padmanabhan, and M. R. V. Krishnan. 2007. Extraction of fatty acids from *Syzygium cumini* seeds with different solvents. *Asian J. Chem.* 19 (4): 2779–2782.

Langenheim, J. H. 1994. Higher plant terpenoids: A phytocentric overview of their ecological roles. *J. Chem. Ecol.* 20 (6): 1223–1280.

Li, L., L. S. Adams, S. Chen, C. Killian, A. Ahmed, and N. P. Seeram. 2009a. *Eugenia jambolana* Lam. berry extract inhibits growth and induces apoptosis of human breast cancer but not non-tumorigenic breast cells. *J. Agric. Food Chem.* 57 (3): 826–831.

Li, L., Y. Zhang, and N. P. Seeram. 2009b. Structure of anthocyanins from *Eugenia jambolana* fruit. *Nat. Prod. Commun.* 4 (2): 217–219.

Li, M., Y. Kai, H. Qiang, and J. Dongying. 2006. Biodegradation of gallotannins and ellagitannins. *J. Basic Microb.* 46 (1): 68–84.

Li, S., N. Huang, X. Hao, L. Li, and S. Li. 2009. Effect of chemical constituents from *Syzygium cumini* (Myrtaceae) on glucose uptake in insulin-resistant L6 cells. *Yunnan Zhiwu Yanjiu* 31 (5): 469–473.

Linde, H. 1983. Two phenols from Java plum blossoms. *Arch. Pharm.* 316 (11): 971–972.

Lorenz, M. M., L. Alkhafadji, E. Stringano, S. Nilsson, I. Mueller-Harvey, and P. Uden. 2014. Relationship between condensed tannin structures and their ability to precipitate feed proteins in the rumen. *J. Sci. Food Agric.* 94 (5): 963–968.

Lustosa, A. K. M., S. M. D. S. S. Da, A. M. d. G. L. Cito, L. J. A. Dantas, M. H. Chaves, O. E. H. De, A. M. V. C. De, and L. F. N. N. De. 1999. Essential oil of *Syzygium cumini* (L.) Skeels (Myrtaceae): Chemical composition and antimicrobial activity. *An. Assoc. Bras. Quim.* 48 (2): 95–97.

Machado, R. R. P., D. F. Jardim, A. R. Souza, E. Scio, R. L. Fabri, A. G. Carpanez, R. M. Grazul, J. P. R. F. de Mendonça, B. Lesche, and F. M. Aarestrup. 2013. The effect of essential oil of *Syzygium cumini* on the development of granulomatous inflammation in mice. *Rev. Bras. Farmacogn.* 23 (3): 488–496.

Mahmoud, I. I., M. S. A. Marzouk, F. A. Moharram, M. R. El-Gindi, and A. M. K. Hassan. 2001. Acylated flavonol glycosides from *Eugenia jambolana* leaves. *Phytochemistry* 58 (8): 1239–1244.

Mahmoud, I. I., F. A. Moharram, M. R. El-Gindi, M. S. A. Marzouk, J. Nolte, and A. M. K. Hassan. 2002. Essential oils from two species of *Eugenia*. *Bull. Faculty Pharm.* (Cairo University) 40 (1): 123–127.

Migliato, K. F., R. R. D. Moreira, J. C. P. Mello, L. V. S. Sacramento, M. A. Correa, and H. R. N. Salgado. 2007. Quality control of *Syzygium cumini* (L.) Skeels fruits. *Rev. Bras. Farmacogn.* 17 (1): 94–101.

Mir, Q. Y., M. Ali, and P. Alam. 2009. Lignan derivatives from the stem bark of *Syzygium cumini* (L.) Skeels. *Nat. Prod. Res.* 23 (5): 422–430.

Mohamed, A. A., S. I. Ali, and F. K. El-Baz. 2013. Antioxidant and antibacterial activities of crude extracts and essential oils of *Syzygium cumini* leaves. *PLoS One* 8 (4): e60269.

Mura, K., H. Shiramatsu, and W. Tanimura. 2000. A substance inhibiting the growth of lactic acid bacteria in duhat (*Syzygium cumini* Skeels) bark. *Biocontrol Sci.* 5 (1): 33–38.

Nair, A. G. R., and S. S. Subramanian. 1962. Chemical examination of the flowers of *Eugenia jambolana*. *J. Sci. Ind. Res.* 21B: 457–458.

Nakatsubo, T., L. Li, T. Hattori, S. Lu, N. Sakakibara, V. L. Chiang, M. Shimada, S. Suzuki, and T. Umezawa. 2007. Roles of 5-hydroxyconiferylaldehyde and caffeoyl CoA O-methyltransferases in monolignol biosynthesis in *Carthamus tinctorius*. *Cell. Chem. Technol.* 41 (9–10): 511–520.

Nazif, N. M. 2007. The anthocyanin components and cytotoxic activity of *Syzygium cumini* (L.) fruits growing in Egypt. *Nat. Prod. Sci.* 13 (2): 135–139.

Nuengchamnong, N., and K. Ingkaninan. 2009. On-line characterization of phenolic antioxidants in fruit wines from family Myrtaceae by liquid chromatography combined with electrospray ionization tandem mass spectrometry and radical scavenging detection. *LWT Food Sci. Technol.* 42 (1): 297–302.

Omar, R., L. Li, T. Yuan, and N. P. Seeram. 2012. α-Glucosidase inhibitory hydrolyzable tannins from *Eugenia jambolana* seeds. *J. Nat. Prod.* 75 (8): 1505–1509.

Owolabi, O. A., D. B. James, E. B. Adejor, N. Q. Nwaozuzu, T. Oloba, and C. D. Luca. 2013. Phytochemical constituents and effect on haematological parameters and lipid profile of aqueous extracts of *Eugenia jambolana* leaves, stem bark and root bark in normal albino rats. *Res. J. Appl. Sci. Eng. Technol.* 6 (10): 1846–1850.

Packiyasothy, E. V., and S. Kyle. 2002. Antimicrobial properties of some herb essential oils. *Food Aust.* 54 (9): 384–387.

Pouysegu, L., D. Deffieux, G. Malik, A. Natangelo, and S. Quideau. 2011. Synthesis of ellagitannin natural products. *Nat. Prod. Rep.* 28 (5): 853–874.

Power, F. B., and T. Callan. 1912. Chemical examination of jambul seeds. *Pharmaceut. J.* 88: 414–417.

Rajasekaran, M., J. S. Bapna, S. Lakshmanan, A. G. R. Nair, A. J. Veliath, and M. Panchanadam. 1988. Antifertility effect in male rats of oleanolic acid, a triterpene from *Eugenia jambolana* flowers. *J. Ethnopharmacol.* 24 (1): 115–121.

Ramaiah, M., and S. S. Nigam. 1969. Chemical study of essential oil of *Eugenia jambolona* (syn.: *Syzygium cumini*, *E. fructicosa*). *Riechstoffe Aromen Koerperpflegemittel* 19 (10): 433–434, 436, 438, 440, 443–434.

Ramos, M. F. S., A. C. Siani, M. C. Souza, E. C. Rosas, and M. G. M. O. Henriques. 2006. Evaluation of the antiinflammatory activity of essential oils from five Myrtaceae species. *Rev. Fitos* 2 (2): 58–66.

Rao, G. V., K. S. Rao, M. S. L. Madhavi, T. Mukhopadhyay, and S. Lavakumar. 2012. Cell migration composition from *Syzygium cumini*. Indian Patent Application IN 2010CH02712 A 20120629.

Ravichandiran, V., and T. Ilango. 2012. Immunomodulatory effect of a herbal formulation tablet comprising equal mixture of Indian herbals namely *Piper nigrum, Zingiger officinale,* and *Syzygium jambolanum.* Indian Patent Application IN 2011CH02015 A 20120127.

Saeed, M. T., A. Rauf, and S. M. Osman. 1987. Fatty acid composition of seed oils of four plant families. *J. Oil Technol. Assoc. India* 19 (4): 86–88.

Sapana, S. K., V. M. Jadhav, and V. J. Kadam. 2009. Development and validation of HPTLC method for determination of 3-hydroxy androstane [16,17-C](6′methyl, 2′-1-hydroxy-isopropene-1-yl) 4,5,6 H pyran in jambul seed (*Syzygium cumini*). *Int. J. PharmTech Res.* 1 (4): 1129–1135.

Sarita, M., M. Bhagya, and T. Shivanandappa. 2012. Post-coital contraceptive activity and teratogenicity of *Eugenia jambolana* Lam. seed. *J. Pharm. Res.* 5 (8): 4295–4298.

Sengupta, P., and P. B. Das. 1965. Terpenoids and related compounds. IV. Triterpenoids from the stem-bark of *Eugenia jambolana. J. Indian Chem. Soc.* 42 (4): 255–258.

Severo, J., R. S. dos Santos, J. Casaril, A. Tiecher, J. A. Silva, and C. V. Rombaldi. 2010. Detannization and preservation of jambolao fruits. *Cienc. Rural* 40 (4): 976–982.

Shafi, P. M., M. K. Rosamma, K. Jamil, and P. S. Reddy. 2002. Antibacterial activity of *Syzygium cumini* and *Syzygium travancoricum* leaf essential oils. *Fitoterapia* 73 (5): 414–416.

Shankar, M. B., J. R. Parikh, R. Geetha, R. S. Mehta, and A. K. Saluja. 2007. Anti-diabetic activity of novel androstane derivatives from *Syzygium cumini* Linn. *J. Nat. Remedies* 7 (2): 214–219.

Sharma, A., V. K. Patel, and P. Ramteke. 2009. Identification of vibriocidal compounds from medicinal plants using chromatographic fingerprinting. *World J. Microbiol. Biotechnol.* 25 (1): 19–25.

Sharma, M., L. Li, J. Celver, C. Killian, A. Kovoor, and N. P. Seeram. 2010. Effects of fruit ellagitannin extracts, ellagic acid, and their colonic metabolite, urolithin A, on Wnt signaling. *J. Agric. Food Chem.* 58 (7): 3965–3969.

Simoes-Pires, C. A., S. Vargas, A. Marston, J. R. Ioset, M. Q. Paulo, A. Matheeussen, and L. Maes. 2009. Ellagic acid derivatives from *Syzygium cumini* stem bark: Investigation of their antiplasmodial activity. *Nat. Prod. Commun.* 4 (10): 1371–1376.

Sobral-Souza, C. E., N. F. Leite, F. A. B. Cunha, A. I. Pinho, R. S. Albuquerque, J. N. P. Carneiro, I. R. A. Menezes, J. G. M. Costa, J. L. Franco, and H. D. M. Coutinho. 2014. Cytoprotective effect against mercury chloride and bioinsecticidal activity of *Eugenia jambolana* Lam. *Arab. J. Chem.* 7 (1): 165–170.

Srivastava, S., and D. Chandra. 2013. Pharmacological potentials of *Syzygium cumini*: A review. *J. Sci. Food Agric.* 93 (9): 2084–2093.

Subramanian, S. S., and A. G. R. Nair. 1972. Flavonoids of the flowers of *Eugenia jambolana. Curr. Sci.* 41 (19): 703–704.

Teixeira, C. C., L. P. Pinto, F. H. Kessler, L. Knijnik, C. P. Pinto, G. J. Gastaldo, and F. D. Fuchs. 1997. The effect of *Syzygium cumini* (L.) Skeels on post-prandial blood glucose levels in non-diabetic rats and rats with streptozotocin-induced diabetes mellitus. *J. Ethnopharmacol.* 56 (3): 209–213.

Teixeira, C. C., C. A. Rava, D. S. P. Mallman, R. Melchior, R. Argenta, F. Anselmi, C. R. Almeida, and F. D. Fuchs. 2000. Absence of antihyperglycemic effect of jambolan in experimental and clinical models. *J. Ethnopharmacol.* 71 (1–2): 343–347.

Teixeira, C. C., L. S. Weinert, D. C. Barbosa, C. Ricken, J. F. Esteves, and F. D. Fuchs. 2004. *Syzygium cumini* (L.) Skeels in the treatment of type 2 diabetes: Results of a randomized, double-blind, double-dummy, controlled trial. *Diabetes Care* 27 (12): 3019–3020.

Timbola, A. K., B. Szpoganicz, A. Branco, F. D. Monache, and M. G. Pizzolatti. 2002. A new flavonol from leaves of *Eugenia jambolana. Fitoterapia* 73 (2): 174–176.

Ultee, A., M. H. J. Bennik, and R. Moezelaar. 2002. The phenolic hydroxyl group of carvacrol is essential for action against the foodborne pathogen *Bacillus cereus. Appl. Environ. Microbiol.* 68 (4): 1561–1568.

Vaishnav, M. M., and P. Jain. 2004. Phytochemical examination of *Syzygium cumini* Linn. *Orient. J. Chem.* 20 (1): 201–204.

Vaishnav, M. M., and D. P. Sahu. 2005. Flavonol glycoside from *Syzygium cumini* root. *J. Inst. Chem.* (India) 77 (1): 26–32.

Vaishnava, M. M., and K. R. Gupta. 1990. Isorhamnetin 3-O-rutinoside from *Syzygium cumini* Linn. *J. Indian Chem. Soc.* 67 (9): 785–786.

Vaishnava, M. M., A. K. Tripathy, and K. R. Gupta. 1992. Flavonoids from *Syzygium cumini* roots. *Fitoterapia* 63 (3): 259–260.

Veigas, J. M., M. S. Narayan, P. M. Laxman, and B. Neelwarne. 2007. Chemical nature, stability and bioefficacies of anthocyanins from fruit peel of *Syzygium cumini* Skeels. *Food Chem.* 105 (2): 619–627.

Veigas, J. M., R. Shrivasthava, and B. Neelwarne. 2008. Efficient amelioration of carbon tetrachloride induced toxicity in isolated rat hepatocytes by *Syzygium cumini* Skeels extract. *Toxicol. In Vitro* 22 (6): 1440–1446.

Venkateswarlu, G. 1952. Nature of the coloring matter of the jambul fruit. *J. Indian Chem. Soc.* 29: 434–437.

Vijayanand, P., L. J. M. Rao, and P. Narasimham. 2001. Volatile flavor components of jamun fruit (*Syzygium cumini* L). *Flavour Fragr. J.* 16 (1): 47–49.

Wormald, M. R., R. J. Nash, A. A. Watson, B. K. Bhadoria, R. Langford, M. Sims, and G. W. J. Fleet. 1996. Casuarine-6-α-D-glucoside from *Casuarina equisetifolia* and *Eugenia jambolana*. *Carbohydr. Lett.* 2 (3): 169–174.

Xu, R., G. C. Fazio, and S. P. T. Matsuda. 2004. On the origins of triterpenoid skeletal diversity. *Phytochemistry* 65 (3): 261–291.

Zhang, J., J. Chen, Z. Liang, and C. Zhao. 2014. New lignans and their biological activities. *Chem. Biodiv.* 11 (1): 1–54.

Zhang, J., and S. Diao. 2012a. Technology of extraction of proanthocyanidins from fruit of *Syzygium cumini* (L.) Skeels. *Shipin Yu Fajiao Gongye* 38 (1): 233–237.

Zhang, J., and S. P. Diao. 2012b. In vitro antioxidant activity of proanthocyanidins in *Syzygium cumini* fruits. *Shipin Kexue* 33 (17): 101–105.

Zhang, L. L., and Y. M. Lin. 2009. Antioxidant tannins from *Syzygium cumini* fruit. *Afr. J. Biotechnol.* 8 (10): 2301–2309.

7 Pharmacognosy and Pharmacopoeial Standards for *Syzygium cumini*

Madan Mohan Pandey and A. K. S. Rawat

CONTENTS

INTRODUCTION

Natural bioactive compounds play an important role in the development of new drugs. The number is still high for natural product–derived compounds for infectious diseases, as well as anti-inflammatory drugs, representing 75% and 26%, respectively. An estimate of the new drugs approved for all diseases shows that out of the 1031 compounds, about 690 are not of purely synthetic origin. These are either

119

from natural sources, modified from natural sources, or based on pharmacophore-derived bioactive compounds from a natural product (Newman et al. 2003). Various edible fruits having colors like orange, red, purple, black, and dark purple produced by the subfamily Myrtoideae of the Myrtaceae are mainly known by their bright anthocyanin colors. These fruits are aromatic and sweet to tart, and many are somewhat astringent in taste, indicating the presence of tannins, phenols, and flavonoids. Their taste is often described as somewhat acrid. Various known antioxidant flavonoids have been isolated from the members of the Myrtaceae family. It was found that leaf extracts of 17 *Eugenia/Syzygium* species of neotropics and paleotropics contained 71% myricetin and quercetin and 24% kaempferol. Some other important phenolics and flavonoids, like methylellagic acid, ellagic acid, prodelphinidin, and procyanidin, have also been found in many other species of Myrtaceae (Hegnauer 1990; Nair et al. 1999).

Syzygium cumini has been used in various indigenous systems of medicine since ancient times. It is commonly known as jamun, jam, and jambul in India. A variety of therapeutic properties are attributed to this plant in the classical traditional medicine systems of India and elsewhere. Among a large number of herbal drugs stated to possess antidiabetic activity, *S. cumini* finds a prominent place in the Ayurveda as well as the Unani system of medicine. It is also an ancient medicinal plant with an illustrious medical history and has been the subject of classical review for more than 100 years. In Indian folk medicine, it is used in the treatment of diabetes mellitus (Kirtikar and Basu 1975). *S. cumini* is one of the most often recommended ingredients in more than 100 formulations used for antihyperglycemic activity. Historically, *S. cumini* has been used in the treatment of diabetes mellitus even before insulin was discovered (Helmstaedter 2007).

CONVENTIONAL MEDICINE AND ETHNOBOTANICAL USES

The plant parts, such as seed, fruit, leaves, flower, and bark, of *S. cumini* are used in traditional medicine. The traditional medicare practitioners in India use different parts of this plant in the treatment of diabetes, blisters in mouth cancer, diarrhea, dysentery, piles, digestive complaints, and stomachache (Jain 1991). Caraka described the use of seeds, leaves, and fruits in decoctions for diarrhea and the bark as an astringent. According to Susruta, the fruits are used in obesity, vaginal discharges, menstrual disorders, and cold infusion in intrinsic hemorrhage (Khare 2004).

FRUITS

The fruits of *S. cumini* have been used the world over for a broad variety of ailments, including cough, diabetes, dysentery, and inflammation (Reynertron et al. 2005). The fruit has a very long history of use for various medicinal purposes and currently has a large market for the treatment of chronic diarrhea and other enteric disorders (Veigas et al. 2007). Its fruits are being used in the Unani system of medicine as a liver tonic. It also enriches blood, gives strength to teeth and gums, and forms good lotion for removing ringworm infection of the head. The vinegar prepared from the fruit is tonic, astringent, carminative, and useful in spleen diseases. Juice of black jamun and mangoes in equal parts relieves thirst and has been found to be very effective in diabetes (Nadkarni 1954). The fruits are acrid and sweet, cooling and astringent to bowels, and increase "vata" and remove bad smell from the mouth. Fresh, ripe fruits are consumed directly or made into tarts, sauces, jams, or other value-added products. Jambolan juice is made from its fresh fruits and is considered excellent for sherbet, syrup, and squashes. The pulp of the fruit is also used in making jams, jellies, and puddings. In the Philippines, the fruits are used in the preparation of wine. Vinegar is prepared by using the juice of unripe fruits. The juice or decoction of *Syzygium* fruits is used to cure spleen enlargement, diarrhea, and urine retention, and it is also diluted with water and used as a gargle for sore throat and as a lotion for ringworm of the scalp. The fruits of *S. cumini* are used not only for medicinal purposes but also in numerous food products. These fruits are fit for human consumption, and they contain various components, like gallic acid, tannins, vitamin C, and anthocyanins (Martinez and Del Valle 1981).

SEEDS

The seeds of *S. cumini* are sweet, are astringent to bowels, have a diuretic property, stop urinary discharge, and are considered good for diabetes. The sprouts are said to be carminative and astringent to bowels (Sharma and Mehta 1969; Kirtikar and Basu 1975). The seed powder is used as a remedy for diabetes (Nadkarni 1954). Seed powder used in combination with mango kernels and curd helps overcome the problem of diarrhea and dysentery, and enlargement of the spleen, and also is used as a diuretic in patients suffering from scanty or suppressed urine. The extract of the seed is used to treat cough, cold, fever, and skin problems such as rashes, and the mouth, throat, intestines, and genitourinary tract ulcers caused by *Candida albicans* (Chandrasekaran and Venkatesalu 2004).

LEAVES

Traditionally, the leaf ash has been used to strengthen the teeth and gums. Oil obtained from the leaves is useful in various skin diseases. Juice of the tender leaves of *S. cumini* and mango leaves mixed with myrobalan is administered, along with goat's milk and honey, to cure dysentery with bloody discharge (chakardata), whereas the juice of tender leaves alone or in combination with carminatives such as cardamom or cinnamon is given, along with goat's milk, to cure diarrhea in children (Nadkarni 1954). In India, the leaves of *S. cumini* are used as a food for tasar silkworms. The leaves of *S. cumini* yield essential oils that are used in soaps and perfumes (Morton 1987).

BARK

According to Ayurveda, *S. cumini* bark is acrid, sweet, digestive, and astringent to the bowels; anthelmintic; and good remedy for sore throat, bronchitis, asthma, thirst, biliousness, dysentery, blood impurities, and curing ulcers, especially as a mouth wash for the astringent effect on mouth ulcerations (Kirtikar and Basu 1975).

It is also a good blood purifier. The bark alone or with other astringents, like cardamom and cinnamon, is used as a decoction in the case of chronic diarrhea and dysentery. It also acts as a gargle in sore throat and spongy gums. The external application of bark has good wound healing properties (Nadkarni 1954; Sharma and Mehta 1969).

PHARMACOGNOSTICAL INVESTIGATION OF SEED

AYURVEDIC PROPERTIES (ANONYMOUS 1999; ROSS 2003; JOSHI 2004)

Rasa: Kasaya, Madhura, Amla
Virya: Sita
Guna: Laghu, Ruksa
Vipala: Madhura, Katu
Karma: Vatala, Pittahara, Kaphahara, Vistambhi, Grahi

VERNACULAR NAMES IN INDIA

The drug jamun (*S. cumini*) is known in different names in different states and languages of India (Anonymous 2008):

Sanskrit: Mahajambu, ksudrajambu
Assam: Jam
Bengali: Jaam, kalajam, badjam
Gujarati: Gambu, jamun, jambuda
Hindi: Jamuna, jamun, jomuna
English: Black plum
Kannada: Jambu nerale, merale, neralamara
Marathi: Jambul
Malayalam: Njaval, naval
Punjabi: Jammu
Tamil: Kottainaval, naaval, navval sambu
Telugu: Nesedu
Orissa: Jamu
Urdu: Jamun

VERNACULAR NAMES IN OTHER COUNTRIES (SOWJANYA ET AL. 2013)

Malaysia: Jambulana, jambulan
Sri Lanka: Jambola
Indonesia: Jamblang, duwet (Java)
Philippines: Duhat (Tagalog, Bisoya), lomboi (ilokana)
Burma: Thabyay-hpyoo
Cambodia: Pring bai
Laos: Va
Thailand: Wa (central), hakhiphae (Chiang Rai)

MACROSCOPIC AND MICROSCOPIC DESCRIPTIONS OF SEED

Macroscopic Characters

The dried fruits contain one or two seeds that are compressed together into a mass resembling a single seed. Seeds are oval to roundish, brownish-black, and about 1.5–2 cm long and 1–0.5 cm wide. The whole seed is enclosed in a cream-colored coriaceous covering. The testa is light brownish in color. It can be easily separated from the kernel, which consists of irregularly folded dark violet to brownish cotyledons. Near one end of the seed is the hilum, which forms a circular slightly elevated scar on the testa and a grooved micropyle adjacent to it. The embryo consists of two large cotyledons enclosing a radicle and plumule (Figure 7.1b and c) (Anonymous 2008).

(a)

(b)

(c)

FIGURE 7.1   *Syzygium cumini*: (a) leaves, (b) fruits, and (c) dried seeds.

Microscopic Characters

Transversely cut sections under microscopic observation show thin-walled testa, followed by two- to three-layered tegmen. The embryo is semicircular in shape with a lot of oleoresin canals. A longitudinally cut section of the seed shows the outer narrow testa and tegmen, and the inner wide and thick embryo. The major area of the section consists of a small radicle in the middle of the seed and two thick cotyledons on either side. Cotyledons contain a lot of oleoresin canals and starch grains (Figures 7.2 and 7.3). A detailed transversely cut section under microscopic observation shows a layer of the epidermis of the testa, often found to be attached externally with a few brownish remnants of the parenchymatous tissue of the mesocarp. One to three rows of parenchymatous subepidermis lie underneath the mesocarp. The remaining tissue is the tegmen traversed with bands of sclereids and isolated or groups of stone cells. Obliquely cut vascular strands are located toward the peripheral region. Tannin cells and cluster crystals of calcium oxalates are present. The innermost hyaline layer of the testa contains tangentially elongated, thin-walled, and parenchymatous cells (Figures 7.4 and 7.5) (Anonymous 2008).

Powder and Organoleptic Characters

The dried seed powder is brown in color. Under the microscope, the powder of the seed shows fragments of colored testa in different views, stone cells and sclereids in different shapes and sizes, simple triangular or oval-shaped starch grains in isolated and grouped manners, tannin cells, cluster crystals of calcium oxalate, oleoresin canals, fragments of scalariform tracheids, and fibers. The taste of the dried powder is astringent, and odor is not characteristic (Figure 7.6) (Anonymous 2008).

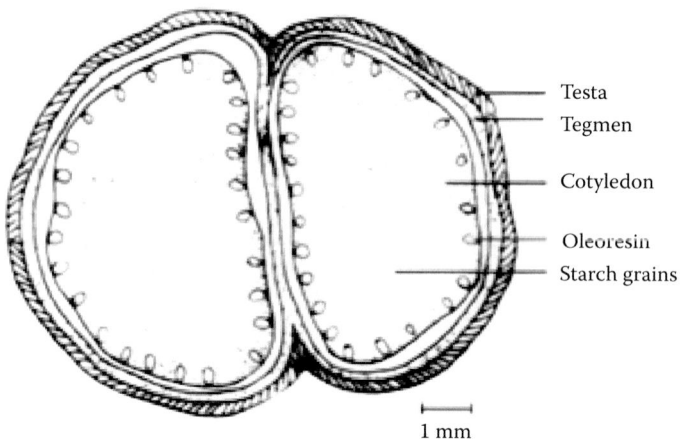

Testa
Tegmen

Cotyledon

Oleoresin
Starch grains

1 mm

FIGURE 7.2   Transverse section (TS) of the *S. cumini* seed (diagrammatic).

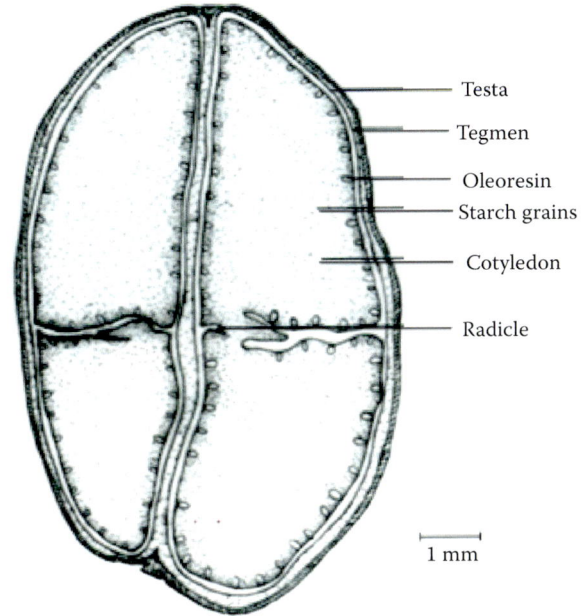

FIGURE 7.3   Longitudinal section (LS) of the *S. cumini* seed (diagrammatic).

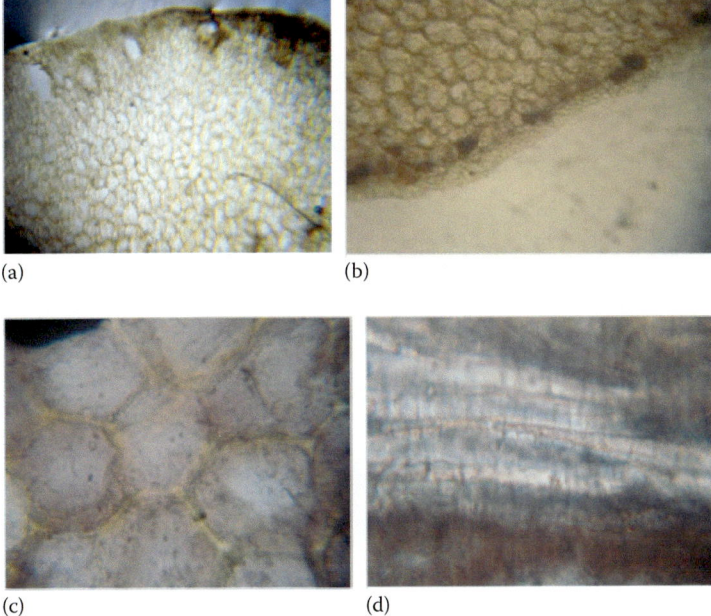

FIGURE 7.4   Microphotographs of the fruit in transverse section (TS): (a) cotyledon showing schizogenous cavities, (b) cells containing starch, (c) mesophyll cells, and (d) testa.

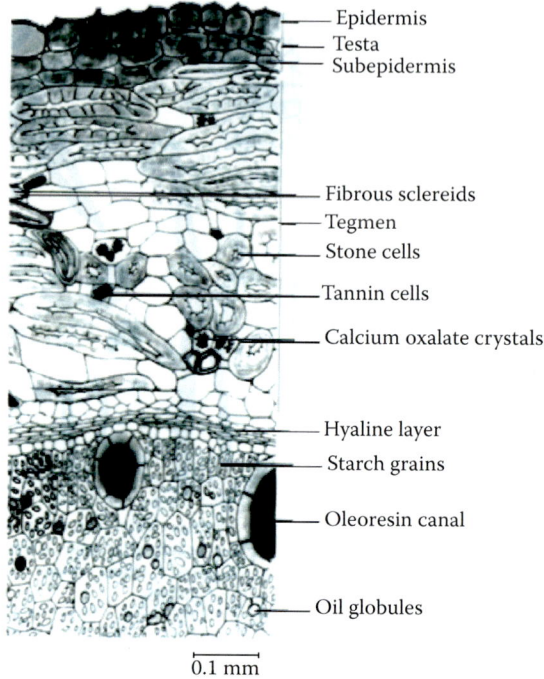

Epidermis
Testa
Subepidermis

Fibrous sclereids
Tegmen
Stone cells

Tannin cells

Calcium oxalate crystals

Hyaline layer
Starch grains

Oleoresin canal

Oil globules

0.1 mm

FIGURE 7.5   Detailed anatomical (microscopical) structure of the seed in transverse section (TS).

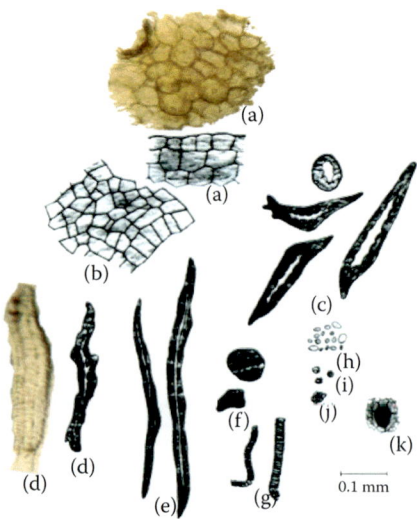

FIGURE 7.6   Powder microscopic characters of *S. cumini* seed: (a) fragment of testa in section view, (b) fragments of testa in surface view, (c) stone cells, (d) sclereids, (e) fibers, (f) tannin cells, (g) tracheids with scalariform thickening, (h) starch grains, (i) cluster crystals of calcium oxalate, (j) group of starch grains, and (k) oleoresin cavity with oil globules.

Physicochemical Standards

The quality control parameters include foreign material, extractive values, and ash values of the dried raw herbal drug. The physicochemical parameters for *S. cumini* seed are as follows (Anonymous 1999, 2008):

Foreign matter:	Not more than 1.0%
Total ash:	Not more than 1.9%
Acid-insoluble ash:	Not more than 0.4%
Alcohol- or ethanol-soluble extract:	Not less than 5.7%
Water-soluble extract:	Not less than 14.5%
Petroleum ether–soluble extract:	Not less than 15%w/w
Chloroform-soluble extract:	Not less than 25%w/w

PHARMACOGNOSTIC STANDARDIZATION OF LEAVES

MACROSCOPIC CHARACTERS

The leaves measure about 10–15 cm long and 4–6 cm wide. These are entire, ovate-oblong, and sometimes lanceolate, and also acuminate, coriaceous, tough, smooth, and shining with numerous nerves uniting within the margin (Figure 7.1a).

ANATOMICAL CHARACTERS

The leaf of *S. cumini* shows a two- to three-layered epidermis. Just beneath the epidermis is the mesophyll tissue that is composed of isodiametric, thin-walled parenchymatous ground tissues and cells. These cells are filled with simple starch grains. In the midrib region, the vascular bundle shows xylem toward the upper epidermis and phloem on the lower side. Starch grains, oil globules, tannin cells, and stone cells are also visible. The leaf contains the upper and lower epidermis, palisade cells, spongy parenchyma, stomata, xylem, phloem, and trichomes (Figure 7.7).

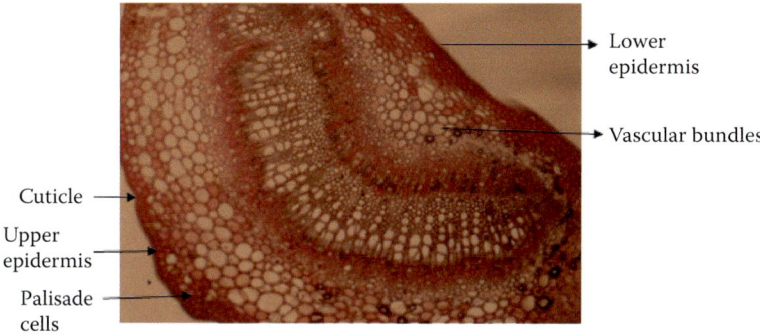

FIGURE 7.7 Transverse section of the leaf through the midrib region of *S. cumini*.

CHEMICAL CONSTITUENTS

Being a member of the family Myrtaceae, *S. cumini* records a number of bioactive molecules, namely, maleic acid, oxalic acid, gallic acid, betulic acid, tannins, protein, flavonoids, and essential oil (Table 7.1). It is also enriched with a number of other bioactive components, like anthocyanins, ellagic acid, glucoside, isoquercetin, myricetin, and kaemferol, in all parts of *S. cumini* (Wilt et al. 1999). Various

TABLE 7.1
Phytochemical Constituents in Different Parts of *Syzygium cumini*

Parts	Chemical Constituents	References
Edible pulp	Vitamin C, vitamin A, riboflavin, nicotinic acid, choline, folic acid, raffinose, glucose, fructose, maleic acid, gallic acid, cyanidin glycoside, cyanidin diglycoside, petunidin, malvidin, delphinidin-3-gentiobioside, and malvidin-3-laminaribioside	Sharma and Seshadri (1955) Anonymous (2002) Teixeira et al. (2006)
Seed	Gallic acid, glycoside jamboline, jambosine, triterpenoid B, pentacyclic triterpenoid (friedelin), tannins, ellagic acid, β-sitoterol, gallitanins, terpenes (1-limonene, dipentene), corilagin, 3,6-hexahydroxydiphenoylglucose, 4,6-hexahydroxydiphenoylglucose, 1-galloylglucose, 3-galloylglucose, quercetin, and elements such as zinc, chromium, vanadium, potassium, and sodium	Chopra et al. (1956) Gupta and Agrawal (1970) Bhatia and Bajaj (1972, 1975) Anonymous (2002) Jagetia and Baliga (2002) Ravi et al. (2004) Evans (2007) Kokate et al. (2008)
Bark	Gallic acid, *Eugenia* triterpenoid B, ellagic acid, pentacyclic triterpenoid (betulinic acid), pentacyclic triterpenoid (friedelin), resin (myricetine), phytosterol (B-sitosterol, myricyl alcohol), eugenin, epi-friedanlol, friedelanol, quercetin, kaempferol, myricetin, gallic acid, ellagic acid, flavonoids, and tannins	Sengupta and Das (1965) Bhatia and Bajaj (1972, 1975) Bhargava et al. (1974) Anonymous (2002) Yogeswar and Sriram (2005) Ivan (2006) Evans (2007) Kokate et al. (2008)
Flower	Oleanolic acid, *Eugenia* triterpenoid A, *Eugenia* triterpenoid B, pentacyclic triterpenoid (Friedelin), tannins, ellagic acid, flavanols (isoquercetin, quercetin, kaempferol, myricetin, myricetin), 3-L-arabinoside, dihydromyricetin, and quercetin galactosides	Bhatia and Bajaj (1972) Subramanian and Nair (1972) Nair and Subramanian (1974) Anonymous (2002) Kokate et al. (2008)
Leaves	Gallitanins, terpenes (1-limonene, dipentene), sesquiterpenes (cadalane type, azulene type), acylated flavanol glycosides, myricetin, myricetin 3-O-4-acetyl-L-rhamnopyranoside, galloyl carboxylase, esterase, betulinic acid, crategolic (maslinic) acid, n-hepatcosane, n-nonacosane, n-hentriacontane, n-octacosanol, n-triacontanol, and n-dotricontanol	Bhatia and Bajaj (1972) Gupta and Sharma (1974) Anonymous (2002) Jagetia and Baliga (2002) Timbola et al. (2002) Yogeswari and Sriram (2005) Kokate et al. (2008)
Roots	Flavonoid glycosides, isorhamnetin 3-O-rutinoside	Vaishnava et al. (1992) Udayan et al. (2006)

minerals and vitamins have also been present, like Ca, Mg, P, Fe, Na, K, Cu, S, Cl, vitamin C, vitamin A, riboflavin, nicotinic acid, choline, and folic acid. Glucose and fructose are the principal source of sweeteners in ripe fruit with no trace of sucrose. Maleic acid is the major acid (0.59% of the weight of fruit (Anonymous 2002; Ivan 2006). A small quantity of oxalic acid has been also reported (Anonymous 2002). Tannins, mainly gallic acid, are responsible for the astringent effect of the fruits (Ivan 2006; Evans 2007). The astringency in fruit is due to the efficiency in combining tissues and proteins and their precipitate. The purple color of the fruit is due to the presence of one or two cyanidin diglycosides (Anonymous 2002; Ivan 2006). The seeds have a greater number of flavonoids and phenolics.

The waxy component of the fleshy pericarp contains a sterol (m.p. 135°C) essential oil. The major component appears to be triterpene hydroxyl acid or oleanolic acid ($C_{30}H_{48}O_3$, m.p. 298°C–300°C) (Anonymous 2002; Ivan 2006). Oleanolic acid is classified into the β-amyrin group of triterpenoids (Evans 2007). The flowers contain oleanolic acid, acetyl oleanolin acid (0.3%, m.p. 260°C–262°C), *Eugenia* triterpenoid A (0.5%), and *Eugenia* triterpenoid B (0.3%). Flowers also contain ellagic acid (0.01%) (Anonymous 2002). Ellagic acid is formed from lactonization of hexahydroxdiphenic acid during chemical hydrolysis of tannins (Kokate et al. 2008). The plant seeds are rich in protein and calcium. The seeds contain tannins (19%), ellagic acid, and gallic acid (1%–2%). A glycoside (jamboline), starch, myricyl alcohol in the unsaponified fraction of seeds, and a small quantity (0.05%) of pale yellow essential oil (specific gravity 0.926 (20°c) are also present (Bhatia and Bajaj 1972, 1975; Anonymous 2002; Ivan 2006; Evans 2007; Kokate et al. 2008). The fresh leaves of the plant contain an essential oil with pleasant odor. The oil contains α-pinene, camphene, β-pinene, myrcene, limonene, cis-ocimene, γ-terpinene, terpinolene, bornyl acetate, α-copaene, β-caryophyllene, α-humulene, γ-cadinene, δ-cadinene, trans-ocimene, cis-ocimene, β-myrcene, α-terpineol, dihydrocarvyl acetate, geranyl butyrate, terpinyl valerate, β-selinene, calacorene, α-murolol, α-santalol, cis-farnesol, lauric acid, myristic acid, palmitic acid, stearic acid, oleic acid, linoleic acid, malvalic acid, sterculic acid, vernolic acids, and unsaponifiable matter of the seed fat (Sowjanya et al. 2013).

The yield and physical characteristics of the leaf oil vary according to the season of collection. The essential oil is reported to be responsible for the antibacterial activity of the leaves of *S. cumini* (Anonymous 2002; Jagetia and Baliga 2002). Stem bark contains pentacyclic triterpenoid betulinic acid (m.p. 306°C–310°C) (Anonymous 2002; Ivan 2006). Betulinic acid is a naturally occurring triterpenoid, which has shown selective cytotoxicity against a number of specific tumors and promising activity against HIV, malaria, and demonstrated immunomodulatory as well as inflammatory actions. A plant sterol, β-sitosterol, is found in almost all parts of the plant. It has the same chemical structure as cholesterol. It has very beneficial pharmacological activity, for example, as an anti-inflammatory and in lowering blood cholesterol (Wang and Ng 1999; Yun et al. 2003). Friedelin ($C_{30}H_{50}O$, m.p. 256°C–260°C) is also a pentacyclic triterpenoid found in plants (Anonymous 2002; Ivan 2006). Plant bark also contains a substance that is an ester of epi-friedelanol ($C_{30}H_{51}OH$) with a fatty acid ($C_{27}H_{55}COOH$). It also contains tannins (10%–12%), gallic acid, ellagic acid, and the resin myricetin (Bhatia and Bajaj 1972).

CONCLUSION

Jamun (*S. cumini*) is widely used in the indigenous systems of medicine for the treatment of various diseases, especially for the management of diabetes. This plant has an extensive range of biological activities. In the many recognized systems of medicine like Ayurveda, it is already mentioned in the herbal remediation for a number of human ailments having various pharmacological activities, such as antidiarrheal, astringent, digestive, antibacterial, antioxidant, antiviral, anti-inflammatory, antitumor, and radioprotective, but the most important activity is antidiabetic. There are many herbal formulations for the management of diabetes, like diabecon, jambalsava, and madhunasini, which contain *S. cumini* as a main ingredient. All parts of this plant are enriched with anthocyanins, ellagic acid, glucoside, isoquercetin, myrecetin, and kaemferol. Various studies of *S. cumini* as an antidiabetic agent, with its possible mechanism of action and delaying complications of diabetes, such as cataract and neuropathy, have been reported. Most of the pharmacological evaluation has been carried out using the seeds of *S. cumini*, but the pharmacological potential of other parts of the plant is of utmost importance and needs to be worked out in the future. The metabolic pathway of the phytochemical constituents needs to be understood, and clinical trials of bioactive fractions or extracts and compounds from jamun also need to be carried out.

REFERENCES

Anonymous. 1999. *The Ayurvedic Pharmacopoeia of India*. Part I, 1st ed., vol. II. New Delhi: Controller of Publications, 54–57.

Anonymous. 2002. *Wealth of India: A Dictionary of Indian Raw Materials and Industrial Products*. Raw Materials 10. New Delhi: National Institute of Science Communication, Council of Scientific and Industrial Research, 100–107.

Anonymous. 2008. *Quality Standards of Indian Medicinal Plants*. Vol. 7. New Delhi: Indian Council of Medical Research, 295.

Bhargava, K. K., R. Dayal, and T. R. Seshadri. 1974. Chemical component of *Syzygium cumini* stem bark. *Curr. Sci.* 43: 645–646.

Bhatia, I. S., and K. L. Bajaj. 1972. Tannins in black-plum (*Syzygium cumini* L.) seeds. *Biochem. J.* 128: 56.

Bhatia, I. S., and K. L. Bajaj. 1975. Chemical constituents of the seeds and bark of *Syzygium cumini*. *Planta Med.* 28: 346–352.

Chandrasekaran, M., and V. Venkatesalu. 2004. Antibacterial and antifungal activity of *S. jambolanum* seeds. *J. Ethnopharmacol.* 91: 105–108.

Chopra, R. N., S. I. Nayar, and I. C. Chopra. 1956. *Glossary of Indian Medicinal Plants*. New Delhi: CSIR-Publication and Information Department, 238.

Evans, W. C. 2007. *Trease and Evans, Pharmacognosy*. 15th ed. New Delhi: Elsevier Indian Pvt. Ltd., 420–421.

Gupta, D. R., and S. K. Agrawal. 1970. Chemical examination of the unsafonicable matter of the seed fat of *Syzygium cumini*. *Sci. Cult.* 36: 298.

Gupta, G. S., and D. P. Sharma. 1974. Triterpenoid and other constituents of *Eugenia jambolana* leaves. *Phytochemistry* 13: 2013–2014.

Hegnauer, R. 1990. *Chemotaxonomie der Pflanzen*. Vol. 9. Berlin: Birkhäuser Verlag, 119–129.

Helmstaedter, A. 2007. Antidiabetic drugs used in Europe prior to the discovery of insulin. *Pharmazie* 62: 717–720.

Ivan, A. R. 2006. *Medicinal Plants of the World: Chemical Constituents, Traditional Uses and Modern Medicinal Uses*. Totowa, NJ: Human Press, 283–289.

Jagetia, G. C., and M. S. Baliga. 2002. *Syzygium cumini* (jamun) reduces the radiation-induced DNA damage in the cultured human peripheral blood lymphocytes: A preliminary study. *Toxicol. Lett.* 132: 19–25.

Jain, S. K. 1991. *Dictionary of Indian Folk Medicine and Ethnobotany*. New Delhi: Deep Publications.

Joshi, S. G. 2004. *Medicinal Plants*. New Delhi: Oxford and IBH Publications, 294.

Khare, C. P. 2004. *Encyclopedia of Indian Medicinal Plants*. New York: Springer-Verlag, 207–208.

Kirtikar, K. R., and B. D. Basu. 1975. *Indian Medicinal Plants*. Vol. II. New Delhi: Jayyed Press, 1052–1053.

Kokate, C. K., A. P. Purohit, and S. B. Gokhale. 2008. *Pharmacognosy*. 14th ed. Pune: Nirali Prakashan, 257–258.

Martinez, S. B., and M. J. Del Valle. 1981. Storage stability and sensory quality of duhat (*Syzygium cumini* Linn.) anthocyanins as food colorant. *UP Home Econ. J.* 9: 1.

Morton, J. 1987. Jambolan. In *Fruits of Warm Climates*, 375–378. Miami: Julia Morton.

Nadkarni, K. M. 1954. *Indian Materia Medica*. Vol. I. Bombay: Popular Book Depot, 516–518.

Nair, A. G. R., S. Krishnan, C. Ravikrishna, and K. P. Madhusudanan. 1999. New and rare flavonol glycosides from leaves of *Syzygium samarangense*. *Fitoterapia* 70: 148–151.

Nair, A. G. R., and S. S. Subramanian. 1974. Chemical examination of the flowers of *Eugenia jambolana*. *J. Sci. Indust. Res.* 21B: 457–458.

Newman, D. J., G. M. Cragg, and K. M. Snader. 2003. Natural products as sources of new drugs over the period 1981–2002. *J. Nat. Prod.* 66: 1022–1037.

Ravi, K., B. Ramchandran, and S. Subramanian. 2004. Protective effect of *Eugenia jambolana* seed kernel on tissue antioxidants in streptozotocin induced diabetic rats. *Biol. Pharm. Bull.* 78: 1212–1217.

Reynertron, K. A., M. J. Basile, and E. J. Kennelly. 2005. Antioxidant potential of seven myrtaceous fruits. *Ethnobot. Res. Appl.* 3: 25–35.

Ross, I. A. 2003. *Medicinal Plants of the World*. 2nd ed., vol. 1. Totowa, NJ: Humana Press, 445–451.

Sengupta, P., and P. B. Das. 1965. Terpenoids and related compounds. Part IV. Triterpenoids from the stem bark of *Syzygium cumini*. *Ind. Chem. Soc.* 42: 255–258.

Sharma, J. N., and T. R. Seshadri. 1955. Survey of anthocyanins from Indian sources: Part II. *J. Sci. Ind. Res.* 14: 211–214.

Sharma, P., and P. M. Mehta. 1969. *Dravyaguna Vignyan*. Parts II and III. Varanasi: Chowkhamba Vidyabhawan, 586.

Sowjanya, K. M., J. Swathi, K. Narendra, and A. K. Satya. 2013. A review on phytochemical constituents and bioassay of *Syzygium cumini*. *Int. J. Nat. Prod. Sci.* 3: 1–11.

Subramanian, S. S., and A. G. R. Nair. 1972. Flavonoids of the flowers of *Eugenia jambolana*. *Curr. Sci.* 41: 703–704.

Teixeira, C. C., F. D. Fuchs, L. S. Weinert, and J. Esteves. 2006. The efficacy of folk medicines in the management of type 2 diabetes mellitus: Results of a randomized controlled trial of *Syzygium cumini* (L) Skeels. *J. Clin. Pharmacol. Ther.* 31: 1–5.

Timbola, A. K., B. Szpoganicz, A. Branco, F. D. Monache, and M. G. Pizzolatti. 2002. A new flavonoid from leaves of *Syzygium cumini*. *Fitoterapia* 73: 174–176.

Udayan, P. S., G. Satheesh, K. V. Tushar, and I. Balachandran. 2006. Medicinal plants used by the Malayali tribe of Shevaroy hills, Yercaud, Salem district, Tamil Nadu. *Zoos Print J.* 21: 2223–2224.

Vaishnava, M. M., A. K. Tripathy, and K. R. Gupta. 1992. Flavonoid glycosides from roots of *Syzygium cumini. Fitoterapia* 63: 259–260.

Veigas, J. M., M. S. Narayan, and B. N. Beelwarne. 2007. Chemical nature stability and bioefficacies of anthocyanins from fruit peel of *S. cumini* Skeels. *Food Chem.* 105: 619–627.

Wang, H. X., and T. B. Ng. 1999. Natural products with hypoglycemic, hypotensive, hypocholesterolemic, antiatherosclerotic and antithrombotic activities. *Life Sci.* 65: 2663–2677.

Wilt, T. J., R. MacDonald, and A. Ishani. 1999. β-Sitosterol for the treatment of benign prostatic hyperplasia: A systematic review. *Br. J. Urol. Int.* 83: 976–983.

Yogeswari, P., and D. Sriram. 2005. Betulinic acid and its derivatives: A review on their biological properties. *Curr. Med. Chem.* 12: 657–666.

Yun, Y., S. Han, E. Park, D. Yim, S. Lee, K. Cho, and K. Kim. 2003. Immunomodulatory activity of betulinic acid by producing pro-inflammatory cytokines and activation of macrophages. *Arch. Pharm. Res.* 26: 1087–1095.

8 Biological Activities of *Syzygium cumini* and Allied Species

Varughese George and Palpu Pushpangadan

CONTENTS

INTRODUCTION

The genus *Syzygium* is of commercial importance with timber, fruit yielding, and medicinally utilized plants (Anonymous 1976). The most economically important species is *Syzygium aromaticum* (clove), of which the unopened flower buds are an important spice. Clove is well known in food preparation and as a traditional remedy for asthma (Kim et al. 1998), digestive system disorders (Baytop 1999), dental disorders, respiratory disorders, headaches, and sore throat in Asian countries (Chaieb et al. 2007). It is used as an anticarcinogenic agent (Zheng et al. 1992). *Syzygium cumini* has been used against dysentery (Shafi et al. 2002), to treat inflammation (Chaudhari et al. 1990) and diabetes mellitus (Bhattarai 1992). *S. cumini* is a traditional medicinal plant in Brazil for its antileishmanial and antifungal activity (Fernada et al. 2008). *Syzygium zeylanicum* is reported to be a stimulant and antirheumatic (Anonymous 1976). The bark of *Syzygium guineense* is used in traditional medicine to treat gastrointestinal upsets and diarrhea (Noudogbessi et al. 2008). *Syzygium samarangense* is cultivated in many parts of India for its edible fruit (Anonymous 1976). *Syzygium jambos* is used traditionally in sub-Saharan Africa to treat infectious diseases (Djipa et al. 2000). In India, it has been used, in a mix with honey or milk, to treat diabetes and digestive diseases, and the fresh fruits have been taken orally to treat stomachache (Duraipandiyan et al. 2006). It is also used to treat illnesses caused by bacterial, fungal, and viral pathogens (Kusumoto et al. 1995); ulcers in the genitourinary tract caused by *Candida albicans*; and cold, cough, fever, and skin problems (Chandrasekaran and Venkatesalu 2004). Acetone and aqueous extracts from the bark of *S. jambos* showed effective antimicrobial activity against *Staphylococcus aureus*, *Yersinia enterocolitica*, and coagulase-negative staphylococci (Duraipandiyan et al. 2006). Antinociceptive activity of leaf extracts of *S. jambos* has also been studied (Ávila-Peña et al. 2007). The ripe fruit may be eaten raw or used for flavoring purposes, but its decoction is a reputed febrifuge (Morton 1987). The plant is used in Polynesian traditional medicine for the treatment of infectious diseases and has been found to elicit antiviral, antifungal, and antibacterial activities (Locher et al. 1995).

MEDICINAL AND PHARMACOLOGICAL IMPORTANCE OF ESSENTIAL OILS FROM THE GENUS *SYZYGIUM*

Essential oil from *S. aromaticum* is reported to possess anthelmintic, analgesic, antibacterial, antifungal, and anticancer properties (Webb and Tanner 1944; Bullerman et al. 1977; Zheng et al. 1992). The essential oil of *S. aromaticum* shows anti-inflammatory, cytotoxic and anesthetic activities (Chaieb et al. 2007). Cai and Wu (1996) reported that the clove oil possessed antimicrobial activity against oral bacteria commonly associated with dental caries and periodontal disease. Therefore, *S. aromaticum* is used as an ingredient in toothpaste and mouth fresheners in India. Clove oil has been reported by many researchers as a good source of antifungal compounds (Delespaul et al. 2000). The essential oil extracted from clove is used as a topical application to relieve pain and promote healing in herbal medicine. It also finds use in the fragrance and flavoring industries. In South Korea, the oil is used as a medicine for the treatment of asthma and various allergic disorders (Kim et al. 1998). It is reported that clove oil could be used as a potential chemopreventive agent due to its antioxidant properties (Lee and Shibamoto 2001). Clove oil was found to be cytotoxic even at low concentrations, with up to 73% of this effect attributable to eugenol (Prashar et al. 2006). An antimicrobial activity study on leaf oil of *Syzygium travancoricum* showed that it was found to be more effective against fungi than bacteria (Radha et al. 2002).

BIOLOGICAL ACTIVITY OF *SYZYGIUM CUMINI* AND ALLIED SPECIES

S. cumini and other members of the genus *Syzygium* are credited with a host of biological activities (Sagrawat et al. 2006). Pharmacological investigations carried out on *S. cumini* and related species confirmed the therapeutic potential of these plants and their importance in traditional, classical, and modern medicine (Ayyanar et al. 2013; Costa et al. 2013; Noor et al. 2013; Srivastava and Chandra 2013). *S. cumini* and other members of the genus *Syzygium* possess antimicrobial, antiviral, antidiabetic, antifertility, antioxidant, antihelminthic, hypolipidemic, antiasthmatic, anti-inflammatory, hepatoprotective, antinephrotoxic, anticytotoxic, central nervous system (CNS) protective, antinociceptive, cytotoxic, antimutagenic, spasmolytic, antifeedant, cosmeceutical, anticancer, and antimite activities. They also act against fluoride-induced toxicity. A brief review on the biological activity and therapeutic potential of *S. cumini* and other important members of the genus *Syzygium* is presented in this chapter.

ANTIMICROBIAL AND ANTIVIRAL ACTIVITIES

Oliveira et al. (2007) reported the antimicrobial activity of *S. cumini* leaf extract. The crude hydroalcoholic extract was active against *Candida krusei* (inhibition zone of 14.7 ± 0.3 mm and minimum inhibitory concentration [MIC] = 70 µg/ml) and multiresistant strains of *Pseudomonas aeruginosa*, *Klebsiella pneumoniae*, and *S. aureus*.

Kothari et al. (2011) studied the antibacterial activity of *S. cumini* seed extracts prepared in methanol and ethanol by disc diffusion and broth dilution assays. Both extracts exerted a broad spectrum of bacteriostatic action against different gram-positive and gram-negative bacteria. Their MIC against susceptible organisms ranged from 154 to 656 µg/ml. The highest total activity was registered by the ethanol extract against *Staphylococcus epidermidis*.

The antibacterial activity of the essential oil of *S. cumini* was assayed using MICs. The main oil constituents were α-pinene (17.53%), α-terpineol (16.67%), and allo-ocimene (13.55%). The oil of *S. cumini* demonstrated strong inhibition activity against the tested bacterial strains, such as gram-positive bacteria, *Bacillus subtilis* ATCC 6633, *S. aureus* ATCC 6538, and *Sarcina lutea* ATCC 9341, and gram-negative bacteria, *Escherichia coli* ATCC 8739, *P. aeruginosa* ATCC 9027, *Agrobacterium tumefaciens* ATCC 1593-2, and *Pectobacterium carotovorum* subsp. *carotovorum* ATCC 39048 (Elansary et al. 2012).

The aqueous extract of the fruits of *S. cumini* was found to be active against the growth of *Fusarium oxysporum*, while the aqueous extract of the bark inhibited the growth of *Alternaria alternata* (Gupta and Bhadauria 2012).

The aqueous, methanolic, hexane, and ethyl acetate extracts of the leaves of *S. cumini* exhibited antimicrobial activity against dental caries causing strains such as *Streptococcus viridans*, *S. mutans*, *E. coli*, *P. aeruginosa*, *S. aureus*, and *B. sub-tilis*. The largest zone of inhibition was obtained with the methanolic extract against *E. coli* (20 mm). The MIC of extract was also determined against the selected micro-organisms showing a zone of inhibition of ≥8 mm. The study concluded that the leaves of *S. cumini* possessed very good antibacterial activity against dental caries–causing microorganisms (Tahir et al. 2012).

Javaid and Samad (2012) reported the antifungal activity of the methanolic extract of the leaves of *S. cumini* against two strains of *Alternaria alternata*, isolated from dying-back trees of two *Eucalyptus* spp., namely, *E. citriodora* and *E. globulus*. All the concentrations (1%, 2%, and 5% w/v) of the methanolic extract significantly reduced the fungal biomass. There were reductions in the ranges of 82%–88% due to different concentrations of the leaf extracts of *S. cumini*.

Vasavi et al. (2014) screened the ethanol extract of *S. cumini* for anti-quorum-sensing (QS) activity using a *Chromobacterium violaceum* CV026 biosensor bioassay and reported that the ethyl acetate fraction (EAF) inhibited C6-*N*-acyl homoserine lactone (AHL)-mediated violacein production in *C. violaceum* CV026.

Gupta et al. (2014) synthesized silver nanoparticles (AgNPs) using the metha-nolic extract of *S. cumini* leaf, and these AgNPs exhibited significant antimicro-bial activity against *S. aureus*, *E. coli*, *B. subtilis*, *K. pneumoniae*, *P. aeruginosa*, *Mycobacterium smegmatis*, *Trichophyton rubrum*, *Aspergillus* sp., and *C. albicans*. AgNPs also inhibited biofilm formation in a wide range of AgNP concentrations and showed good compatibility of AgNPs with human embryonic kidney cells (HEK 293). The results indicated that AgNPs can be used for designing novel antibacterial, antifungal, and antibiofilm agents.

J. P. Singh et al. (2016) examined the antioxidant and antimicrobial activities of jambolan fruit polyphenols against reference pathogenic strains (*S. aureus*, methicillin-resistant *Staphylococcus aureus* [MRSA], *E. coli*, *K. pneumoniae*, and *C. albicans*) and compared them with those of polyphenol standards (gallic acid, quercetin, caffeic acid, sinapic acid, and delphinidin chloride). The study showed gallic acid and quercetin with higher antioxidant activities (2,2-diphenyl-1-picrylhydrazyl [DPPH] and 2,2′-azino-bis(3-ethylbenzthiazoline-6-sulfonic acid) [ABTS]) than the other standards, and the jambolan fruit polyphenol extract exhibited a broad-spectrum antimicrobial activity against reference pathogenic strains with a zone of inhibition and MIC in the range of 14.3–23.0 mm and 0.5–2.5 mg/ml, respectively.

R. Singh et al. (2016) tested silver, gold (AuNPs), and gold-silver bimetallic (Au-AgNPs) nanoparticles synthesized from medicinal plants, such as *Barleria prio-nitis*, *Plumbago zeylanica*, and *S. cumini*, against *Mycobacterium tuberculosis* and *Mycobacterium bovis* Bacillus Calmette–Guérin (BCG). Au-AgNPs synthesized from *S. cumini* showed profound efficiency, specificity, and selectivity to kill myco-bacteria, with their selectivity index in the range of 94–108.

Voigt et al. (2013) tested the antimicrobial activity of hydroalcoholic extracts of *S. cumini* leaves against bacteria related to bovine mastitis and found that the hydroalcoholic extracts of fresh plant are effective against *Streptococcus uberis*, *S. agalacteae*, and *S. dysgalacteae*.

Saroj et al. (2015) reported the essential oil constituents of *S. cumini*, such as 7-hydroxycalamenene, 7-acetoxycalamenene, 1-epi-cubenol, α-terpineol, and (Z)-β- and (E)-β-ocimene, as potential antifungal agents against the phytopathogenic fungi, *Rhizoctonia solani* AG 4HG-III, and *Choanephora cucurbitarum*.

Rahnama et al. (2012) reported antibacterial effects of *S. aromaticum*. They reported synergetic effects of a combination of essential oils extracted from *Myristica fragrans*, *Zataria multiflora*, *S. aromaticum*, and *Zingiber officinale* against *Listeria monocytogenes* in brain heart infusion broth. The MIC and minimum bacterial concentration (MBC) with essential oils and pH values (5, 6, and 7) alone and in combination with nicin (5 µg/ml) were determined. By decreasing pH, these antibacterial effects were increased.

Cock (2012) reported the antimicrobial activity of methanolic extracts of *Syzygium australe* and *Syzygium leuhmannii* leaves against a wide range of gram-positive and gram-negative bacteria.

Harris et al. (2011) reported the limited screening of tropical plant extracts for inhibitory activity against the essential enzyme peptidyl-tRNA hydrolase (Pth). Initial screening was conducted through an electrophoretic mobility assay and Northern blot detection. The ability of Pth to cleave the peptide-tRNA ester bond was assessed. The ethanol bark extract of *Syzygium johnsonii* showed strong inhibitory potential. Molecular docking studies point to *Syzygium* polyphenolics as the potential source of inhibition.

The antibacterial activity of the essential oils from the leaves of *S. cumini* and *S. travancoricum*, collected from Kerala, India, was evaluated against *Bacillus sphaericus*, *B. subtilis*, *S. aureus*, *E. coli*, *P. aeruginosa*, and *Salmonella typhimurium*. The essential oils showed considerable antibacterial activity, especially against *S. typhimurium* (Shafi et al. 2002).

Gopan et al. (2008) reported the chemical composition and antimicrobial activity of the leaf oil from *Syzygium gardneri*. The major constituent in the oil was caryophyllene oxide (49.6%). Monoterpenes were detected at a low percentage (1.2%). The oil was tested for antimicrobial activity against gram-positive and gram-negative bacteria, as well as against the fungi *C. albicans* and *C. glabrata*.

Chandra et al. (1998) reported the antiviral activity of an extract of *Syzygium megacarpum* against encephalitis causing virus.

ANTIDIABETIC ACTIVITY

Eight known compounds isolated from the roots of *S. cumini* were evaluated by Shi-Fei et al. (2009) for their ability to enhance the glucose consumption in insulin-resistant L6 muscle cells induced by high-concentration insulin and glucose. All the compounds significantly enhanced the glucose consumption of insulin-resistant cells in the presence and absence of insulin. The glucose consumption was increased by 17.35% and 51.11% by friedelin at a concentration of 10 µg/ml without insulin and 5,7,3′,4′,5′-pentahydroxyflavone at a concentration of 0.1 μg/ml with insulin, respectively.

Rajput et al. (2009) investigated the possible antidiabetic effects of alcoholic extract of bark of *S. cumini*, *Ficus benghalensis*, and *Butea monosperma* in alloxan-induced diabetic rats. The acute oral toxicity showed that the polyherbal formulation was safe until 2000 mg/kg body weight, and no macroscopical organ abnormalities were observed in acute oral models. Oral administration of 200 mg/kg body weight of aqueous solution of polyherbal formulation for seven days exhibited significant reduction in the blood glucose level in diabetic rats. A comparison was made between the action of a prepared polyherbal formulation and a known antidiabetic drug, glibenclamide 600 μg/kg body weight. The antidiabetic effect of the polyherbal formulation was nearly comparable to that observed with glibenclamide.

Shafiuddin et al. (2011) reported antidiabetic, antihyperlipidemic, and antioxidant activity of a polyherbal formulation containing *Alpinia galangal*, *Swertia chirata*, *S. cumini* (= *Eugenia jambolana*), *Momordica charantia*, *Gymnema sylvestre*, *Fumaria officinalis*, and *Azadirachta indica* in streptozotocin (STZ)-induced diabetic rats. Diabetes was induced by a single intraperitoneal injection of STZ (50 mg/kg) in male Wistar rats. Rats with fasting blood glucose levels of ≥250 mg/dl, after seven days of STZ administration, were randomized into different groups (six rats in each group). The groups were treated with an oral administration of aqueous extract of formulation (250 and 500 mg/kg body weight) and glibenclamide (2.5 mg/kg body weight) for 21 days. At the end of the study, blood glucose, lipid profiles, and enzymatic and nonenzymatic liver antioxidant levels were estimated. Oral administration of formulation for 21 days significantly reduced the blood glucose level in STZ-induced diabetic rats. Supplementation with formulation showed significant improvement in the lipid levels and glucose tolerance. In addition, formulation supplementation decreased oxidative stress by improving endogenous antioxidant levels.

Saravanan et al. (2011) studied the hypolipidemic activity of the polyherbal drug Zignidd in STZ-induced diabetic rats. Zignidd contains 17 species, including *S. cumini* (= *E. jambolana*) bark (6 g). The hypolipidemic effect of Zignidd (polyherbal formulation) was evaluated in STZ-induced diabetic dyslipidemia in Wistar rats by assaying their triglyceride (TG), total cholesterol (TC), and high-density lipoprotein (HDL) levels. Diabetic Wistar rats were treated orally once a day up to 14 days with doses of 200 and 400 mg/kg body weight of Zignidd. Glibenclamide 4 mg/kg was used as the standard drug. TG, TC, and HDL were estimated. Low-density lipoprotein (LDL) and very low-density lipoprotein (VLDL) levels were calculated from the above measurements by using the Friedewald formula. The anti-atherogenic index (AAI) was calculated from TC and HDL. A significant decrease in TG ($p < 0.05$), TC ($p < 0.05$), VLDL ($p < 0.005$), and LDL ($p < 0.05$) and an increase in HDL ($p < 0.05$) and AAI were observed after 14 days treatment of Zignidd when compared with diabetic control. The results demonstrate that Zignidd possesses a significant hypolipidemic effect against STZ-induced diabetic dyslipidemia in rats.

Middha et al. (2012) studied the hypoglycemic agents used in traditional medicine and showed that *S. cumini* and *Trigonella foenum-graecum* (seed), *Moringa alba* (leaf), *Punica granatum* (peel), *Emblica officinalis*, and *M. charantia* possessed the highest hypoglycemic activity of varying degrees. *S. cumini* and *T. foenum-graecum* showed better activity in neutral and basic media than others.

Alam et al. (2012) studied the putative antidiabetic constituents from *S. cumini* leaves. From the nuclear magnetic resonance (NMR) data, four different compounds, lupeol, 12-oleanen-3-β-acetate, stigmasterol, and β-sitosterol, were identified from the *n*-hexane fraction of the plant extract. These compounds have potential antidiabetic activities that support the traditional use of the leaves for treating diabetes.

Santos et al. (2012) conducted a transversal descriptive study with 158 diabetic patients enrolled in the program HIPERDIA at the Programas de Saúde da Família of Vitoria de Santo Antão–Pernambuco, Brazil, between July 2009 and May 2010, with data collected by means of a structured form. Among interviewees, 36% reported the use of medicinal plants considered hypoglycemic. A total of 35 different plants belonging to 24 families were cited, and the most frequently cited species belonged to Asteraceae (12.5%) and Myrtaceae (9.37%). The most prevalent medicinal plant was "pata-de-vaca" (*Bahuinia* sp.), with 16.8%, followed by "azeitona roxa" (*S. cumini*: syn. *Syzygium jambolanum*) and "insulin" (*Cissus sicyoides*). Most individuals (58%) cultivated the medicinal plant they used, and for those who acquired them, the main source was "raizeiros" (people similar to healers but who only sell medicinal plants) (28.16%).

Srivastava et al. (2012) reported the hypoglycemic and hypolipidemic activity of *S. cumini* (= *E. jambolana*) pulp and seed extract in STZ-induced diabetic albino rats. *S. cumini* pulp and seed extract at a dose of 200 mg/kg body weight showed therapeutic effect in diabetes-induced albino rats.

Three new hydrolyzable tannins (HTs), two gallotannins, jamutannins A and B, and an ellagitannin, iso-oenothein C, along with eight known phenolic compounds were isolated from the seeds of *S. cumini* (= *E. jambolana*). The structures were elucidated on the basis of spectroscopic data analysis. All compounds isolated were evaluated for α-glucosidase inhibitory effects compared with the clinical drug acarbose, and the hydrolysable tannins showed an α-glucosidase inhibitory effect (Omar et al. 2012).

Roy et al. (2014) performed an *in silico* analysis to understand the binding of small molecules from *Syzygium* sp. with α-glucosidase inhibitory potential in human maltase glucoamylase (MGAM), a potent molecular target for controlling postprandial glucose surplus in type 2 diabetes. The study confirmed myricetin as the most potent inhibitor, with a high binding affinity for both N- and C-terminals of MGAM.

In an attempt to provide evidence of the empirically supported benefits of the use of *S. cumini* in diabetes, De Bona et al. (2014) demonstrated that *S. cumini* aqueous leaf extract (ASc) was found to be more effective in preventing an increase in adenosine deaminase (ADA) activity than phenolic compounds, and that ASc might collaborate to improve endothelial dysfunction, antioxidant, anti-inflammatory, and antithrombotic properties of adenosine by affecting its metabolism.

Tripathi and Kohli (2014) reported that daily continuous oral treatment of STZ-induced diabetic Wistar albino rats with ethanolic and aqueous extracts of stem bark of *S. cumini* for three weeks resulted in significant reductions in fasting blood glucose levels compared with diabetic controls.

Tong et al. (2014) demonstrated that *S. cumini* (= *E. jambolana*) contained potent α-amylase inhibitors because of monomeric and polymeric HTs, which could potentially alleviate postprandial hyperglycemia in diabetic patients.

Chinni et al. (2014) reported the inhibitory potential (with a lower K-i value) of *S. cumini* fruits in CYP2C9-mediated diclofenac metabolism in human liver microsomes, suggesting the potential pharmacokinetic and pharmacodynamic interactions of *S. cumini* fruits when concomitantly administered with other drugs.

The highly enriched fractions obtained from a broad ethyl acetate fraction of seed extract of *S. cumini* yielded maslinic acid, 5-(hydroxymethyl) furfural, gallic acid, valoneic acid dilactone, rubuphenol, and ellagic acid. The isolated constituents showed promising *in vitro* antidiabetic activity during aldose reductase (AR) and protein-tyrosine phosphatase 1B (PTP1B) inhibition assays (Sawant et al. 2015).

Mahajan et al. (2015) demonstrated that the daily administration of "GSPF kwath," a polyherbal formulation containing 10 herbs, including *S. cumini* regularly, for six months showed significant reduction of blood glucose and glycosylated hemoglobin levels in patients with type 2 diabetes mellitus. There was also a significant increase in HDL cholesterol levels and concomitant decreases in TC, TG, LDL, and VLDL levels. Patients exhibited a significant improvement in the biochemical markers for oxidative stress.

Bitencourt et al. (2015) evaluated the effect of the aqueous seed extract of *S. cumini* (ASc) on ADA activity, lipoperoxidation (cerebral cortex, kidney, liver, and pancreas), and biochemical (serum) and histopathological (pancreas) parameters in diabetic rats. The study revealed that the short-term treatment with ASc has an important protective role under pathophysiological conditions caused by the early stage of diabetes mellitus.

Sanches et al. (2016) studied the metabolic effects of hydroethanolic extract of *S. cumini* leaf (HESc) on lean and monosodium L-glutamate (MSG)–induced obese rats. Obese rats treated with HESc showed a twofold increase in lipolytic activity in the periepididymal fat pad, as well as brought TG levels in serum, liver, and skeletal muscle back to close to those found in lean animals. HESc also improved hyperinsulinemia and insulin resistance in obese + HESc rats, resulting in partial reversal of glucose intolerance, compared with obese rats. HESc had no effect in lean rats. The data demonstrate that *S. cumini* leaf improved peripheral insulin sensitivity via stimulating or modulating β-cell insulin release, which was associated with improvements in metabolic outcomes in MSG-induced obese rats.

Bitencourt et al. (2016) assessed the protective effects of an aqueous extract (ASc) of polymeric nanoparticles containing ASc (NPASc) prepared from *S. cumini* seeds. Both formulations showed high protection against oxidized LDL (ox-LDL) particles, antifungal activity against *Candida guilliermondii* and *Candida haemulonii*, and no acute toxicity in the Artemia salina lethality assay and in rats. The above findings highlight the possibility of expanding the use of *S. cumini* to ameliorate the chronic complications of diabetes mellitus, and the lack of toxicity of the nanoparticles indicates NPASc could be a safe candidate for drug delivery systems.

Manaharan et al. (2012) isolated six flavonoid compounds from *Syzygium aqueum* leaf extract as potential antihyperglycemic agents. The compounds isolated were 4-hydroxybenzaldehyde, myricetin-3-*O*-rhamnoside, europetin-3-*O*-rhamnoside, phloretin, myrigalone-G, and myrigalone-B. The compounds myricetin-3-*O*-rhamnoside and europetin-3-O-rhamnoside showed high inhibitory activities, with half-maximal effective concentration (EC_{50}) values of 1.1 and 1.9 µM against α-glucosidase and EC_{50} values of 1.9 and 2.3 µM against α-amylase, respectively.

Oral administration of an aqueous extract of *S. jambos* complemented with insulin injection in male C57BLKS/J (db/db) and C57BL/J (ob/ob) genetically obese mice for 10 weeks showed that the C57BL/J ob/ob mice on *S. jambos* treatments showed better blood glucose modulation over time (Gavillan-Suarez et al. 2015).

Chuan et al. (2012) reported that a fraction from *S. samarangense* and Perry fruit extract ameliorates insulin resistance via modulating insulin signaling and the inflammation pathway in tumor necrosis factor α–treated FL83B mouse hepatocytes.

In an experimental study to investigate the glucose uptake activity of the water extracts from the leaves and fruit of edible Myrtaceae plants, including guava (*Psidium guajava*), wax apples (*S. samarangense*), Pu-Tau (*S. jambos*), and Kan-Shi Pu-Tau (*S. cumini*) in the insulin-resistant FL83B mouse hepatocytes, Chang and Shen (2013) inferred that *S. samarangense* fruit extract (SSFE) exhibited the highest glucose uptake activity. They postulated that vescalagin, an active component in *S. samarangense*, might alleviate the insulin resistance in mouse hepatocytes.

ANTIOXIDANT ACTIVITY

Banerjee et al. (2005) reported the antioxidant activity of the fruit skin of *S. cumini*. Zhi-Ping et al. (2008) investigated the antioxidant activity of *S. cumini* leaf extracts using the DPPH free radical scavenging and ferric reducing antioxidant power (FRAP) assays. The methanolic extract and its four-water, ethyl acetate, chloroform, and *n*-hexane fractions were prepared and subjected to antioxidant evaluation. The results showed that the ethyl acetate fraction had stronger antioxidant activity than the other ones. High-performance liquid chromatography (HPLC) data indicated that *S. cumini* leaf extracts contained phenolic compounds, such as ferulic acid and catechin, responsible for their antioxidant activity.

Kheaw-on et al. (2009) studied the antioxidant capacity of flesh and seed from *S. cumini* fruits. The reddish-purple color of the "wa" or black plum (*S. cumini*) fruit is due to the anthocyanins, which are recognized as having strong antioxidant properties. Methanolic and acidified methanolic extracts, which included anthocyanins (crude extracts), were prepared from flesh (pulp and peel) and seeds of green, magenta, and dark purple fruits of both fresh and dried plant materials. The extracts were evaluated for antioxidant capacity, total phenolic content, and total anthocyanin contents. Acidified methanolic extracts gave consistently higher levels of all parameters studied. High levels of antioxidants were found in the fruit, with extracts from the seeds having greater antioxidant capacity and total phenolic content than flesh. Anthocyanin was detected only in the flesh of magenta and dark purple fruits. The good yield of fruit from this tree and the high yield of antioxidants found in this study strongly suggested that this tropical edible fruit can be a potentially rich source of natural antioxidant.

Tannins extracted from *S. cumini* fruit showed a very good DPPH radical scavenging activity and FRAP. HTs were identified as ellagitannins, consisting of a glucose core surrounded by gallic acid and ellagic acid units. Condensed tannins were identified as B-type oligomers of epiafzelechin (propelargonidin) with a degree of polymerization up to 11 (Liang-Liang and Yi-Ming 2009).

Elansary et al. (2012) reported the antioxidant activity of the essential oil obtained from the leaves of *S. cumini* using the DPPH method. The total antioxidant activity (TAA) was 11.13%.

Jayachandra and Devi (2012) examined the antioxidant activity of the methanolic extract of *S. cumini* bark by *in vitro* methods such as DPPH scavenging assay, hydrogen peroxide scavenging assay, and FRAP assay. The half-maximal inhibitory concentration (IC_{50}) values of the methanolic extract of *S. cumini* for DPPH and hydrogen peroxide scavenging activity were found to be 53.3% at a concentration of 600 mg and 42.03% at 1.2 mg/ml, respectively. The FRAP value was found to be 810 μg Fe^{2+}/g. The extract showed significant antioxidant activity in all antioxidant assays when compared with ascorbic acid.

Tobal et al. (2012) evaluated the use of *S. cumini* fruit extract as an antioxidant additive in orange juice and its sensorial impact. The work was undertaken to explore the possibility of consumption of *S. cumini* fruit by adding its extract to orange juice, making good use of its functional (antioxidant) properties. *S. cumini* fruit extract was characterized in terms of its anthocyanin content (2.11 g/100 g expressed in cyanidine-3-glucoside equivalents), total phenolic compounds (360 mg/100 g expressed in gallic acid equivalents [GAE]), and antioxidant capacity evaluated by the DPPH free radical scavenging method. The effects of the addition of *S. cumini* fruit crude extract, as well as its chromatographic fractions on the juice, were assessed chemically by headspace solid-phase microextraction and gas chromatography coupled with a mass spectrometry detector. Only six compounds had their chromatographic peak intensities clearly changed, and the results are discussed in terms of the inhibition of the formation of 2-octanone, hexanol, α-copaene, and α-panasinsene and the conservation of octyl acetate and *p*-menth-1-en-9-ol. Sensory evaluation of orange juice with and without *S. cumini* crude extract addition did not show any significant differences in the sensorial profile, discriminative, and acceptance tests.

El-Anany and Ali (2013) demonstrated that administration of various levels (400, 800, and 1200 ppm) of pomposia extracts (*S. cumini*) as a natural antioxidant in comparison with butylhydroxytoluene (BHT) as a synthetic antioxidant did not cause any significant changes in the biochemical parameters and did not show any adverse effect on the liver and kidney tissues of the tested rats. The results of the study thus suggest using pomposia juice as safe food grade substance.

Kaneria and Chanda (2013) evaluated the antioxidant and antimicrobial activities of different solvent extracts (petroleum ether, toluene, ethyl acetate, acetone, and water) of *S. cumini* leaves and found that the acetone extract had good antioxidant and remarkable antimicrobial activity.

A comparative evaluation of antioxidant activities of different kinds of extracts from *S. cumini* leaves by Mohamed et al. (2013) showed that the methanol extract exhibited a higher antioxidant and antibacterial activities than methylene chloride and essential oil extracts. A higher content of both total phenolics and flavonoids was found in the methanolic extract than in the other extracts. Due to their antioxidant and antibacterial properties, the leaf extracts from *S. cumini* might be used as natural preservative ingredients in the food or pharmaceutical industries.

Eshwarappa et al. (2014) examined the antioxidant activities of leaf gall extracts (aqueous and methanol) of *S. cumini* using DPPH, nitric oxide scavenging, hydroxyl scavenging, and FRAP methods. The presence of phenolics, flavonoids, phytosterols, terpenoids, and reducing sugars was detected in both extracts, and the methanolic extract showed higher antioxidant potential than the standard ascorbic acid.

Shrikanta et al. (2015) detected a higher amount of resveratrol and polyphenol (gallic acid) and high antioxidant activity in jaumun seeds (*S. cumini*).

An aqueous leaf extract of *S. cumini* was shown to have a reduced DPPH radical, partial prevention of lipid peroxidation induced by Fe^{2+}/citrate, and effectiveness against mitochondrial swelling induced by Ca^{2+}, thereby indicating the potential of *S. cumini* leaf extract as a useful therapeutic for the treatment of diseases related to mitochondrial dysfunctions (Ecker et al. 2015).

Mussi et al. (2015) studied the effect of the air temperature and velocity on the drying kinetics in the spouted bed of jambolao residue (peel and seeds) and the consequent changes on the antioxidant activity, anthocyanins, and mineral contents in the dried products. They reported that air temperature showed a positive effect on anthocyanin degradation, and their content was reduced by 60%–70% after drying, whereas the antioxidant activities of dried residue ranged from 93% to 97%. The minerals in 50 g of dried sample product contained, on average, 25% of the daily recommended intake of Mg for adults, while Fe presented 8.7%, and Zn and Ca, 3.2%.

Singh et al. (2015) studied the preparation of antioxidant-rich gluten-free eggless muffins from rice flour blended with varying amounts of *S. cumini*–jambolan fruit pulp (JFP) and xanthan gum (XG). The incorporation of JFP and XG increased batter viscoelasticity, and JFP incorporation increased greenness, cohesiveness, resilience, water activity, total phenolic content, total flavonoid content, DPPH, and ABTS inhibition of the muffins. Sensory analyses revealed that JFP incorporation improved the consumer acceptability of the muffins.

Veber et al. (2015) evaluated the phenolic contents and antiradical activity in *S. cumini* leaves and fruits at different stages of maturation. The study reported a higher average amount of phenolic compounds (237.52 mg GAE/100 g) in leaves, followed by the greenest fruit (109.17 mg GAE/100 g), and a higher antioxidant activity was detected in the green fruits (IC_{50} 2.27 mg/ml) than in the leaves (IC_{50} 23.07 mg/ml).

Coelho et al. (2016) evaluated the composition of phenolic compounds in frozen pulp and the juice of jambolan (*S. cumini*) and their antioxidant properties. The main phenolic compounds in the jambolan were the organic acids (malic and lactic acids), flavonoids (procyanidin B1 and catechin), monoglucoside anthocyanin (cyanidin 3-glucoside), and phenolic acids (gallic and chlorogenic acids). The antioxidant activities of jambolan products were considered acceptable compared with other potentially antioxidant foods.

Geun and Shibamoto (2001) reported that the antioxidant activity of clove bud (*S. aromaticum*) extract and its major aroma components, eugenol and eugenyl acetate, was comparable to that of the natural antioxidant, α-tocopherol (vitamin E).

A methanolic extract of *Syzygium calophyllifolium* stem bark was shown to have effective radical scavenging ability in DPPH, ABTS(+), phosphomolybdenum, FRAP, superoxide, and metal-chelating assays, while the leaf methanol extract was reported to have significant antioxidant activity. *S. calophyllifolium* leaf and bark extracts could therefore be taken as a good source of natural antioxidant supplement in food to defend oxidative stress–related disorders like diabetes (Chandran et al. 2015).

Jayasinghe et al. (2007) reported the isolation of three dihydrochalcones, phloretin 4′-*O*-methyl ether (2′,6′-dihydroxy-4′-methoxydihydrochalcone), myrigalone G (2′,6′-dihydroxy-4′-methoxy-3′-methyldihydrochalcone), and myrigalone B (2′,6′-dihydroxy-4′-methoxy-3,5′-dimethyldihydrochalcone), with radical scavenging properties toward the DPPH radical by a spectrophotometric method from the dichloromethane extract of the leaves of *S. jambos*.

A study was carried out to evaluate the antioxidant and tumor cell suppression potential of methanolic extract of *S. jambos* leaf in *in vitro* model systems. The different radical systems, comprising superoxide radical (O_2), hydroxyl radical (OH), and nitric oxide radical (NO) systems, were studied against a methanolic extract of *S. jambos*. The percentage of inhibition and IC_{50} showed that the *S. jambos* extract has potential antioxidant activity against free radicals. The tumor cell suppression or antiproliferative activity was demonstrated in three different cancer cell lines: MCF7 (breast cancer cell line), HepG2 (hepatocancer cell line), and A549 (lung cancer cell line). The sulforhodamine B (SRB) method was followed for evaluating the tumor cell suppression activity. The growth inhibition of 50% (GI_{50}), total growth inhibition (TGI), and lethal concentration of 50% (LC_{50}) of *S. jambos* extract showed the tumor cell suppression potential of the plant. The efficacy of the extract was highest in MCF7 cells, followed by HepG2 cells and A549 cells (Selvam et al. 2011).

Wei and Ismail (2012) reported that a methanolic extract of *Syzygium polyanthum* leaves showed mild antioxidant activity with an IC_{50} value of 90.85 µg/ml compared with the standard quercetin (24.09 µg/ml) on DPPH radical scavenging assay. Nevertheless, the total phenolics were analyzed via the Folin–Ciocalteau method, in which 111.25 mg of GAE and 312.52 mg of caffeic acid equivalent (CAE)/100 g dry leaves were obtained. Analyses by HPLC and liquid chromatography–mass spectrometry confirmed the presence of gallic acid and caffeic acid as the major phenolic acids in the methanolic *S. polyanthum* leaf extract. The total flavonoid analysis based on the down method and HPLC suggested only a minute percentage of flavonoids in the methanolic extract.

Vasanthi et al. (2012) reported the antioxidant activity of *Syzygium samarangense* fruit extract (SSFE) through *in vitro* models such as the ABTS method, antioxidant capacity by radical scavenging activity using the DPPH assay and FRAP method, and total phenolic and total flavanoid contents. The total phenolic content of SSFE was performed employing the literature method involving the Folin–Ciocalteu reagent and gallic acid as the standard, and it was found to be 162.58 ± 0.51 µg/mg GAE. Total flavanoid content was 310 µg/mg as quercetin equivalents per gram dry weight basis. The IC_{50} value of antioxidant activity of the fruit extract was found to be 140 µg/ml in the standard (L-ascorbic acid), The IC_{50} value for SSFE was found to be 175 µg/ml and 250 µg/ml for DPPH and ABTS scavenging activity, respectively.

The increasing absorbance of the reaction mixture indicated the high reducing antioxidant power of the fruit extract.

Methanol extract obtained from *S. zeylanicum* leaves exhibited potent antioxidant activity. The water extract obtained from this methanol extract by sequential extraction with hexane, chloroform, ethyl acetate, and *n*-butanol also showed the strongest antioxidant activity among extracts. This water extract was further fractionated by column chromatography with various concentrations of methanol solutions. Among the six resultant fractions, the fraction developed with 20% methanol exhibited the most potent antioxidant activity. The one peak among the three major HPLC peaks in this fraction was isolated and purified using a preparative HPLC. The structure of a pure compound was elucidated as a novel macrocyclic ellagitannin using a ^{1}H/^{13}C NMR and a high-resolution electrospray ionization mass spectrometer. This newly isolated compound, which was named zeylaniin A, exhibited potent antioxidant activities in the assays of DPPH, oxygen radical absorbance capacity, and malonadehyde–gas chromatography. *S. zeylanicum* leaves can be a possible source of natural antioxidants (Nomi et al. 2012).

HEPATOPROTECTIVE ACTIVITY

The oral treatment of rats with a methanolic extract of *S. cumini* (= *E. jambolana*) at a dose of 200 mg/kg for seven days resulted in significant hepatoprotection for total protein (6.6 g/dl) only, while at 400 mg/kg, the restoration was significant for enzyme markers, that is, SGOT, SGPT, and ALP (22.5 U/L, 18.5 U/L, and 58.5 U/L) in CCl$_4$-treated rats. The maximum tolerated dose of the extract was found to be greater than 3000 mg/kg taken by mouth. Thus, from the study it was concluded that methanolic extracts of *S. cumini* leaves had a hepatoprotective effect against CCl$_4$-induced liver injury (Mani et al. 2012).

Biochemical and histopathological studies by Islam et al. (2015) on the hepatoprotective effect of seed extract of *S. cumini* in CCl$_4$-induced stressed adult male (Sprague Dawley) rats revealed that the extract in 250 and 500 mg/kg body weight doses protected the liver from CCl$_4$-induced stress.

CARDIOPROTECTIVE ACTIVITY

Atale et al. (2013) studied the cardioprotective properties of methanolic seed extract (MSE) of *S. cumini* in diabetic *in vitro* conditions and confirmed the suppression of ractive oxygen species (ROS) production by MSE in glucose-induced cells, thus protecting the cardiac cells from glucose-induced stress.

R. M. Ribeiro et al. (2014) demonstrated that a hydroalcoholic extract of *S. cumini* leaves reduced the blood pressure and heart rate of spontaneously hypertensive rats, and that the antihypertensive effect was probably due to the inhibition of the arterial tone and extracellular calcium influx.

Herculano et al. (2014) demonstrated that a single oral administration of hydroalcohol extract from the fruits of *S. cumini* (EHSCF) reduced the significant mean arterial pressure in spontaneously hypertensive rats, indicating the antihypertensive effect of the extract.

S. cumini methanolic pulp extract (MPE), a naturally derived gallic acid–enriched antioxidant, was shown to have a significant protective effect against the malathion-mediated oxidative stress in cardiac myocytes compared with *S. cumini* ethanolic and aqueous pulp extracts (Atale et al. 2014).

Chagas et al. (2015) reviewed the cardiometabolic properties (antihyperglyce-mic, hypolipemic, anti-inflammatory, cardioprotective, and antioxidant activities) of *S. cumini* by correlating its already identified phytochemicals with their described mechanisms of action. The data highlighted that some compounds target multiple metabolic pathways, thereby becoming potential pharmacological tools.

ANTI-INFLAMMATORY AND ANTINOCICEPTIVE ACTIVITIES

The ethnolic extract of the bark of *S. cumini* showed significant anti-inflammatory activity in carrageenan (acute)-, kaolin-carrageenan (subacute)-, formaldehyde (subacute)-induced paw edema, and cotton pellet granuloma (chronic) tests in rats. The extract did not induce any gastric lesion in both acute and chronic ulcerogenic tests in rats (Muruganandan et al. 2001).

Tanko et al. (2008) reported the antinociceptive and anti-inflammatory activities of ethanol extract of *S. aromaticum* flower bud using acetic acid–induced abdominal contractions in mice and formalin-induced hind paw edema in Wistar rats. Three doses of the ethanol extract (50, 100, and 200 mg/kg body weight intraperitoneally) were used for both studies. The extract had an LD_{50} of 565.7 mg/kg body weight intraperitoneally in mice.

Modi et al. (2010) studied the anti-inflammatory activity of the methanolic and aqueous extracts of the seeds of *S. cumini* in Wistar rats using the carrageenan-induced left hind paw edema. The methanolic and aqueous extracts at a dose of 250 mg/kg body weight showed moderate to significant anti-inflammatory activity. The methanolic and aqueous extracts of *S. cumini* reduced the edema induced by carrageenan by 48.29% and 68.85%, respectively, on oral adminis-tration of 250 mg/kg body weight, compared with the untreated control group. Diclofenac sodium at 100 mg/kg body weight inhibited the edema volume by 75.08%. The results indicated that the aqueous extract shows more significant anti-inflammatory activity than methanolic extracts when compared with the standard and untreated control.

Jain et al. (2010) evaluated the anti-inflammatory activity of ethyl acetate and methanol extracts of *S. cumini* leaves in carrageenan-induced paw edema in Wistar rats at dose levels of 200 and 400 mg/kg administrated orally. Both extracts exhib-ited significant anti-inflammatory activity, which supports the traditional medicinal utilization of the plant.

Belle et al. (2013) studied the effect of aqueous seed extract of *S. cumini* (ASc) in the activity of enzymes involved in lymphocyte functions in lymphocytes iso-lated from healthy human donors. ASc inhibited the ADA and dipeptidyl peptidase IV (DPP-IV) activities without alteration in the CD26 expression (DPP-IV protein). No alterations were observed in the acetylcholinesterase (AChE) activity or in cell viability. These results indicated that the inhibition of the DPP-IV and ADA activi-ties was dependent on the time of exposition to ASc.

Machado et al. (2013) detected anti-inflammatory and apoptotic activity of the essential oil of *S. cumini in vivo*. The anti-inflammatory action and chronic granulomatous inflammation in BALB/c mice, intravenously infected with *M. bovis* BCG, were judged by measuring and classifying the granulomas formed in the hepatic parenchyma. A reduction in the granulomatous area and a change in the pattern of the granulomas were found. Antimycobacterial activity of the essential oil against *M. bovis* was detected *in vitro* by an interferometric method in liquid culture medium.

Siani et al. (2013) evaluated the anti-inflammatory activity of the essential oils from the leaves of *S. cumini*, as well as some of their terpene-enriched fractions (+V = more volatile; –V = less volatile) obtained by vacuum distillation. Anti-inflammatory activity was assessed in the lipopolysaccharide-induced pleurisy model, by measuring the inhibition of total leukocyte, neutrophil, and eosinophil migration in the mice pleural lavage, after treatment with the oils at 100 mg/kg. Eosinophil migration was inhibited by *S. cumini* (67%) and *S. cumini* (+V) (63%), and this efficacy was correlated with the presence of β-pinene and β-caryophyllene in the oils.

A study on the effects of *S. cumini* aqueous extract (SCC) on indomethacin-induced acute gastric ulceration revealed that SCC acted as an antioxidant, anti-inflammation, and antiulcer against indomethacin (Chanudom and Tangpong 2015).

Latief et al. (2015) investigated the anti-inflammatory and antioxidant effects of a flavonoid glucoside, trimeric myricetin rhamnoside (TMR), isolated from leaves of *S. cumini* (= *E. jambolana*). The TMR was studied for anti-inflammatory activity in carrageenan-induced hind paw edema and antioxidant activity in lung by cecal ligation and puncture (CLP)–induced sepsis in mice. The results of this study concluded that the TMR appears to have potential benefits in diseases that are mediated by both inflammation and oxidative stress and support the pharmacological basis of use of *S. cumini* as traditional herbal medicine for the treatment of inflammatory diseases.

Investigation of the antinociceptive effect of ethanol extracts from *S. cumini* leaves on formalin- and glutamate-induced orofacial nociception in mice resulted in a highly significant reduction in the percentage of paw lick time during a formalin pain test, and a marked inhibition of glutamate-induced orofacial nociception (Quintans et al. 2014).

Kandati et al. (2012) reported that *Syzygium alternifolium* chloroform root extract showed 5-LOX inhibition activity similar to that of the standard drug, zileutin. In *in vivo* anti-inflammatory studies, *S. alternifolium* chloroform root extract exhibited better anti-inflammatory activity than its methanolic root extract. Similar results were observed in the acetic acid–induced writhing model and radiant heat–induced model. Hence, it was concluded that *S. alternifolium* chloroform root extract exhibited a greater degree of efficacy than the *S. alternifolium* methanolic extract, and a good *in vitro* and *in vivo* anti-inflammatory effect.

An aqueous extract of dried flower buds of *S. aromaticum* (1 g/kg body weight) inhibited the formation of edema induced by carrageenan and decreased granuloma in the cotton pellet granuloma model. The extract, when compared with the disease control, is reported to decrease the elevated levels of succinate dehydrogenase ($p < 0.001$), xanthine oxidase ($p < 0.05$), and lipid peroxidation, and increase the activity of catalase (CAT) ($p < 0.001$) and glutathione peroxidase ($p < 0.01$) in the animal models (T. Ahmad et al. 2012).

Ávila-Peña et al. (2007) reported the analgesic potential of a leaf hydroalcoholic extract of *S. jambos* in rats. Hot-plate and formalin tests were used to estimate cutaneous nociception, whereas measurements of forelimb grip force were done to assess muscular nociception under normal and inflammatory conditions. In the hot-plate test, *S. jambos* extract produced a significant increase in the withdrawal response latencies in a dose-dependent manner (10–300 mg/kg intraperitoneally) and with a maximal effect (analgesic efficacy) similar to that of morphine. The extract (100–300 mg/kg intraperitoneally) significantly reduced pain scores in all phases of the formalin test, with an analgesic efficacy higher than that shown by diclofenac. Although the extract (300 mg/kg) did not alter grip force in intact rats, it reversed the reduction in grip force induced by bilateral injection carrageenan in the forelimb triceps. This analgesic effect of the extract on muscle hyperalgesia was not antagonized, but enhanced, by naloxone. Thus, the *S. jambos* extract has remarkable analgesic effects on both cutaneous and deep muscle pain that are not mediated by opioid receptors.

Raga et al. (2011) reported that cycloartenyl stearate, lupenyl stearate, sitosteryl stearate, and 24-methylenecycloartenyl stearate from the air-dried leaves of *S. samarangense* exhibited potent analgesic and anti-inflammatory activities.

ANTIFERTILITY ACTIVITY

Sarita et al. (2012) studied the postcoital contraceptive activity and teratogenecity effect of various extracts of *S. cumini* (= *E. jambolana*) seed. Petroleum ether, ethyl acetate, and ethanol extracts of the seeds were administered orally at dose levels of 200 and 600 mg/kg body weight from 10 to 18 days of pregnancy. A strong abortifacient activity (97.07%) was observed at 600 mg/kg body weight of rats treated with ethyl acetate seed extract, but did not show any developmental toxicity and teratogenicity effect in rats.

The precoital antifertility activity of the petroleum ether, ethyl acetate, and ethanol extracts of seeds of *S. cumini* was tested in female albino rats. Of these, the ethyl acetate extracts were found to be most effective in causing significant anti-implantation activity. The ethyl acetate extract also exhibited a strong antiestrogenic activity when administered alone. It also inhibited an estrogen-induced gain in uterine weight when administered along with ethinyl estradiol (Sarita and Bhagya 2012).

SPASMOLYTIC ACTIVITY

Amor et al. (2005) reported the isolation of spasmolytic flavonoids from *S. samarangense*. Four rare C-methylated flavonoids with a chalcone and a flavanone skeleton were isolated from the hexane extract of *S. samarangense* and were subsequently tested for spasmolytic activity. All flavonoids, identified as 2'-hydroxy-4',6'-dimethoxy-3'-methylchalcone, 2',4'-dihydroxy-6'-methoxy-3',5'-dimethylchalcone, 2',4'-dihydroxy-6'-methoxy-3'-methylchalcone, and 7-hydroxy-5-methoxy-6,8-dimethylflavanone, showed dose-dependent spasmolytic activity in the rabbit jejunum with IC_{50} values of 148.3 ± 69.4, 77.2 ± 43.5, 142.4 ± 58.6, and 178.5 ± 37.5 µg/ml (mean ± SEM), respectively.

ANTIASTHMATIC ACTIVITY

The possible antiasthma activity of macerated and Soxhlet extracted leaves of *S. cumini* on tracheal chains of guinea pigs was evaluated. The relaxant effects of four cumulative concentrations of macerated and Soxhlet extracts (0.25, 0.5, 0.75, and 1.0 w/v) in comparison with saline as a negative control and four cumulative concentrations of theophylline (0.25, 0.5, 0.75, and 1.0 mM) as a positive control were examined on precontracted tracheal chains of two groups of six guinea pigs each: group 1 ($N = 6$) on tracheal chains contracted by 60 mM KCl, group 2 ($N = 6$) on nonincubated tracheal chains contracted by 10 μM methacholine hydrochloride, and group 3 ($N = 4$) on tissues incubated with 1 μM propranolol. A decrease in contractile tone of the tracheal chains was considered a relaxant effect. The isolated guinea pig trachea precontracted with KCl, methacholine, and tissues incubated with propranolol was used to study the relaxation of macerated and Soxhlet extracts of leaves of *S. cumini*. In group 1 experiments, only the 1.0 mM concentration of theophylline and 1.0 W/V of the soxhlet extracts of *Syzygium cumini* leaves showed a significant relaxant effect compared with that of saline ($p < 0.001$ for both concentrations), which were significantly greater than those of macerated extracts. In group 2 experiments, only the last two higher concentrations of theophylline and Soxhlet extract showed a significant relaxant effect compared with that of saline. The effects of the two higher concentrations of theophylline in this group were significantly greater than those of the macerated and Soxhlet extracts ($p < 0.01$), and in group 2 and 3 experiments, both the macerated and Soxhlet extracts showed concentration-dependent relaxant effects compared with that of saline ($p < 0.05$ to $p < 0.001$ for both extracts). The relaxant effects of the macerated and Soxhlet extracts in group 1 were significantly lower than those of groups 2 and 3. In the group 3 experiment, a potent relaxant effect was observed (Mahapatra and Pradhan 2012).

CYTOTOXIC ACTIVITY

Prashar et al. (2006) reported the cytotoxicity of clove (*S. aromaticum*) oil and its major components to human skin cells. The essential oil extracted from clove is used as a topical application to relieve pain and promote healing in herbal medicine and also finds use in the fragrance and flavoring industries. Clove oil has two major components, eugenol and β-caryophyllene, which constitute 78% and 13% of the oil, respectively. Clove oil and these components are generally recognized as safe, but the *in vitro* study demonstrated cytotoxic properties of both the oil and eugenol toward human fibroblasts and endothelial cells. Clove oil was found to be highly cytotoxic at concentrations as low as 0.03% (v/v), with up to 73% of this effect attributable to eugenol. β-Caryophyllene did not exhibit any cytotoxic activity, indicating that other cytotoxic components may also exist within the parent oil.

Cytotoxicity assays on five C-methylated chalcones from *S. samarangense* showed one of these flavonoids, 2′,4′-dihydroxy-6-methoxy-3′,5′-dimethyl chalcone, as a cytotoxic natural product with potential cancer applications, especially in human ovarian and mammary adenocarcenoma (Amor et al. 2007).

Anticytotoxic, Antimutagenic, and Anticancer Activities

Devkar et al. (2012) studied the protective role of *Brassica olerecea* and *S. cumini* (= *E. jambolana*) extracts against H_2O_2-induced cytotoxicity in H9C2 cells. The study suggests that *E. jambolana* seed extract is capable of cardioprotective activity due to the high number of flavonoids in it that are instrumental in lowering intracellular oxidative stress, preventing depletion of cellular antioxidants, and improving cell viability.

Aqil et al. (2012) reported the cancer chemoprotective potential of *S. cumini* anthocyanins and other polyphenolics extracted with acidic ethanol and enriched by amberlite XAD7/HP20 (1:1). The pulp powder was found to contain 0.54% anthocyanins, 0.17% ellagic acid and ellagitannins, and 1.15% total polyphenolics. Jamun seed contained no detectable anthocyanins but had higher amounts of ellagic acid and ellagitannins (0.5%) and total polyphenolics (2.7%) than the pulp powder. Upon acid hydrolysis, the pulp extract yielded five anthocyanidins by HPLC: malvidin (44.4%), petunidin (24.2%), delphinidin (20.3%), cyanidin (6.6%), and peonidin (2.2%). Extracts of both jamun pulp (1445 ± 64 μmol of trolox equivalent [TE]/g) and seeds (3379 ± 151 μM of TE/g) showed a high oxygen radical absorbance capacity. Their high antioxidant potential was also reflected by ABTS and DPPH scavenging and ferrous ion-chelating activities. The hydrolyzed pulp and seed extracts showed significant antiproliferative activity in human lung cancer A549 cells. However, unhydrolyzed extracts showed much less activity.

Tripathi et al. (2013) studied the protective effects of *S. cumini* extract (SCE) (100 and 200 mg/kg) against genotoxicity and oxidative stress induced by cyclophosphamide (CP) in mice. The study revealed that SCE significantly inhibited the frequencies of aberrant metaphases, chromosomal aberrations, micronuclei formation, and cytotoxicity in mouse bone marrow cells induced by CP. SCE also produced a significant reduction of abnormal sperm; antagonized the reduction of CP-induced superoxide dismutase (SOD), CAT, and glutahione (GSH) activities; and inhibited increased malondialdehyde (MDA) content in the liver.

The antimutagenicity of *S. cumini* assayed by *E. coli* rifampicin resistance showed that purified anthocyanin suppressed the mutagenic SOS DNA repair process in *E. coli*, and thus indicated suppression of the error-prone DNA repair pathway as one of the major mechanisms of antimutagenicity of this fruit (Saxena et al. 2013).

Aqil et al. (2016) reported that *S. cumini* (jamun) significantly offset 17β estrogen-mediated alterations in mammary cell proliferation, ER-α, cyclinD1, and candidate miRNAs in female August–Copenhagen Irish rats, and that the modulation of these biomarkers correlated with a reduction in mammary carcinogenicity.

De Bona et al. (2014) evaluated the *in vitro* effect of gallic acid and aqueous *S. cumini* leaf extract (ASc) on ADA and DPP-IV activities, cell viability, and oxidative stress parameters in lymphocytes exposed to 2,2′-azobis-2-amidinopropane dihydrochloride (AAPH). They reported that ASc reduced the AAPH-induced increase in ADA activity, but no effect was observed on DPP-IV activity, and that ASc increased protein sulfhydryl (P-SH) groups and cellular viability and decreased LDH activity, but was not able to reduce the AAPH-induced lipid peroxidation. The study highlighted the immunomodulatory and cytoprotective effects of ASc, which showed more protective effects than gallic acid.

Miyazawa and Hisama (2003) reported the antimutagenic activity of phenylpropanoids from clove (*S. aromaticum*). Phenylpropanoids that possess antimutagenic activity were isolated from the buds of clove. The isolated compounds suppressed the expression of the *umu* gene following the induction of the SOS response in *S. typhimurium* TA1535/pSK1002 treated with various mutagens. The suppressive compounds were mainly localized in the ethyl acetate extract fraction of the processed clove.

Bioassay-guided fractionation of the methanolic extracts of the pulp and seeds of the fruits of *S. samarangense* led to the identification of four cytotoxic compounds and eight antioxidants on the basis of HPLC-photodiode array (PDA) analysis, mass spectrometry, and various NMR spectroscopic techniques. Three *C*-methylated chalcones, 2′,4′-dihydroxy-3′,5′-dimethyl-6′-methoxychalcone, 2′,4′-dihydroxy-3′-methyl-6′-methoxychalcone (stercurensin), and 2′,4′-dihydroxy-6′-methoxychalcone (cardamonin), were isolated and displayed cytotoxic activity (IC_{50} = 10, 35, and 35 μM, respectively) against the SW-480 human colon cancer cell line. Also, a number of known antioxidants were obtained, including six quercetin glycosides (reynoutrin, hyperin, myricitrin, quercitrin, quercetin, and guaijaverin), one flavanone ([*S*]-pinocembrin), and two phenolic acids (gallic acid and ellagic acid) (Simirgiotis et al. 2008).

ANTINEPHROTOXIC ACTIVITY

Adikay et al. (2010) studied the nephroprotector activity of an ethanol extract of fruits of *S. cumini* (250 and 500 mg/kg taken by mouth) on cisplatin-induced nephrotoxicity (6 mg/kg intraperitoneally) in albino rats. The nephroprotector activity of *S. cumini* was assessed by estimating the levels of blood urea nitrogen, serum creatinine, serum total proteins, urinary protein, and lipid peroxidation in the kidney. Cisplatin elevated the serum marker level, increased the protein excretion in urine, reduced the creatinine clearance, and increased the renal MDA level. Animals that received an ethanol extract of fruits of *S. cumini* significantly reversed the effects induced by cisplatin in a dose-dependent manner.

Adikay and Belide (2012) evaluated the effect of an anthocyanin fraction of *S. cumini* on cisplatin-induced nephrotoxicity in male albino rats. The anthocyanin fraction was administered by gastric intubation at two dose levels. Animals were divided into five groups. Group I animals received vehicle. On day 1, Group II animals received cisplatin (6 mg/kg intraperitoneally, single dose). Groups III (7.5 mg/kg) and IV (15 mg/kg) received anthocyanin fractions at 1 h before and 24 and 48 h after cisplatin injection. Group V received only the anthocyanin fraction. After 72 h of cisplatin injection, blood and urine were collected and estimated for serum marker level and creatinine clearance. MDA levels and histological studies were also conducted. Cisplatin caused renal damage characterized by elevation of blood urine nitrogen, serum creatinine, and MDA level, with a marked drop in creatinine clearance; the anthocyanin fraction reversed all the effects induced by cisplatin in dose-dependent manner. Histological studies also substantiated the above-mentioned results.

Against Fluoride-Induced Toxicity

K. R. Ahmad et al. (2012) reported ameliorative effects of *S. cumini* fruit pulp extract in mice against fluoride exposure (from NaF) in relation to histopathological and histometric changes of testis and sperm micrometry. Sperm micrometry revealed a significant decline ($p < 0.05$) in head length (7.93 µ), breadth (3.96 µ), tail length (83.25 µ), and the length (22.79 µ) and diameter (1.06 µ) of the middle part in the NaF group compared with the control group (9.79, 4.12, 101.16, 25, and 1.198 µ, respectively). These latter changes from NaF exposure were significantly reversed on posttreatment with jambul fruit pulp extract. Signs of testicular histopathologies from NaF were effectively reversed with jambul extract treatment.

CNS Protective Activity

Kumar et al. (2007) investigated the ethyl acetate and methanol extracts of the seeds of *S. cumini* for CNS activity on albino mice in rota rod and actophotometer at dose levels of 200 and 400 mg/kg. Both extracts exhibited significantly CNS activity.

Anti-*Leishmania* and Molluscicidal Activity

Dias et al. (2013) examined the chemical composition and biological potential of the essential oil extracted from *S. cumini* leaves collected in Brazil. Gas chromatography–mass spectrometry analyses revealed a high abundance of monoterpenes (87.12%) in the oil. Eleven compounds were identified, with the major components being α-pinene (31.85%), (Z)-ocimene (28.98%), and (E)-ocimene (11.71%). The essential oil showed significant activity against *Leishmania amazonensis* and a molluscicidal effect against *Biomphalaria glabrata*.

Potential antileishmanial activity was detected with the hexanic extract of *S. cumini* (IC_{50} of 31.64 µg/ml) leaves against stationary-phase promastigotes of *L. amazonensis* and murine macrophages (T. G. Ribeiro et al. 2014).

Franca et al. (2015) demonstrated that *S. cumini* essential oil and its major constituent, α-pinene, have significant anti-*Leishmania* activity modulated by macrophage activation, with acceptable levels of cytotoxicity in murine macrophages and human erythrocytes.

Antihelminthic Activity

Dibua and Odo (2012) studied the antihelminthic activity of the ethanolic extract of *S. aromaticum*. The bioassay was carried out on 70 local earthworms (*Pheretima posthuma*): 10 live worms in 20 mg/ml of each ethanolic extract, 20 mg/ml of piperazine and 20 mg/ml of 5% dimethylformamide (DMF). The *S. aromaticum* extract showed antihelminthic activity ($p < 0.05$).

Treatment of Retinitis

Priya et al. (2013) studied the binding affinity of five anthocyanin compounds from *S. cumini* fruit peel with the X-linked retinitis pigmentosa (RP2) gene (a mutant of this gene causes loss of vision in humans) and revealed cyanidin 3,5 diglucoside with lowest G score (−12.62 kcal/mol) as an inhibitor that could be of potential use in the treatment of retinitis pigmentosa in humans.

ANTITERMITE ACTIVITY

In choice and no-choice trials under laboratory and field conditions to study the feeding preferences of *Coptotermes heimi*, a widely present termite species in Pakistan, for 18 different wood species, *S. cumini* was ranked as the most resistant to the termite attack (Rasib and Ashraf 2014).

Mite lethality was estimated using a complete exposure method test with the *S. aromaticum* (clove) oil at different concentrations, and a systemic administration method of oil at different concentrations diluted in syrup was placed in feeders for bees. The LC_{50} for the complete exposure method at 24 h was 0.59 µl/dish. The inferior and superior limits obtained were 0.47×10^{-6} and 1.22 µl/dish, respectively. The LC_{50} estimated at 48 h showed a slight decrease compared with that recorded at 24 h. Ratio selection (LC_{50} of *Apis mellifera*/LC_{50} of *Varroa destructor*) for the complete exposure method was 26.46 and 13.35 for 24 and 48 h, respectively. Regarding the systemic administration method, the mite LC_{50} at 24 h was 12,300 ppm. The inferior and superior limits calculated were 9,214 and 15,178 ppm, respectively. The LC_{50} estimated at 48 h showed a slight decrease compared with that recorded at 24 h. Ratio selection for the systemic administration method was 3.05 and 2.22 for 24 and 48 h, respectively. *S. aromaticum* oil was found to be an attractant for *V. destructor* at 4.8% (w/w) concentration. The results showed that oil toxicity against *V. destructor* differed depending on its administration (Maggi et al. 2010).

ANTIFEEDANT ACTIVITY

Extracts of *S. aromaticum* (= *Syzygium caryophyllatum*) were examined for antifeedant activity against the fourth instar larvae of the Mexican bean beetle. The extracts showed strong antifeedant activity (Jayasinghe et al. 2003).

AS NATURAL COLORANT AND COSMECEUTICAL

Anthocyanins are potential candidates for use as natural food colorant. Chaudhary and Mukhopadhyay (2013) identified all six major types of anthocyanins in *S. cumini* fruit skin peel by ultraperformance liquid chromatography studies. They suggested that the highest anthocyanin yield (763.80 mg, 100 ml[−1]), highest chroma, and hue angle in the red color range could be obtained when 20% ethanol was used in combination with 1% acetic acid. The optimized solvent can be used to extract anthocyanins from the *S. cumini* fruits and used as natural colorants in the food industries. Santiago et al. (2016) demonstrated that the anthocyanin from the fruit peel powder of *S. cumini* can be an alternative source of a stable natural colorant for commercial use in the food industry. The peel powder of *S. cumini* also proved to be rich in dietary fibers and thus a good ingredient for low-calorie diets, as well as having low lipid content.

Palanisamy et al. (2011) found *S. aqueum* leaf extracts to have a significant composition of phenolic compounds, protective activity against free radicals, and low pro-oxidant capability. Its ethanolic extract, in particular, is characterized by its excellent radical scavenging activity with EC_{50} values of 133 µg/ml DPPH, 65 µg/ml ABTS, and 71 µg/ml (galvinoxyl); low pro-oxidant capabilities; and a phenolic content of 585–670 mg GAE/g extract. The extract also displayed other activities, deeming it an ideal cosmetic ingredient. A substantial tyrosinase inhibition activity with an IC_{50} of about 60 µg/ml was observed. In addition, the extract was also found to have anticellulite activity tested for its ability to cause 98% activation of lipolysis of adipocytes (fat cells) at a concentration of 25 µg/ml. In addition, the extract was not cytotoxic to Vero cell lines up to a concentration of 600 µg/ml.

CONCLUSION AND FUTURE PERSPECTIVES

The history of the use of *S. cumini* goes back several millennia to ancient written Ayurvedic documents such as *Charaka Samhita* and *Bhavaprakasha*. The prophylactic and curative properties of *S. cumini* and related species from the same genus were known to ancient physicians. This group of plants became an important component in the curative arsenal of several ancient cultures. The fruit of *S. cumini* enriched the fruit baskets of the village folk, while the flower buds of *S. aromaticum* added flavor to the cuisine of different cultures and has been an important component of the spice trade of several nations. The essential oil of *S. aromaticum* has been used in several perfumery and cosmetic products.

Since *S. cumini* and allied species are used in a number of ethnomedical preparations, they have been the subject of intensive studies of a large number of research workers across the globe. These studies have confirmed the ethnomedical claims on the therapeutic potential of *Syzygium* trees. A good number of powerful antioxidants have been isolated and characterized from these plants. Essential oils, as well as the extracts prepared from different parts of the plants, have been demonstrated to possess antibacterial, antifungal, and antiviral properties. Some of the extracts exhibited anti-inflammatory and nociceptive properties. Besides, several parts of *S. cumini*, particularly the seed, are shown to possess powerful antidiabetic properties. Thus, the plant *S. cumini* and related species from the same genus have both prophylactic and curative properties. Future research on these species should focus on the development of affordable and alternate drugs for the treatment of chronic diseases such as diabetes and inflammation.

REFERENCES

Adikay, S., and P. Belide. 2012. Effect of anthocyanin fraction on Cispaltin-induced nephro-toxicity. *Int. J. Res. Ayurveda Pharm.* 3: 587–590.

Adikay, S., P. Belide, and B. Koganti. 2010. Protective effect of fruits of *Syzygium cumini* against Cisplatin-induced acute renal failure in rats. *J. Pharm. Res.* 3: 2756–2758.

Ahmad, K. R., T. Nouroze, K. Raees, T. Abbas, M. A. Kanwal, S. Noor, and S. Jabeen. 2012. Protective role of jambul (*Syzygium cumini*) fruit-pulp extract against fluoride-induced toxicity in mice testis: A histopathological study. *Fluoride* 45: 281–289.

Ahmad, T., T. S. Shinkafi, I. Routray, A. Mahmood, and S. Ali. 2012. Aqueous extract of dried flower buds of *Syzygium aromaticum* inhibits inflammation and oxidative stress. *J. Basic Clin. Pharm.* 3: 323–327.

Alam, R. M., A. B. Rahman, M. Moniruzzaman, M. F. Kadir, M. A. Haque, M. R. H. Alvi, and M. Ratan. 2012. Evaluation of antidiabetic phytochemicals in *Syzygium cumini* (L.) Skeels (family: Myrtaceae). *J. Appl. Pharm. Sci.* 2: 94–98.

Amor, E. C., M. Irene, M. Villasenor, R. Antemano, Z. Perveen, P. Gisela, P. Concepcion, and M. I. Choudahry. 2007. Cytotoxic C-methylated chalcones from *Syzygium samarangense*. *Pharm. Biol.* 45: 777–783.

Amor, E. C., I. M. Villasenor, M. N. Ghayur, A. H. Gilani, and M. I. Choudhary. 2005. Spasmolytic flavonoids from *Syzygium samarangense* (Blume) Merr. L. M. Perry. *Z. Naturforsch. C* 60: 67–71.

Anonymous. 1976. *The Wealth of India, Raw Materials*. Vol. 10. New Delhi: Council of Scientific and Industrial Research, 93–107.

Aqil, F., A. Gupta, R. Munagala, J. Jeyabalan, H. Kausar, R. J. Sharma, I. P. Singh, and R. C. Gupta. 2012. Antioxidant and antiproliferative activities of anthocyanin/ellagitannin-enriched extracts from *Syzygium cumini* L. (jambu, the Indian blackberry). *Nutr. Cancer* 64: 428–438.

Aqil, F., Jeyabalan, J., Munagala, R., Singh, I. P., and Gupta, R. C. 2016. Prevention of hormonal breast cancer by dietary jamun. *Mol. Nutr. Food Res.* 60: 1470–1481.

Atale, N., M. Chakraborty, S. Mohanty, S. Bhattacharya, D. Nigam, M. Sharma, and V. Rani. 2013. Cardioprotective role of *Syzygium cumini* against glucose-induced oxidative stress in h9c2 cardiac myocytes. *Cardiovasc. Toxicol.* 13: 278–289.

Atale, N., K. Gupta, and V. Rani. 2014. Protective effect of *Syzygium cumini* against pesticide-induced cardiotoxicity. *Environ. Sci. Pollution Res.* 21: 7956–7972.

Ávila-Peña, D., N. Pena, L. Quinteros, and H. Suarez-Roca. 2007. Antinociceptive activity of *Syzygium jambos* leaves extracts on rats. *J. Ethnopharmacol.* 112: 380–385.

Ayyanar, M., P. Subash-Babu, and S. Ignacimuthu. 2013. *Syzygium cumini* (L.) Skeels, a novel therapeutic agent for diabetes: Folk medicinal and pharmacological evidences. *Complement. Ther. Med.* 21: 232–243.

Banerjee, A., N. Dasgupta, and B. De. 2005. In vitro study of antioxidant activity of *Syzygium cumini* fruit. *Food Chem.* 90: 727–733.

Baytop, T. 1999. *Therapy with Medicinal Plants in Turkey (Past and Present)*. No. 3255, 2nd ed. Istanbul: Istanbul Publications of the Istanbul University, 334–335.

Belle, L. P., P. E. R. Bitencourt, F. H. Abdalla, K. S. de Bona, A. Peres, L. D. K. Maders, and M. M. Moretto. 2013. Aqueous seed extract of *Syzygium cumini* inhibits the dipeptidyl peptidase IV and adenosine deaminase activities, but it does not change the CD26 expression in lymphocytes in vitro. *J. Physiol. Biochem.* 69: 119–124.

Bhattarai, N. K. 1992. Folk herbal remedies of Sinhupalchok district, central Nepal. *Fitoterapia* 63: 145–155.

Bitencourt, P. E. R., K. S. D. Bona, L. O. Cargnelutti, G. Bonfanti, A. Pigatto, A. Boligon, M. L. Athayde, F. Pierezan, R. A. Zanette, and M. B. Moretto. 2015. *Syzygium cumini* seed extract ameliorates adenosine deaminase activity and biochemical parameters but does not alter insulin sensitivity and pancreas architecture in a short-term model of diabetes. *J. Complement. Integr. Med.* 12: 187–193.

Bitencourt, P. E. R., L. M. Ferreira, L. O. Cargnelutti, L. Denardi, A. Boligon, M. Fleck, R. Brandao, M. L. Athayde, L. Cruz, R. A. Zanette, S. H. Alves, and M. B. Moretto. 2016. A new biodegradable polymeric nanoparticle formulation containing *Syzygium cumini*: Phytochemical profile, antioxidant and antifungal activity and in vivo toxicity. *Ind. Crop. Prod.* 83: 400–407.

Bullerman, L. B., F. Y. Lieu, and S. A. Seier. 1977. Inhibition of growth and aflatoxin production of cinnamon and clove oils: Cinnamic aldehyde and eungenol. *J. Food Sci.* 42: 1107.

Cai, L., and C. D. Wu. 1996. Compounds from *Syzygium aromaticum* possessing growth inhibitory activity against oral pathogens. *J. Nat. Prod.* 59: 987–990.

Chagas, V. T., L. M. Franca, S. Malik, P. de Andrade, and M. Antonio. 2015. *Syzygium cumini* (L.) Skeels: A prominent source of bioactive molecules against cardiometabolic diseases. *Front. Pharmacol.* 6: 259.

Chaieb, K., H. Hajlaoui, T. Zamantar, A. B. Kahla-Nakbi, M. Rouabhia, K. Mahdouani, and F. Bakhrouf. 2007. The chemical composition and biological activity of clove essential oil, *Eugenia caryophyllata* (*Syzygium aromaticum* L. Myrtaceae): A short review. *Phytother. Res.* 21: 501–506.

Chandra, K., P. Gupta, K. L. Singh, and J. S. Tandon. 1998. Antiviral activity of an extract of *Syzygium megacarpa* against encephalitis causing virus. *Indian J. Virol.* 14: 31–35.

Chandran, R., S. Sathyanarayanan, M. Rajan, M. Kasipandi, and T. Parimelazhagan. 2015. Anti-oxidant, hypoglycemic and anti-hyperglycemic properties of *Syzygium calophyllifolium*. *Bangladesh J. Pharmacol.* 10: 672–680.

Chandrasekaran, M., and V. Venkatesalu. 2004. Antibacterial and antifungal activity of *Syzygium jambolanum* seeds. *J. Ethnopharmacol.* 91: 105–108.

Chang, W. C., and S. C. Shen. 2013. Effect of water extracts from edible Myrtaceae plants on uptake of 2-(n-(7-nitrobenz-2-oxa-1,3-diazol-4-yl)amino)-2-deoxyglucose in TNF—Treated FL83B mouse hepatocytes. *Phytother. Res.* 27: 236–243.

Chanudom, L., and J. Tangpong. 2015. Anti-inflammation property of *Syzygium cumini* (L.) Skeels on indomethacin-induced acute gastric ulceration. *Gastroenterol. Res. Pract.* 2015: 1–12.

Chaudhari, A. K. N., S. Pal, A. Gomes, and S. Bhattacharya. 1990. Antiinflammatory and related actions of *Syzygium cumini* seed extract. *Phytother. Res.* 4: 5–10.

Chaudhary, B., and K. Mukhopadhyay. 2013. Solvent optimization for anthocyanin extraction from *Syzygium cumini* L. Skeels using response surface methodology. *Int. J. Food Sci. Nutr.* 64: 363–371.

Chinni, S., A. Dubala, A. Kosaraju, R. B. Khatwal, M. N. S. Kumar, and E. Kannan. 2014. Effect of crude extract of *Eugenia jambolana* Lam. on human cytochrome P450 enzymes. *Phytother. Res.* 28: 1731–1734.

Chuan, S. S., C. W. Chang, and C. C. Li. 2012. Fraction from wax apple [*Syzygium samarangense* (Blume) Merrill and Perry] fruit extract ameliorates insulin resistance via modulating insulin signaling and inflammation pathway in tumor necrosis factor α-treated FL83B mouse hepatocytes. *Int. J. Mol. Sci.* 13: 8562–8577.

Cock, I. E. 2012. Antimicrobial activity of *Syzygium australe* and *Syzygium leuhmannii* leaf methanolic extracts. *Pharmacogn. Commun.* 2: 71–77.

Coelho, E. M., A. De, C. Luciana, L. C. Correa, M. T. Bordignon-Luiz, and M. D. S. Lima. 2016. Phenolic profile, organic acids and antioxidant activity of frozen pulp and juice of the jambolan (*Syzygium* cumini). *J. Food Biochem.* 40: 211–219.

Costa, A. G. V., D. F. Garcia-Diaz, P. Jimenez, and P. I. Silva. 2013. Bioactive compounds and health benefits of exotic tropical red-black berries. *J. Function. Foods* 5: 539–549.

De Bona, K. S., G. Bonfanti, P. E. R. Bitencourt, L. O. Cargnelutti, P. S. da Silva, T. P. da Silva, R. A. Zanette, A. S. Pigatto, and M. B. Moretto. 2014. *Syzygium cumini* is more effective in preventing the increase of erythrocytic ADA activity than phenolic compounds under hyperglycemic conditions in vitro. *J. Physiol. Biochem.* 70: 321–330.

Delespaul, Q., V. G. D. Billerbeck, C. G. Roques, G. Michel, C. M. Vinuales, and J. M. Bessiere. 2000. The antifungal activity of essential oils as determined by screening methods. *J. Essent. Oil Res.* 12: 256–266.

Devkar, R. V., A. V. Pandya, and N. H. Shah. 2012. Protective role of *Brassica olerecea* and *Eugenia jambolana* extracts against H_2O_2 induced cytotoxicity in H9C2 cells. *Food Funct.* 3: 837–843.

Dias, C. N., K. A. F. Rodrigues, F. A. A. Carvalho, S. M. P. Carneiro, J. G. S. Maia, E. H. A. Andrade, and D. F. C. Moraes. 2013. Molluscicidal and leishmanicidal activity of the leaf essential oil of *Syzygium cumini* (L.) Skeels from Brazil. *Chem. Biodiv.* 10: 1133–1141.

Dibua, U. M. E., and G. E. Odo. 2012. *In vitro* antimicrobial and antihelminthic activity of the ethanolic extracts of *Allium sativum, Allium cepa, Lantana camara, Averrhoa carambola* and *Syzygium aromaticum. J. Med. Plants Res.* 6: 5059–5068.

Djipa, C. D., D. Delmee, and Q. J. Laclercg. 2000. Antimicrobial activity of bark extracts of *Syzygium jambos* (L.) Alston (Myrtaceae). *J. Ethnopharmacol.* 71: 307–313.

Duraipandiyan, V., M. Ayyanar, and S. Ignacimuthu. 2006. Antimicrobial activity of some ethnomedicinal plants used by Paliyar tribe from Tamil Nadu, India. *BMC Complement Altern. Med.* 6: 35.

Ecker, A., F. A. Vieira, A. S. Prestes, M. M. dos Santos, A. Ramos-Ferreira, D. Dias, G. T. de Macedo, C. V. Klimaczewski, R. L. Seeger, R. J. B. Teixeira, and B. Nilda. 2015. Effect of *Syzygium cumini* and *Bauhinia forficata* aqueous-leaf extracts on oxidative and mitochondrial parameters in vitro. *EXCLI J.* 14: 1219–1231.

El-Anany, A. M., and R. F. M. Ali. 2013. Biochemical and histopathological effects of administration various levels of Pomposia (*Syzygium cumini*) fruit juice as natural antioxidant on rat health. *J. Food Sci. Technol. Mysore* 50: 487–495.

Elansary, H. O., M. Z. M. Salem, N. A. Ashmawy, and M. M. Yacout, 2012. Chemical composition, antibacterial and antioxidant activities of leaves essential oils from *Syzygium cumini* L., *Cupressus sempervirens* L. and *Lantana camara* L. from Egypt. *J. Agric. Sci.* (Toronto) 4: 144–152.

Eshwarappa, R. S. B., R. S. Iyer, S. R. Subbaramaiah, S. A. Richard, and B. L. Dhananjaya. 2014. Antioxidant activity of *Syzygium cumini* leaf gall extracts. *Bioimpacts* 4: 101–107.

Fernada, G. B., M. L. Bouzada, R. Fabri, M. O. Matos, F. O. Moreira, E. Scio, and E. S. Coimbra. 2008. Antileishmanial and antifungal activity of plants used in traditional medicine in Brazil. *J. Med. Plants Res.* 2: 246–249.

Franca, K. A. R., L. V. Amorim, C. N. Dias, D. F. C. Moraes, F. Denise, S. M. C. Portela, and F. A. A. de Carvalho. 2015. *Syzygium cumini* (L.) Skeels essential oil and its major constituent alpha-pinene exhibit anti-*Leishmania* activity through immunomodulation in vitro. *J. Ethnopharmacol.* 160: 32–40.

Gavillan-Suarez, J., A. Aguilar-Perez, N. Rivera-Ortiz, K. Rodriguez-Tirado, W. Figueroa-Cuilan, Morales- L. Santiago, G. Maldonado-Martinez, L. A. Cubano, and M. M. Martinez-Montemayor. 2015. Chemical profile and in vivo hypoglycemic effects of *Syzygium jambos, Costus speciosus* and *Tapeinochilos ananassae* plant extracts used as diabetes adjuvants in Puerto Rico. *BMC Complement. Altern. Med.* 15: 244.

Geun, L. K., and T. Shibamoto. 2001. Antioxidant property of aroma extract isolated from clove buds [*Syzygium aromaticum* (L.) Merr. et Perry]. *Food Chem.* 74: 443–448.

Gopan, R., V. George, N. S. Pradeep, and M. G. Sethuraman 2008. Chemical composition and antimicrobial activity of the leaf oil from *Syzygium gardneri* Thw. *J. Essent. Oil Res.* 20: 72–74.

Gupta, K., S. Barua, S. N. Hazarika, A. K. Manhar, D. Nath, N. Karak, N. D. Namsa, R. Mukhopadhyay, V. C. Kalia, and M. Mandal. 2014. Green silver nanoparticles: Enhanced antimicrobial and antibiofilm activity with effects on DNA replication and cell cytotoxicity. *RSC Adv.* 4: 52845–52855.

Gupta, M., and R. Bhadauria. 2012. Evaluation of anti-fungal potential of aqueous extract of *Syzygium cumini* Linn. against *Alternaria alternata* Nees. and *Fusarium oxysporum* Schle. *Int. J. Pharma Bio Sci.* 3: 571–577.

Harris, S. M., H. McFeeters, I. V. Ogungbe, L. R. Cruz-Vera, W. N. Setzer, B. R. Jackes, and R. L. McFeeters. 2011. Peptidyl-tRNA hydrolase screening combined with molecular docking reveals the antibiotic potential of *Syzygium johnsonii* bark extract. *Nat. Prod. Commun.* 6: 1421–1424.

Herculano, E. A., C. D. F. da Costa, A. K. B. F. Rodrigues, J. X. Araujo-Junior, A. E. G. Santana, P. H. B. Franca, E. A. Gomes, M. J. Salvador, F. B. P. Moura, and E. A. N. Ribeiro. 2014. Evaluation of cardiovascular effects of edible fruits of *Syzygium cumini* (L) Skeels Myrtaceae in rats. *Trop. J. Pharm. Res.* 13: 1853–1861.

Islam, M., K. Hussain, A. Latif, F. K. Hashmi, H. Saeed, N. I. Bukhari, S. S. Hassan, M. Z. Danish, and B. Ahmad. 2015. Evaluation of extracts of seeds of *Syzygium cumini* L. for hepatoprotective activity using CCl_4-induced stressed rats. *Pak. Vet. J.* 35: 197–200.

Jain, A., S. Sharma, M. Goyal, S. Dubey, S. Jain, J. Sahu, A. Sharma, and A. Kaushik. 2010. Anti-inflammatory activity of *Syzygium cumini* leaves. *Int. J. Phytomed.* 2: 124–126.

Javaid, A., and S. Samad. 2012. Screening of allelopathic trees for their antifungal potential against *Alternaria alternata* strains isolated from dying-back *Eucalyptus* spp. *Nat. Prod. Res.* 26: 1697–1702.

Jayachandra, K., and V. S. Devi. 2012. *In-vitro* antioxidant activity of methanolic extract of *Syzygium cumini* Linn. bark. *Asian J. Biomed. Pharm. Sci.* 2: 45–49.

Jayasinghe, U. L. B., B. M. M. Kumarihamy, A. G. D. Bandara, J. Waiblinger, and W. Kraus. 2003. Antifeedant activity of some Sri Lankan plants. *Nat. Prod. Res.* 17: 5–8.

Jayasinghe, U. L. B., R. M. S. Ratnayake, M. M. W. S. Medawala, and Y. Fujimoto 2007. Dihydrochalcones with radical scavenging properties from the leaves of *Syzygium jambos*. *Nat. Prod. Res.* 21: 551–554.

Kandati, V., P. Govardhan, C. S. Reddy, A. R. Nath, and R. R. Reddy. 2012. *In-vitro* and *in-vivo* anti-inflammatory activity of *Syzygium alternifolium* (Wt.) Walp. *J. Med. Plants Res.* 6: 4995–5001.

Kaneria, M., and S. Chanda. 2013. Evaluation of antioxidant and antimicrobial capacity of *Syzygium cumini* L. leaves extracted sequentially in different solvents. *J. Food Biochem.* 37: 168–176.

Kheaw-on, N., R. Chaisuksant, O. Suntornwat, S. Kanlayanarat, Y. Desjardins, and V. Srilaong. 2009. Antioxidant capacity of flesh and seed from *Syzygium cumini* fruits. *Acta Hortic.* 837: 73–78.

Kim, H. M., E. H. Lee, S. H. Hong, H. J. Song, M. K. Shin, S. H. Kim, and T. Y. Shin. 1998. Effect of *Syzygium aromaticum* extract on immediate hypersensitivity in rats. *J. Ethnopharmacol.* 60: 125–131.

Kothari, V., S. Seshadri, and P. Mehta. 2011. Fractionation of antibacterial extracts of *Syzygium cumini* (Myrtaceae) seeds. *Res. Biotechnol.* 2: 53–63.

Kumar, A., N. Padmanabhan, and M. R. V. Krishnan. 2007. Central nervous system activity of *Syzygium cumini* seed. *Pak. J. Nutr.* 6: 698–700.

Kusumoto, I. T., T. Nakabayashi, H. Kida, H. Miyashirol, M. Hattori, T. Namba, and K. Shimotohno. 1995. Screening of various plant extracts used in Ayuvedic medicine for inhibitory effects on human immunodeficiency virus type 1 (IIIV-1) protease. *Phytother. Res.* 9: 180–184.

Latief, N., S. Anand, M. C. Lingaraju, V. Balaganur, N. N. Pathak, J. Kalra, D. Kumar, B. K. Bhadoria, and S. K. Tandan. 2015. Effect of trimeric myricetin rhamnoside (TMR) in carrageenan-induced inflammation and caecal ligation and puncture-induced lung oxidative stress in mice. *Phytother. Res.* 29: 1798–1805.

Lee, K. G., and T. Shibamoto. 2001. Antioxidant property of aroma extract isolated from clove buds [*Syzygium aromaticum* (L.) Merr. et Perry]. *Food Chem.* 74: 443–448.

Liang-Liang, Z., and L. Yi-Ming. 2009. Antioxidant tannins from *Syzygium cumini* fruit. *Afr. J. Biotechnol.* 8: 2301–2309.

Locher, O. P., M. T. Burch, H. F. Mower, J. Berestecky, H. Davis, B. Van Poel, A. Lasure, D. A. Vanden Berghe, and A. J. Vlietinck. 1995. Anti-microbial activity and anti-complement activity of extracts obtained from selected Hawaiian medicinal plants. *J. Ethnopharmacol.* 49: 23–32.

Machado, R. R. P., D. F. Jardim, A. R. Souza, E. Scio, R. L. Fabri, A. G. Carpanez, R. M. Grazul, J. P. R. F. de Mendonca, B. Lesche, and F. M. Aarestrup. 2013. The effect of essential oil of *Syzygium cumini* on the development of granulomatous inflammation in mice. *Braz. J. Pharmacogn.* 23: 488–496.

Maggi, M. D., S. R. Ruffnengo, L. B. Gende, E. G. Sarlo, M. J. Eguaras, P. N. Bailac, and M. I. Ponzi. 2010. Laboratory evaluations of *Syzygium aromaticum* (L.) Merr. et Perry essential oil against *Varroa destructor. J. Essent. Oil Res.* 22: 119–122.

Mahajan, S., P. Chauhan, S. K. Subramani, A. Anand, D. Borole, H. Goswamy, and G. B. K. S. Prasad. 2015. Evaluation of "GSPF kwath": A *Gymnema sylvestre*-containing polyherbal formulation for the treatment of human type 2 diabetes mellitus. *Eur. J. Integr. Med.* 7: 303–311.

Mahapatra, P. K., and D. Pradhan. 2012. Relaxant effects of *Syzygium cumini* leaves on guinea pig tracheal chains and its possible mechanism(s). *J. Biomed. Pharm. Res.* 1: 11–16.

Manaharan, T., D. Appleton, C. H. Ming, and U. D. Palanisamy. 2012. Flavonoids isolated from *Syzygium aqueum* leaf extract as potential antihyperglycaemic agents. *Food Chem.* 132: 1802–1807.

Mani, S., G. Narayanaswamylaxmanan, D. Chellappan, A. Mohammad, P. Padugu, R. Garvandula, S. Ande, U. Anu, and U. Vallapu. 2012. Screening of methanolic extract of *Eugenia jambolana* leaves for its hepatoprotective activity in carbon tetrachloride induced rats. *Int. J. Appl. Res. Nat. Prod.* 5: 14–18.

Middha, S. K., T. Usha, P. Tripathi, K. Y. Marathe, T. Jain, B. Bhatt, Y. P. Masurkar, and V. Pande. 2012. An *in vitro* studies on indigenous ayurvedic plants, having hypoglycemic activity. *Asian Pac. J. Trop. Dis. Suppl.* 1: S46–S49.

Miyazawa, M., and M. Hisama. 2003. Antimutagenic activity of phenylpropanoids from clove (*Syzygium aromaticum*). *J. Agric. Food Chem.* 51: 6413–6422.

Modi, D. C., J. K. Patel, B. N. Shah, and B. S. Nayak. 2010. Antiinflammatory activity of seeds of *Syzygium cumini* Linn. *J. Pharm. Educ. Res.* 1: 68–70.

Mohamed, A. A., S. I. Ali, and F. K. El-Baz. 2013. Antioxidant and antibacterial activities of crude extracts and essential oils of *Syzygium cumini* leaves. *PLoS One* 8: e60269.

Morton, J. 1987. Malay apple. In *Fruits of Warm Climates*, ed. J. F. Morton and F. Julia. Miami: Morton Publishing, 378–381.

Muruganandan, S., K. Srinivasan, S. Chandra, S. K. Tandan, J. Lal, and V. Raviprakash. 2001. Anti-inflammatory activity of *Syzygium cumini* bark. *Fitoterapia* 72: 369–375.

Mussi, L. P., A. O. Guimaraes, K. S. Ferreira, and N. R. Pereira. 2015. Spouted bed drying of jambolao (*Syzygium cumini*) residue: Drying kinetics and effect on the antioxidant activity, anthocyanins and nutrients contents. *Food Sci. Technol.* 61: 80–88.

Nomi, Y., S. Shimizu, Y. Sone, M. T. Tuyet, T. P. Gia, M. Kamiyama, T. Shibamoto, K. Shindo, and Y. Otsuka. 2012. Isolation and antioxidant activity of zeylaniin A, a new macrocyclic ellagitannin from *Syzygium zeylanicum* leaves. *J. Agric. Food Chem.* 60: 10263–10269.

Noor, A., V. S. Bansal, and M. A. Vijayalakshmi. 2013. Current update on anti-diabetic bio-molecules from key traditional Indian medicinal plants. *Curr. Sci.* 104: 721–727.

Noudogbessi, J. P., P. Yedomonhan, D. C. K. Sohounhlou, J. C. Chalchat, and G. Figueredo. 2008. Chemical composition of essential oil of *Syzygium guineense* (Willd.) DC. var. *guineense* (Myrtaceae) from Benin. *Rec. Nat. Prod.* 2: 33–38.

Oliveira, G. F., N. A. J. C. Furtado, A. A. S. Filho, C. H. G. Martins, J. K. Bastos, W. R. Cunha, and M. L. Andrade e Silva. 2007. Antimicrobial activity of *Syzygium cumini* (Myrtaceae) leaves extract. *Braz. J. Microbiol.* 38: 381–384.

Omar, R., L. Y. Li, T. Yuan, and N. P. Seeram. 2012. α-Glucosidase inhibitory hydrolyzable tannins from *Eugenia jambolana* seeds. *J. Nat. Prod.* 75: 1505–1509.

Palanisamy, U. D., L. T. Ling, T. Manaharan, V. Sivapalan, T. Subramaniam, M. H. Helme, and T. Masilamani. 2011. Standardized extract of *Syzygium aqueum*: A safe cosmetic ingredient. *Int. J. Cosmetic Sci.* 33: 269–275.

Prashar, A., I. C. Locke, and C. S. Evans. 2006. Cytotoxicity of clove (*Syzygium aromaticum*) oil and its major components to human skin cells. *Cell Prolif.* 39: 241–248.

Priya, S. S. L., P. R. Devi, and A. Madeswaran. 2013. In silico docking studies of RP2 (X-linked retinitis pigmentosa) protein using anthocyanins as potential inhibitors. *Bangladesh J. Pharmacol.* 8: 292–299.

Quintans, J. S. S., R. G. Brito, P. G. V. Aquino, P. H. B. Franca, P. S. Siqueira-Lima, A. E. G. Santana, E. A. N. Ribeiro, M. J. Salvador, J. X. Araujo-Junior, and L. J. Quintans-Junior. 2014. Antinociceptive activity of *Syzygium cumini* leaves ethanol extract on orofacial nociception protocols in rodents. *Pharm. Biol.* 52: 762–766.

Radha, R., R. Latha, and M. S. Swaminathan. 2002. Chemical composition and bioactivity of essential oil from *Syzygium travancoricum* Gamble. *Flav. Fragr. J.* 17: 352–354.

Raga, D. D., C. L. C. Cheng, K. C. I. C. Lee, W. Z. P. Olaziman, V. J. A. Guzman, S. C. Chang, F. C. Franco, and C. Y. Ragasa. 2011. Bioactivities of triterpenes and a sterol from *Syzygium samarangense*. *Z. Naturforsch. C* 66(5/6): 235–244.

Rahnama, M., M. Najimi, and S. Ali 2012. Antibacterial effects of *Myristica fragrans*, *Zataria multiflora* Boiss, *Syzygium aromaticum*, and *Zingiber officinale* Rosci essential oils, alone and in combination with nisin on *Listeria monocytogenes*. *Comp. Clin. Pathol.* 21: 1313–1316.

Rajput, R., M. B. Patil, M. S. Sikarwar, H. Rajput, M. Prasad, S. Tondon, and R. S. Yadav. 2009. Antidiabetic effect of polyherbal formulation in alloxan-induced diabetes mellitus. *Plant Arch.* 9: 365–367.

Rasib, K. Z., and H. Ashraf. 2014. Feeding preferences of *Coptotermes heimi* (Isoptera: Termitidae) under laboratory and field conditions for different commercial and non-commercial woods. *Int. J. Trop. Insect Sci.* 34: 115–126.

Ribeiro, R. M., V. F. P. Neto, K. S. Ribeiro, D. A. Vieira, I. C. Abreu, S. N. Silva, M. S. S. Cartagenes, S. M. F. Freire, A. C. R. Borges, and M. O. R. Borges. 2014. Antihypertensive effect of *Syzygium cumini* in spontaneously hypertensive rats. *Evid. Based Complement. Alternat. Med.* 2014: 605452.

Ribeiro, T. G., M. A. Chavez-Fumagalli, D. G. Valadares, J. R. Franca, P. S. Lage, M. C. Duarte, P. H. R. Andrade et al. 2014. Antileishmanial activity and cytotoxicity of Brazilian plants. *Exp. Parasitol.* 143: 60–68.

Roy, D., V. Kumar, K. K. Acharya, and K. Thirumurugan. 2014. Probing the binding of *Syzygium*-derived alpha-glucosidase inhibitors with N- and C-terminal human maltase glucoamylase by docking and molecular dynamics simulation. *Appl. Biochem. Biotechnol.* 172: 102–114.

Sagrawat, H., A. S. Mann, and M. D. Kharya. 2006. Pharmacological potential of *Eugenia jambolana*: A review. *Pharmacogn. Mag.* 2: 96–104.

Sanches, J. R., L. M. Franca, V. T. Chagas, R. S. Gaspar, K. A. dos Santos, L. M. Goncalves, D. M. Sloboda et al. 2016. Polyphenol-rich extract of *Syzygium cumini* leaf dually improves peripheral insulin sensitivity and pancreatic islet function in monosodium L-glutamate-induced obese rats. *Front. Pharmacol.* 7: 48.

Santiago, M. C. P. A., A. C. M. S. Gouvea, F. M. Peixoto, R. G. Borguini, R. L. O. Godoy, S. Pacheco, L. S. M. Nascimento, and R. I. Nogueira. 2016. Characterization of jamelao (*Syzygium cumini* (L.) Skeels) fruit peel powder for use as natural colorant. *Fruits* 71: 3–8.

Santos, M. M., M. G. S. Nunes, and R. D. Martins. 2012. Empirical use of medicinal plants for diabetes treatment. *Rev. Bras. Plantas Med.* 14: 327–334.

Saravanan, G., B. Kalaiselvi, and G. P. Reddy. 2011. Hypolipidemic effect of Zignidd (poly-herbal formulation) in streptozotocin-induced diabetic rats. *Int. J. Drug Formul. Res.* 2: 151–160.

Sarita, M., and M. Bhagya. 2012. Antiimplantation and anti-estrogenic activity of *Eugenia jambolana* Lam. seed. *J. Pharm. Res.* 5: 2607–2609.

Sarita, M., M. Bhagya, and T. Shivanandappa. 2012. Post-coital contraceptive activity and teratogenicity of *Eugenia jambolana* Lam. seed. *J. Pharm. Res.* 5: 4295–4298.

Saroj, A., V. S. Pragadheesh, Palanivelu, A. Yadav, S. C. Singh, A. Samad, A. S. Negi, and C. S. Chanotiya. 2015. Anti-phytopathogenic activity of *Syzygium cumini* essential oil, hydrocarbon fractions and its novel constituents. *Ind. Crop. Prod.* 74: 327–335.

Sawant, L., V. K. Singh, S. Dethe, A. Bhaskar, J. Balachandran, D. Mundkinajeddu, and A. Agarwal. 2015. Aldose reductase and protein tyrosine phosphatase 1B inhibitory active compounds from *Syzygium cumini* seeds. *Pharm. Biol.* 53: 1176–1182.

Saxena, S., S. Gautam, and A. Sharma. 2013. Comparative evaluation of antimutagenicity of commonly consumed fruits and activity-guided identification of bioactive principles from the most potent fruit, Java plum (*Syzygium cumini*). *J. Agric. Food Chem.* 61: 10033–10042.

Selvam, N. T., V. Venkatakrishnan, S. D. Kumar, and S. Murugesan. 2011. Antioxidant and tumour cell suppression potential of methanolic extract of *Syzygium jambos* (Linn) leaf. *Asian J. Exp. Biol. Sci.* 2: 673–678.

Shafi, P. M., M. K. Rosamma, K. Jamil, and P. S. Reddy. 2002. Antibacterial activity of *Syzygium cumini* and *Syzygium travancoricum* leaf essential oils. *Fitoterapia* 73: 414–416.

Shafiuddin, M., J. Parmar, D. K. Suresh, S. A. Hussain, and M. Shinde. 2011. Antidiabetic, antihyperlipidemic and antioxidant activity of polyherbal formulation. *Int. J. Drug Formul. Res.* 2: 280–302.

Shi-Fei, L., H. Nian-Xu, H. Xiao-Jiang, L. Ling, and L. Shun-Lin. 2009. Effect of chemical constituents from *Syzygium cumini* (Myrtaceae) on glucose uptake in insulin-resistant L6 cells. *Acta Bot. Yunnan.* 31: 469–473.

Shrikanta, A., A. Kumar, and V. Govindaswamy. 2015. Resveratrol content and antioxidant properties of underutilized fruits. *J. Food Sci. Technol. Mysore* 52: 383–390.

Siani, A. C., M. C. Souza, M. G. M. O. Henriques, and M. F. S. Ramos. 2013. Anti-inflammatory activity of essential oils from *Syzygium cumini* and *Psidium guajava*. *Pharm. Biol.* 51: 881–887.

Simirgiotis, M. J., S. Adachi, S. To, H. Yang, K. A. Reynertson, M. S. Basile, R. R. Gil, I. B. Weinstein, and E. J. Kennelly. 2008. Cytotoxic chalcones and antioxidants from the fruits of *Syzygium samarangense* (wax jambu). *Food Chem.* 107: 813–819.

Singh, J. P., A. Kaur, K. Shevkani, and N. Singh. 2015. Influence of jambolan (*Syzygium cumini*) and xanthan gum incorporation on the physicochemical, antioxidant and sensory properties of gluten-free eggless rice muffins. *Int. J. Food Sci. Tech.* 50: 1190–1197.

Singh, J. P., A. Kaur, N. Singh, L. Nim, K. Shevkani, H. Kaur, and D. S. Arora. 2016. In vitro antioxidant and antimicrobial properties of jambolan (*Syzygium cumini*) fruit polyphe-nols. *Food Sci. Technol.* 65: 1025–1030.

Singh, R., L. Nawale, M. Arkile, S. Wadhwani, U. Shedbalkar, S. Chopade, D. Sarkar, B. Chopade, and A. Balu. 2016. Phytogenic silver, gold, and bimetallic nanoparticles as novel antitubercular agents. *Int. J. Nanomed.* 11: 1889–1897.

Srivastava, B., A. K. Sinha, S. Gaur, Y. Barshiliya, and S. B. Singh. 2012. Study of hypo-glycaemic and hypolipidemic activity of *Eugenia jambolana* pulp and seed extract in streptozotocin induced diabetic albino rats. *Asian J. Pharm. Life Sci.* 2: 10–19.

Srivastava, S., and D. Chandra. 2013. Pharmacological potentials of *Syzygium cumini*: A review. *J. Sci. Food Agric.* 93: 2084–2093.

Tahir, L., S. Ahmed, N. Hussain, I. Perveen, and S. Rahman. 2012. Effect of leaves extract of indigenous species of *Syzygium cumini* on dental caries causing pathogens. *Int. J. Pharma Bio Sci.* 3: 1032–1038.

Tanko, Y., A. Mohammed, M. A. Okasha, A. H. Umar, and R. A. Magaji. 2008. Antinociceptive and anti-inflammatory activities of ethanol extract of *Syzygium aromaticum* flower bud in Wistar rats and mice. *Afr. J. Trad. Complement. Altern. Med.* 5: 209–212.

Tobal, T. M., R. Silva, E. Gomes, H. M. A. Bolini, and M. Boscolo. 2012. Evaluation of the use of *Syzygium cumini* fruit extract as an antioxidant additive in orange juice and its sensorial impact. *Int. J. Food Sci. Nutr.* 63: 273–277.

Tong, W. Y., H. Wang, V. Y. Waisundara, and D. Huang. 2014. Inhibiting enzymatic starch digestion by hydrolyzable tannins isolated from *Eugenia jambolana*. *Food Sci. Technol.* 59: 389–395.

Tripathi, A. K., and S. Kohli. 2014. Pharmacognostical standardization and antidiabetic activity of *Syzygium cumini* (Linn.) barks (Myrtaceae) on streptozotocin-induced diabetic rats. *J. Complement. Integr. Med.* 11: 71–81.

Tripathi, P., R. K. Patel, R. Trpathi, and N. R. Kanzariya. 2013. Investigation of antigenotoxic potential of *Syzygium cumini* extract (SCE) on cyclophosphamide-induced genotoxicity and oxidative stress in mice. *Drug Chem. Toxicol.* 36: 396–402.

Vasanthi, S., M. Sasikumar, A. Y. Sangilimuthu, S. Venkatachalapathi, and V. K. Gopalakrishnan. 2012. *In vitro* antioxidant activity of *Syzygium samarangense* Merr. et Perry. fruit extract. *J. Pharm. Res.* 5: 3426–3430.

Vasavi, H. S., A. B. Arun, and P. D. Rekha. 2014. Inhibition of quorum sensing in *Chromobacterium violaceum* by *Syzygium cumini* L. and *Pimenta dioica* L. *Asian Pac. J. Trop. Biomed.* 3: 954–959.

Veber, J., L. A. Petrini, L. B. Andrade, and J. Siviero. 2015. Determination of phenolic compounds and antioxidant capacity of aqueous and ethanolic extracts of jambul (*Syzygium cumini*). *Rev. Bras. Pl. Med.* 17: 267–273.

Voigt, M. F., L. F. Damé Schuch, C. L. Gonçalves, A. Faccin, D. B. A. Schiavon, B. C. Bohm, and L. L. Ferreira. 2013. Antibacterial activity of extracts of *Syzygium cumini* (L.) Skeels (jambolão), against microorganisms associated with bovine mastitis. *Rev. Cubana Pl. Med.* 18: 495–501.

Webb, A. H., and F. W. Tanner. 1944. Effect of spices and flavouring materials on growth of yeast. *Food Res.* 10: 273.

Wei, H. L., and I. S. Ismail. 2012. Antioxidant activity, total phenolics and total flavonoids of *Syzygium polyanthum* (Wight) Walp leaves. *Int. J. Med. Aroma. Plants* 2: 219–228.

Zheng, G. Q., P. M. Kenney, and K. T. Lam. 1992. Sesquiterpenes from clove (*Eugenia caryophyllata*) as potential anticarcinogenic agents. *J. Nat. Prod.* 55: 999–1003.

Zhi-Ping, R., Z. Liang-Liang, and L. Yi-Ming. 2008. Evaluation of the antioxidant activity of *Syzygium cumini* leaves. *Molecules* 13: 2545–2556.

9 Cancer Immunology of *Syzygium cumini*

Sabira Mohammed and K. B. Harikumar

CONTENTS

INTRODUCTION

Syzygium cumini, known as "Jambu" or "Brahaspati" in Sanskrit, is an important medicinal plant. It has been widely used as a cure in Indian traditional systems of medicine, as well as in common folklore. Various parts of the plant (fruit, seeds, leaves, and bark) are used in the treatment of different diseases as diverse as digestive ailments, diabetes mellitus, gingivitis, renal problems, and so forth. The recent trend in utilizing natural products as therapeutic agents has brought out novel uses for the different parts of this plant. In this chapter, we have compiled the studies that show the anti-inflammatory and anticarcinogenic potential of this plant.

ANTI-INFLAMMATORY ACTIVITY OF *S. CUMINI*

There were several reports on the anti-inflammatory potential of *S. cumini*. Ethanolic (70%) extract of stem bark of the plant was studied in different rat models of inflammation, and no cytotoxicity of the extract was observed at the tested doses. The stem bark extract (300 and 1000 mg/kg body weight and, for some experiments, 100 mg/kg body weight) inhibited carrageenan-, kaolin-carrageenan-, and formaldehyde-induced paw edema formation. A similar effect was observed in cotton pellet granuloma formation (a model of chronic inflammation) in rats, as seen from reduced granuloma formation in treated groups (Muruganandan et al. 2001).

Modi et al. (2010) and Kumar et al. (2008) reported the anti-inflammatory potential of seeds of *S. cumini*. Both methanolic and aqueous extracts of the seeds inhibited the paw edema formation in rats. Their findings also confirmed that aqueous extract was found to be more potent than methanolic extracts. The anti-inflammatory activity of the leaf extracts was shown by Kumar et al. (2010) in acute and chronic models of inflammation, namely, the carrageenan paw model and the granuloma pouch model, in albino rats. The extract was shown to contain hydroxyl, ester, olefin, and carbonyl components that might be involved in causing this effect.

The anti-inflammatory activity of ethyl acetate and methanol extracts of leaves of *S. cumini* was reported by Jain et al. (2010). Oral administration of 200 and 400 mg/kg body weight of the extracts significantly reduced the paw edema formation in rats. Siani et al. (2013) demonstrated the anti-inflammatory effect of the essential oils from the leaves of *S. cumini* in a lipopolysaccharide-induced pleurisy model. The possible mode of action is hinted to be through inhibition of eosinophil migration. They also suggest the possibility of employing these essential oils for the treatment of inflammatory diseases. The anti-inflammatory effect of the essential oils from the leaves of the plant was yet again demonstrated by Machado et al. (2013). The anti-inflammatory effect was shown on chronic granulomatous inflammation in mice.

Lima and colleagues from Brazil took a different approach for their study. They collected the leaves of the plant every month for a year, and the efficacy of the extracts and tannin-free fractions was checked for carrageenan-induced paw edema formation in rats. High-performance liquid chromatography (HPLC) analysis revealed that extracts were rich in phenolics and flavonoids. The phenolic-rich fraction was most potent, compared with other fractions, for anti-inflammatory activity (Lima et al. 2007).

The extract and different fractions of the root bark are potent in bringing in an anti-inflammatory effect, as shown by Saha et al. (2013). Of the various fractions tested, the petroleum ether fraction was the most effective in inhibiting carrageenan-induced rat paw edema. It was also observed that the petroleum ether fraction of root bark inhibited paw edema significantly in rats. We further characterized the active components of this fraction and identified three major compounds: β-sitosterol, stigmasterol, and lupeol. Oral administration of the ethanolic extract of *Eugenia jambolana* in gum acacia was capable of inhibiting carrageenan-induced paw edema formation in albino rats, as demonstrated by Kariyil and Sujarani (2011). A similar effect of anti-inflammatory action of *S. cumini* in carrageenan-induced inflammatory models was shown by Mathur et al. (2011) using the methanolic extracts of the seeds.

The anti-inflammatory activity of the ethanolic fraction of bark extract was also tested against inflammation induced by histamine, 5-hydroxytryptamine (5-HT), bradykinin, and Prostaglandin E2 (PGE2) in rats. *S. cumini* extract significantly decreased the paw edema caused by all the agents except bradykinin (Muruganandan et al. 2002).

Using an *in vitro* antiallergic model where rat peritoneal exudate cells are treated with the calcium ionophore A23187, the release of histamine secretion was measured. An ethanol extract of *S. cumini* inhibited the histamine release, further supporting its anti-inflammatory potential (Hossain et al. 2008).

Brito et al. (2007) reported a detailed study of the antiallergic function of a water extract of *S. cumini* collected from Brazil. Preliminary HPLC studies demonstrated that the extract is rich in hydrolyzable tannins and flavonoids. Oral administration extract significantly inhibited the mouse paw edema formation induced by compound 48/80, histamine, and 5-HT but had no effect on platelet-aggregating factor-induced mouse paw edema formation. Further, they tested the effect of the extract in a mouse model of allergic pleurisy and found that there was a significant decrease in the accumulation of eosinophil in the extract-treated group. Analysis of cytokines in pleural lavage fluid using enzyme-linked immu-nosorbent assay (ELISA) revealed that there is a decrease in interleukin 5 (IL-5) and CCL11 production. This could be one of the mechanism actions of *S. cumini* (Brito et al. 2007).

The potential of the fruit extract in protecting from hepatic inflammation was shown (Donepudi et al. 2012) when C57 Bl/6 mice that underwent bile duct liga-tion surgery exhibited good recovery following the administration of the fruit extract.

ANTICARCINOGENIC EFFECT OF *S. CUMINI*

IN VITRO STUDIES

Fruit extracts of *S. cumini* rich in anthocynanis were tested for their antiproliferative potential in a series of breast cancer cells. *In vitro* studies revealed that the extracts inhibited the proliferation of MCF-7 and MDA-MB-231 breast cancer cells, but were least effective for MCF-10A, a type of normal breast cells (Li et al. 2009). Yadav et al. (2011) also confirmed this observation when they found that fruit extract inhibited the growth of MCF-1, PC-3, and A2780 cells.

Another study from Egypt, using freshly ripened fruits, demonstrated antileuke-mic effects of *S. cumini* (Afify et al. 2011). Six successive extracts were made in the order of hexane, chloroform, diethyl ether, ethyl acetate, ethanol, and distilled water, followed by purification of six active principles from *S. cumini* fruit extracts. The authors observed that the ethanolic fraction possesses the highest activity against cells derived from acute myeloid leukemia patient samples (Afify et al. 2011). The interesting component of this study was that the authors used cancer cells isolated from patients.

The pulp, seed, and peel extract of the plant were found to induce apoptosis in human breast (MCF-7) and prostate cancer (PC-3) cells, and out of the three extracts, pulp extract was found to be more potent. Yadav et al. (2011) also reported a similar observation where an ethanolic extract of seeds of *S. cumini* induced apoptosis in different human cancer cell lines.

A methanolic extract of partially ripened fruit skin was also tested for its activity in comparison with crude extract in two cervical cancer cell lines, HeLa and SiHa, *in vitro*. The crude extract was found to be more effective in inducing apoptosis (Barh and Viswanathan 2008).

The cancer chemoprotective ability of *S. cumini* was further shown by Aqil et al. (2012), where the anthocyanins and polyphenols extracted with acidic ethanol showed growth inhibition of the human lung cancer cell line, A549. The effect was attributed to the presence of 5-anthocyanidins and ellagic acid and ellagitannins.

IN VIVO STUDIES

There were several studies reported on the anticancer potential of different parts of *S. cumini*. Goyal et al. (2010) studied the effect of water extract of dried fruit in a benzo[*a*]pyrene (BaP)-induced gastric carcinogenesis model in Swiss albino mice. They observed that tumor incidence, tumor yield, tumor burden, and cumulative number of papillomas were reduced in mice treated with extract compared with untreated control.

Goyal and his group also reported the anticarcinogenic effect of seed extract of *S. cumini* in 7,12-dimethyl benz(*a*)anthracene (DMBA)–induced two-stage skin carcinogenesis in mice. Oral administration of seed extract was found to inhibit the skin papilloma formation, decrease the levels of lipid peroxides, and increase the levels of antioxidant enzymes like superoxide dismutase and catalase, compared with groups treated with carcinogen only. Histological analysis confirmed that administration of extract reduced epidermal hyperplasia (Parmar et al. 2010, 2011).

The antiproliferative effect of the petroleum ether fraction of the plant's root bark extract was shown by Saha et al. (2013), where there was a considerable inhibition of cotton pellet–induced granuloma formation.

IMMUNOMODULATORY ACTIVITY OF *S. CUMINI*

Seed extract of *S. cumini* was tested for its immunomodulatory activity. It was found to increase the delayed-type hypersensitivity (DTH) reaction and humoral antibody titers in rats in a dose-dependent manner. Similarly, the treatment also increased the total number of white blood cells, neutrophils, and lymphocytes in rats (Mastan et al. 2008). This study pointed out that *S. cumini* seed extract has the potential to stimulate the hematopoietic system of the body, which, in turn, indicated that the plant has treatment potential in immune-deficient conditions arising during radiation therapy or chemotherapy.

Bellé et al. (2013) evaluated the effect of an aqueous extract of seeds of *S. cumini* for enzymes involved in lymphocyte functions. Lymphocytes isolated from donors were treated with extract, and activities of adenosine deaminase (ADA), dipeptidyl peptidase IV (DPP-IV), and acetylcholinesterase (AChE) and CD26 expression were analyzed. *S. cumini* inhibited the activities of ADA and DPP-IV but not that of AchE. Moreover, there was no effect on CD26 expression (Bellé et al. 2013). This could be one of the possible mechanisms of action of *S. cumini*.

The immunomodulatory effects of the aqueous extracts of both bark and seed of *S. cumini* were studied in an alloxan-induced diabetic model in mice (Daisy et al. 2004). The total numbers of white blood cells, red blood cells, and lymphocyte (B and T) populations were significantly increased in *S. cumini* extract–treated animals compared with the control group, and correlated with a decrease in serum glucose levels (Daisy et al. 2004).

RADIOPROTECTIVE EFFECT OF *S. CUMINI*

Radiation therapy is one of the chief modalities in cancer therapy. However, one of the major drawbacks of this therapy is the damage caused to adjacent normal tissues. An ideal radioprotectant should be one that protects the normal tissues without compromising the anticancer potential of radiation. Such agents have a great demand in radiation and nuclear medicine. Several groups evaluated the radioprotective effect of *S. cumini* using different systems.

The leaves of *S. cumini* were tested as a radioprotectant using a micronucleus assay. The lymphocytes were isolated from blood samples of healthy donors and exposed to γ-irradiation in the presence and absence of leaf extract. *S. cumini* was found to reduce the formation of micronuclei in lymphocytes (Jagetia and Baliga 2002).

Arun et al. (2011) also confirmed that seed extract *S. cumini* inhibited the micronuclei formation in mouse bone marrow cells induced by genotoxic stress. The micronuclei formation was also assessed in the isolated splenocytes of irradiated mice. Similar to previous observations, *S. cumini* extract reduced the incidence of micronuclei, and a concentration-dependent inhibition of lipid peroxides was also observed in mice brain extract prepared after irradiation (Jagetia et al. 2012).

S. cumini leaf extract was also found to be effective in protecting mice from a lethal dose of radiation (Shetty and Vidyasagar 2008). Animals were sacrificed postirradiation, and damage in the intestinal system was studied. Irradiation caused significant reduction in the height of villi and crypt number, and increased the number of goblet cells and dead cells. Treatment with extract prevented these changes and protected the mouse gastrointestinal system from radiation-induced deleterious damage. The same group also reported the radioprotective potential of seed extract of *S. cumini* (Jagetia et al. 2005).

POSSIBLE MECHANISM OF ACTION

There were several studies on various pharmacological activities of *S. cumini* in the literature. However, very limited studies were reported for isolation of active principles from this plant. The bark extract has shown the presence of betulinic acid, which is a known anticancer and antiviral agent. Seeds were reported to contain different tannins and compounds, such as corilagin, jambosine, quercetin, gallic acid, kaempferol, and myricetin. Many of these compounds are known for their antioxidant potential and might be contributing to the radioprotective activity of *S. cumini*. Quercetin is a good suppressor of NF-kB signaling, which is involved in various inflammatory cascades leading to cancer initiation and progression. *S. cumini* is a rich source of various components having antioxidant, free radical scavenging, and anti-inflammatory activity, which might contribute to the observed effects reported in this chapter. Further studies are required to completely characterize the lead compounds in this medicinal plant.

REFERENCES

Afify, A. M. R., S. A. Fayed, E. A. Shalaby, and H. A. El-Shemy. 2011. *Syzygium cumini* (pomposia) active principles exhibit potent anticancer and antioxidant activities. *Afr. J. Pharm. Pharmacol.* 5: 948–956.

Aqil, F., A. Gupta, R. Munagala, J. Jeyabalan, H. Kausar, R. J. Sharma, I. P. Singh, and R. C. Gupta. 2012. Anti-oxidant and anti-proliferative activities of anthocyanin/ellagitannin-enriched extracts from *Syzygium cumini* L. (jamun, the Indian blackberry). *Nutr. Cancer* 64: 428–438.

Arun, R., M. V. Prakash, S. K. Abraham, and K. Premkumar. 2011. Role of *Syzygium cumini* seed extract in the chemoprevention of in vivo genomic damage and oxidative stress. *J. Ethnopharmacol.* 134: 329–333.

Barh, D., and G. Viswanathan. 2008. *Syzygium cumini* inhibits growth and induces apoptosis in cervical cancer cell lines: A primary study. *Ecancermedicalscience* 2: 83.

Bellé, L. P., P. E. Bitencourt, F. H. Abdalla, K. S. Bona, A. Peres, L. D. Maders, and M. B. Moretto. 2013. Aqueous seed extract of *Syzygium cumini* inhibits the dipeptidyl peptidase IV and adenosine deaminase activities, but it does not change the CD26 expression in lymphocytes in vitro. *J. Physiol. Biochem.* 69: 119–124.

Brito, F. A., L. A. Lima, M. F. S. Ramos, M. J. Nakamura, S. C. Cavalher-Machado, A. C. Siani, M. G. Henriques, and A. L. F. Sampaio. 2007. Pharmacological study of antiallergic activity of *Syzygium cumini* (L.) Skeels. *Braz. J. Med. Biol. Res.* 40: 105–115.

Daisy, P., N. N. Priya, and M. Rajathi. 2004. Immunomodulatory activity of *Eugenia jambolana*, *Clitoria ternatea*, and *Phyllanthus emblica* on alloxan-induced diabetic rats. *J. Exp. Zool.* 7: 269–278.

Donepudi, A. C., L. M. Aleksunes, M. V. Driscoll, N. P. Seeram, and A. L. Slitt. 2012. The traditional Ayurvedic medicine, *Eugenia jambolana* (Jamun fruit), decreases liver inflammation, injury and fibrosis during cholestasis. *Liver Int.* 32: 560–573.

Goyal, P. K., P. Verma, P. Sharma, J. Parmar, and A. Agarwal. 2010. Evaluation of anti-cancer and anti-oxidative potential of *Syzygium cumini* against benzo[a]pyrene (BaP) induced gastric carcinogenesis in mice. *Asian Pac. J. Cancer Prev.* 11: 753–758.

Hossain, S. J., I. Tsujiyama, M. Takasugi, M. A. Islam, R. S. Biswas, and H. Aoshima. 2008. Total phenolic content, antioxidative, anti-amylase, anti-glucosidase, and antihistamine release activities of Bangladeshi fruits. *Food Sci. Technol. Res.* 14: 261–268.

Jagetia, G. C., and M. S. Baliga. 2002. *Syzygium cumini* (jamun) reduces the radiation-induced DNA damage in the cultured human peripheral blood lymphocytes: A preliminary study. *Toxicol. Lett.* 132: 19–25.

Jagetia, G. C., M. S. Baliga, and P. Venkatesh. 2005. Influence of seed extract of *Syzygium cumini* (jamun) on mice exposed to different doses of gamma-radiation. *J. Radiat. Res.* 46: 59–65.

Jagetia, G. C., P. C. Shetty, and M. S. Vidyasagar. 2012. Inhibition of radiation-induced DNA damage by jamun, *Syzygium cumini*, in the cultured splenocytes of mice exposed to different doses of γ-radiation. *Integr. Cancer Ther.* 11: 141–153.

Jain, A., S. Sharma, M. Goyal, S. Dubey, S. Jain, J. Sahu, A. Sharma, and A. Kaushik. 2010. Anti-inflammatory activity of *Syzygium cumini* leaves. *Int. J. Phytomed.* 2: 124–126.

Kariyil, B. J., and S. Sujarani. 2011. Anti-inflammatory activity of *Eugenia jambolana* and *Trigonella foenum graecum* in experimental animal model. *J. Adv. Vet. Res.* 1 (2): 65–68.

Kumar, A., R. Illavarasan, T. Jayachandran, M. Deecaraman, R. M. Kumar, P. Aravindan, N. Padmanabhan, and M. R. V. Krishnan. 2008. Anti-inflammatory activity of *S. cumini* seeds. *Afr. J. Biotechnol.* 7: 941–943.

Kumar, K. P., P. D. Prasad, A. N. Rao, P. D. Reddy, and G. Abhinay. 2010. Anti-inflammatory activity of *Eugenia jambolana* in albino rats. 2010. *Int. J. Pharma Bio Sci.* 1: 435–438.

Li, L., L. S. Adams, S. Chen, C. Killian, A. Ahmed, and N. P. Seeram. 2009. *Eugenia jambolana* Lam. berry extract inhibits growth and induces apoptosis of human breast cancer but not non-tumorigenic breast cells. *J. Agric. Food Chem.* 57: 826–831.

Lima, L. A., A. C. Siani, F. A. Brito, A. L. F. Sampaio, M. Henriques, and C. A. Riehl. 2007. Correlation of anti-inflammatory activity with phenolic content in the leaves of *Syzygium cumini* (L.) Skeels (Myrtaceae). *Quím. Nova* 30: 860–864.

Machado, R. R. P., D. F. Jardim, A. R. Souza, E. Scio, R. L. Fabri, A. G. Carpanez, R. M. Grazul, J. P. R. F. de Mendonca, B. Lesche, and F. M. Aarestrup. 2013. The effect of essential oil of *Syzygium cumini* on the development of granulomatous inflammation in mice. *Rev. Bras. Farmacogn.* 23: 488–496.

Mastan, S. K., A. Saraseeruha, V. Gourishankar, G. Chaitanya, N. Raghunandan, G. A. Reddy, and K. E. Kumar. 2008. Immunomodulatory activity of methanolic extract of *Syzygium cumini* seeds. *Pharmacologyonline* 3: 895–903.

Mathur, A., R. Purohit, D. Mathur, G. B. K. S. Prasad, and V. K. Dua. 2011. Pharmacological investigation of methanol extract of *S. cumini* seeds and *Crateva nurvula* bark on the basis of antimicrobial, anti-oxidant and anti-inflammatory properties. *Der Chem. Sin.* 2: 174–181.

Modi, D. C., J. K. Patel, B. N. Shah, and B. S. Nayak. 2010. Antiinflammatory activity of seeds of *Syzygium cumini* Linn. *J. Pharm. Educ. Res.* 1: 68–70.

Muruganandan, S., S. Pant, K. Srinivasan, S. Chandra, S. K. Tandan, J. Lal, and R. V. Prakash. 2002. Inhibitory role of *Syzygium cumini* on autacoid-induced inflammation in rats. *Indian J. Physiol. Pharmacol.* 46: 482–486.

Muruganandan, S., K. Srinivasan, S. Chandra, S. K. Tandan, J. Lal, and V. Raviprakash. 2001. Anti-inflammatory activity of *Syzygium cumini* bark. *Fitoterapia* 72: 369–375.

Parmar, J., P. Sharma, P. Verma, and P. K. Goyal. 2010. Chemopreventive action of *Syzygium cumini* on DMBA-induced skin papillomagenesis in mice. *Asian Pac. J. Cancer Prev.* 11: 261–265.

Parmar, J., P. Sharma, P. Verma, P. Sharma, and P. K. Goyal. 2011. Modulation of DMBA-induced biochemical and histopathological changes by *Syzygium cumini* seed extract during skin carcinogenesis. *Int. J. Cur. Biomed. Pharmaceut. Res.* 1: 24–30.

Saha, S., E. V. S. Subrahmanyam, C. Kodangala, S. C. Mandal, and S. C. Shastry. 2013. Evaluation of antinociceptive and anti-inflammatory activities of extract and fractions of *Eugenia jambolana* root bark and isolation of phytoconstituents. *Rev. Bras. Farmacogn.* 23 (4): 651–661.

Shetty, P. C., and M. S. Vidyasagar. 2008. Treatment of mice with leaf extract of jamun (*Syzygium cumini* Linn. Skeels) protects against the radiation induced damage in the intestinal mucosa of mice exposed to different doses of γ-radiation. *Pharmacologyonline* 1: 169–195.

Siani, A. C., M. C. Souza, M. G. M. O. Henriques, and M. F. S. Ramos. 2013. Anti-inflammatory activity of essential oils from *Syzygium cumini* and *Psidium guajava*. *Pharm. Biol.* 51 (7): 881–887.

Yadav, S. S., G. A. Meshram, S. Shinde, R. C. Patil, S. M. Manohar, and M. V. Upadhye. 2011. Antibacterial and anticancer activity of bioactive fraction of *Syzygium cumini* L. seeds. *Hayati J. Biosci.* 18: 118–122.

10 The Use of *Syzygium cumini* in Nanotechnology

Avnesh Kumari, Vineet Kumar,
and Sudesh Kumar Yadav

CONTENTS

INTRODUCTION

Nanotechnology deals with the design and production of nanoparticles (NPs) by controlled manipulation of size and shape at the nanoscale. Potential benefits of NPs in biomedical and industrial applications for human health are well accepted in literature (David et al. 2005). Recent reports focus on the interaction of NPs with cells and biological milieu (Kumari and Yadav 2011). Metallic NPs are important due to their applications in emerging areas of nanotechnology (Freeman et al. 1995; Ullman 1996; Zhao et al. 1998; Chang et al. 1999; Link et al. 1999; Manna et al. 2000; Jin et al. 2001; Li et al. 2001; Wang et al. 2001; Rao et al. 2002; Zheng 2002). Size, shape, and surface chemistry play an important role in controlling the physical, chemical, optical, and electronic properties of these NPs. However, the development of a suitable process or mechanism of NP synthesis, as well as full control of size and monodispersity, is still a challenge to current nanotechnology. Various chemical methods are used for the synthesis of NPs (Dahl et al. 2007). However, the chemically synthesized NPs may not be safe to use for biological and therapeutic purposes due to utilization of toxic and hazardous chemicals for their synthesis.

Green nanotechnology involves the application of green chemistry principles for the fabrication of nanoscale materials (Dahl et al. 2007; Iravani 2011). Green nanotechnology strives to discover synthetic methods that eliminate harmful reagents and enhance the efficiency of existing methods. For these reasons, the green chemistry approach is preferred nowadays for the synthesis of nanomaterials (Ravindran et al. 2003; Albrecht et al. 2006). Such nanomaterials would be no less harmful to human beings. Thus, more emphasis is now being given to green synthesis of NPs. The major advantages of green synthesis are the choice of environmentally benign solvent medium, benign reducing agent, and nontoxic material for the stabilization of NPs (capping agent).

Biological synthesis of NPs is preferred over chemical synthesis due to less capital and energy being involved in it. Synthesis of NPs using bioresources has captured significant attention from the scientific community. The use of bioresources such as bacteria, yeast, and fungi has been described for the formation of NPs (Mohanpuria et al. 2008). The rate of NP formation, and therefore the size of the NPs, could be manipulated by controlling processing parameters. Efforts are also being made for the synthesis of NPs of different chemical composition, sizes, and controlled monodispersity. Among bioresources, plants have also been extensively explored for the synthesis of NPs.

The extracts of various plants provide environment-friendly multifunctional materials that can act as reducing and capping agents for the synthesis and stabilization of NPs. These functional molecules are compatible with the green chemistry principles and synthesis. Synthesis of NPs by plant extracts is advantageous over other biological processes (Klaus et al. 1999; Nair and Pradeep 2002; Willner et al. 2006; Konishi et al. 2007; Kumar et al. 2009, 2010; Kumar and Yadav 2011a, 2011b, 2011c) because it does not need a cell culture process (Shankar et al. 2004a). The huge amount of phytochemicals in the plant extracts act as reducing as well as capping agents (Shankar et al. 2003, 2004a; Ankamwar et al. 2005; Chandran et al. 2006; Gericke and Pinches 2006) that reduce metal salts to NPs with subsequent surface coating (Li et al. 2007). Many plant extracts have shown potential to synthesize NPs extracellularly (Table 10.1).

Syzygium cumini has been widely used for the treatment of various ailments in traditional and folk medicine. Many pharmacological activities of *S. cumini* have been reported (Ayyanar and Subhash Babu 2012; see Chapter 8 in this volume for a review). Recently, this plant has been explored for the synthesis of metallic NPs (Kumar et al. 2010). This plant has also shown potential for the synthesis of polymeric NPs (Kumari et al. 2012; Kumar et al. 2014). In this chapter, we discuss the pharmaceutical importance of *S. cumini* and its use in nanotechnology.

PHARMACEUTICAL IMPORTANCE OF *S. CUMINI*

Bark, leaves, and fruits of *S. cumini* have been used in traditional medicine since ancient times (Ayyanar and Subhash Babu 2012; Sagrawat et al. 2006). It has been valued in the Ayurveda and Unani medicine systems for possessing a variety of therapeutic properties (Kumar et al. 2009). In Ayurveda, bark is used for the treatment of sore throat, bronchitis, asthma, thirst, dysentery, and blood impurities (Kirtikar

TABLE 10.1
List of Plants Screened for the Synthesis of Nanoparticles

Name of Plant	References
Aegle marmelos	Lal and Nayak 2012
Alfalfa	Gardea-Torresdey et al. 2003
Allium cepa	Parida et al. 2011
Allium sativum	Rastogi and Arunachalam 2012
Aloe vera	Chandran et al. 2006
Anacardium occidentale	Sheny et al. 2011
Ananas comosus	Lal and Nayak 2012
Andrographis paniculata	Sulochana et al. 2012
Astragalus gummifer	Kora and Arunachala 2012
Azadirachta indica	Shankar et al. 2004b
Brassica juncea	Haverkamp and Marshall 2009
Calendula officinalis	Lal and Nayak 2012
Capsicum annuum	Li et al. 2007
Carica papaya	Lal and Nayak 2012
Catharanthus roseus	Lal and Nayak 2012
Centella asiatica	Palaniselvam et al. 2012
Cinnamomum verum	Lal and Nayak 2012
Citrullus colocynthis	Satyavani et al. 2011
Citrus limonium	Lal and Nayak 2012
Citrus sinensis	Lal and Nayak 2012
Coffea arabica and *Camellia sinensis*	Nadagouda and Varma 2008
Coriandrum sativum	Sathyavathi et al. 2010
Datura metel	Lal and Nayak 2012
Desmodium triflorum	Ahmad et al. 2011
Diopyros kaki	Song et al. 2010
Dioscorea bulbifera	Ghosh et al. 2012
Dioscorea oppositifolia	Maheswari et al. 2012
Elettaria cardamomom	Gnanajobitha et al. 2012; Lal and Nayak 2012
Emblica officinalis	Ankamwar et al. 2005
Ficus benghalensis	Lal and Nayak 2012
Gardenia jasminoides	Jia et al. 2009
Glycine max	Petla et al. 2012
Glycyrrhiza glabra	Dinesh et al. 2012
Helianthus annus	Leela and Vivekanandan 2008
Hibiscus cannabinus	Bindhu and Umadevi 2012
Hibiscus rosasinensis	Lal and Nayak 2012
Hydrilla verticillata	Sable et al. 2012
Jatropha curcas	Hudlikar et al. 2012; Joglekar et al. 2011
Justicia gendarussa	Fazaludeen et al. 2012
Lantana camara	Sivakumar et al. 2012
Macrotyloma uniflorum	Aromal et al. 2012
Mentha arvensis	Lal and Nayak 2012

(Continued)

TABLE 10.1 (CONTINUED)
List of Plants Screened for the Synthesis of Nanoparticles

Name of Plant	References
Mentha piperita	U.K. Parashar et al. 2009
Mirabilis jalapa	Vankar and Bajpai 2010
Morinda pubescens	Mary and Inbathamizh 2012
Murraya koenigii	Philip et al. 2011
Nyctanthes arbor-tristis	Lal and Nayak 2012
Ocimum sanctum	Ramteke et al. 2013; Soundarrajan et al. 2012
Parthenium hysterophorus	Kumar et al. 2012; V. Parashar et al. 2009
Pedilanthus tithymaloides	Sundaravadivelan and Nalini 2011
Pelargonium graveolens	Shankar et al. 2003
Piper betle	Mallikarjuna et al. 2012
Piper nigrum	Lal and Nayak 2012
Plumeria rubra	Patil et al. 2012
Sesuvium portulacastrum	Nabikhan et al. 2010
Solanum xanthocarpum	Amin et al. 2012
Sorghum	Njagi et al. 2011
Swietenia mahogany	Mondal et al. 2011
Syzygium aromaticum	Deshpande 2010
Syzygium aromaticum	Lal and Nayak 2012
Tamarindus indica	Lal and Nayak 2012
Terminalia catappa	Ankamwar 2010
Trianthema decandra	Geethalakshmi and Sarada 2010
Tridax procumbens	Gopalakrishnan 2012
Vitis vinifera	Pavani et al. 2012
Zingiber officinale	Singh et al. 2011

and Basu 1918). In Unani medicine, the ash of leaves is used for strengthening gums and teeth, seeds are used as a remedy for diabetes, and bark has good wound healing properties.

S. cumini is rich in anthocyanins, glucoside, ellagic acid, isoquercetin, kaemferol, and myrcetin. Leaves contain acylated flavonol glycosides (Mahmoud et al. 2011), quercetin, myricetin, myricitin, triterpenoids (Gupta and Sharma 1974), esterase, galloyl carboxylase (Bhatia et al. 1974), and tannins (Morton 1987). Leaves of *S. cumini* are also rich in polyphenols, flavonoids, and essential oils (Mukherjee et al. 1998; Timbola et al. 2002; Lima et al. 2007). Seeds are rich in alkaloids, amino acids, flavonoids, glycosides, phytosterols, saponins, steroids, tannins, and triterpenoids. The stem also contains many phytochemicals, like betulinic acid, friedelin, epifriedelanol, β-sitosterol, eugenin, fatty acid ester of epi-friedelanol (Sengupta and Das 1965), quercetin, kaemferol, myricetin, gallic acid, ellagic acid (Bhargava et al. 1974), bergenins (Kopanski and Schnelle 1988), flavonoids, and tannins (Bhatia and Bajaj 1975). The astringent property of stem bark is due to gallo- and ellagitannins.

Flowers of *S. cumini* contain dihydromyricetin (Nair and Subramanian 1962), quercetin-3-D-galactoside, kaemferol, quercetin, myricetin, isoquercetin, myricetin-3-L-arbinose, oleanolic acid, acetyl oleanolic acid (Rastogi and Mehrotra 1990), and eugenol triterpenoid B (Nair and Subramanian 1962). Roots are rich in flavonoid glycosides (Vaishnava et al. 1992) and isorhamnetin-3-O-rutinoside (Vaishnava and Gupta 1990). Gallic acid imparts sourness, and anthocyanins give color to the fruits (Venkateswarlu 1952). The fruits are rich in raffinose, glucose, fructose (Srivastava 1953), citric acid, mallic acid (Lewis et al. 1956), gallic acid, anthocyanins (Jain and Seshadri 1975), delphinidin-3-gentibioside (Venkateswarlu 1952), cyanidin diglycoside, petunidin, and malvidin (Sharma and Sheshadri 1955). Most of these compounds have been reported to possess antioxidant and free radical scavenging activities (Ruan et al. 2008). These phytochemicals are responsible for the medicinal properties of seeds of *S. cumini*.

The bark of this plant has been used in medicine due to its astringent, sweet, refrigerant, carminative, diuretic, digestive, antihelminthic, febrifuge, constipating, stomachic, and antibacterial properties. The fruits and seeds are also used in treating diabetes, pharyngitis, spleenopathy, urethrorrhea, and ringworm infection. The leaves have been used to treat diabetes, constipation, leucorrhea, stomachalgia, fever, gastropathy, strangury, and dermopathy, and to inhibit blood discharges in the feces (Bhandary et al. 1995; Glyphis and Puttick 1998; Warrier et al. 1996).

Ayurveda mentions the use of *S. cumini* for the treatment of diabetes mellitus. Many traditional practitioners use different parts of this plant for the treatment of diabetes, cancer, colic, digestive problems, piles, and pimples (Jain 1991). Unani medicine uses different parts of this plant for ringworm infections and gum and teeth problems and as a liver tonic.

Almost all parts of *S. cumini* extract show pharmacological activities (Figure 10.1). Different parts of *S. cumini*, like fruits, seeds, and bark extracts, have exhibited

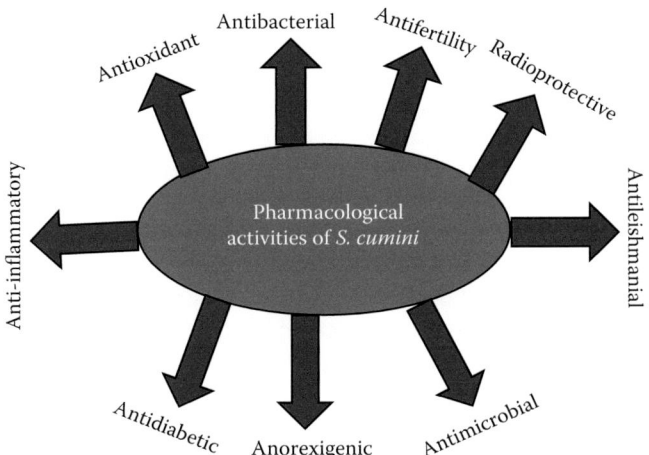

FIGURE 10.1   Pharmacological activity of different parts of *S. cumini*.

promising activity against diabetes mellitus. Several earlier studies have documented antibacterial (Bhuiyan et al. 1996), radioprotective, antifertility, anti-inflammatory, anti-leishmanial, and anorexigenic (Rajasekaran et al. 1988) activities of *S. cumini* extracts (Damasceno et al. 2002; Lima et al. 2007; Ruan et al. 2008). *S. cumini* plant parts contain compounds that are responsible for its medicinal and pharmacological properties. But many of these have poor aqueous solubility, permeability, and bioavailability.

USE OF NANOTECHNOLOGY IN IMPROVING THE EFFICACY OF BIOACTIVES

Nanotechnology is the most important field for generating novel biomedical applications. Nanomedicine is the application of nanotechnology to medicine. It exploits novel physical, chemical, and biological properties of materials at nanoscale. Nanotechnology has made great progress in improving the performance of many therapeutic payloads. Nanotechnology has a great impact on biomedical technology, significantly improving the performance of drugs in terms of efficacy, safety, and patient compliance (Kumari et al. 2010). Nanotechnology provides a variety of nanovectors (NVs) that can be smartly synthesized and used for the delivery of therapeutic cargos (Gelperina et al. 2005). NVs for drug delivery are defined as particles in the range of 1–1000 nm. The advantages of NVs as drugs are high stability, possibility of incorporation of hydrophilic and hydrophobic drugs, and possibility of administration by various routes (Gelperina et al. 2005). These NVs can also be designed for controlled and sustained release of drugs. Drug is adsorbed, dissolved, or dispersed throughout the matrix, or the drug is confined to an aqueous or oily core surrounded by a shell-like wall (Gelperina et al. 2005). Alternatively, drug can be chemically conjugated to the surface of NVs. NVs protect therapeutic cargos from premature degradation in the biological milieu, enhance bioavailability, and prolong presence in blood and cellular internalization. Size and size distribution of NVs are important to determine their cellular internalization and permeation across biological barriers (Kumari et al. 2011). Size and surface chemistry of NVs determine their *in vitro* and *in vivo* performance (Suri et al. 2007).

Therapeutic cargos are either conjugated or adsorbed with NVs. Many NVs, such as liposomes (Costin et al. 2002; Sinico et al. 2005; Xiong et al. 2005), micelles (Yokoyama et al. 1998; Liggins and Burt 2002; Marin et al. 2002), dendrimers (Chauhan et al. 2003; Liu et al. 2008; Niu et al. 2013), quantum dots (Misra 2008), magnetic NPs (Rouhollah et al. 2013), polymeric NPs (Kumar et al. 2014), and metallic NPs (Cheng et al. 2013), have been used for improving efficacy, bioavailability, and therapeutic activity. These NVs have been used to improve the performance of many therapeutic cargos, like doxorubicin, paclitaxel, and camptothecin. Different types of NVs are synthesized using various methods. The choice of NVs depends on the nature of the drug and type of application. Critical parameters considered during the fabrication of NVs are size, surface chemistry, drug loading, drug stability, and drug release kinetics (Kumari et al. 2010). The surface of NVs can be smartly designed to tag many biological entities, like ligands and receptors, on their surface for targeted delivery of payloads (Kumari et al. 2011). Smart designing of NVs with respect to target site and route of administration will solve the problems faced by therapeutic cargos. Therapeutic cargos that act on a disease target must be able to accumulate at the target site in sufficient concentration and for sufficient time. The chemical structure of therapeutic cargos and their physicochemical properties determine their biodistribution. Biodistribution of therapeutic cargos can be changed by chemically modifying their surface (Solomon and D'Souza 2011). However, the incorporation of therapeutic cargos into NVs increases their biodistribution without the need for chemical modification. This is a distinct advantage for therapeutic drugs that cannot be chemically modified without loss of pharmacological activity (Solomon and D'Souza 2011). NVs can also increase the performance of many therapeutic payloads extracted from *S. cumini* plant.

TECHNIQUES USED FOR NANOPARTICLE CHARACTERIZATION

The size and shape of NVs are responsible for their unique properties. NVs cannot be characterized by ordinary instruments. Highly sophisticated instruments with high resolution are required for characterization of NVs. Characterization of NVs by more than one technique is essential for the authenticity and validity of NV size and shape. Many of the analytical techniques that have been used for NV characterization are transmission electron microscopy (TEM), scanning electron microscopy (SEM), atomic force microscopy (AFM), dynamic light scattering (DLS) (Figure 10.2), x-ray based methods, and spectroscopic techniques.

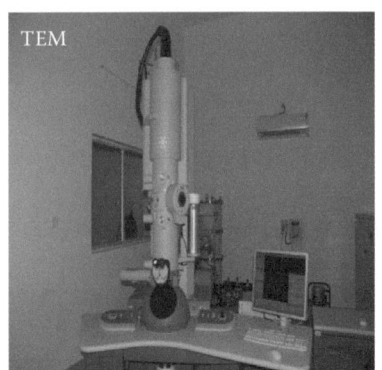
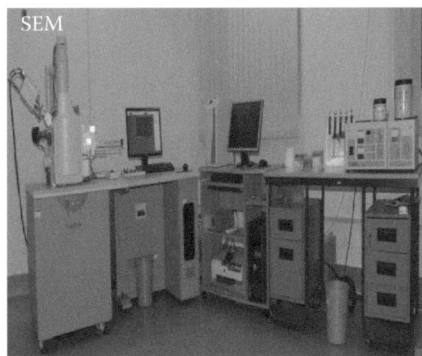
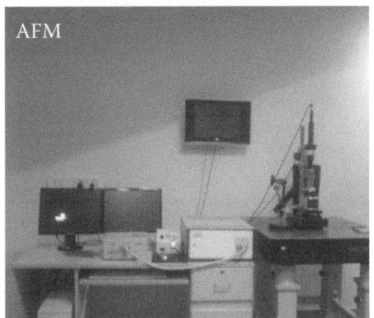
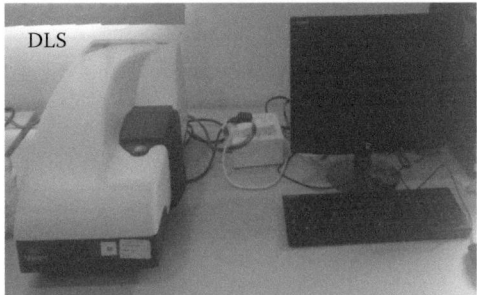

FIGURE 10.2　Instruments for the characterization of nanomaterials.

Electron microscopy techniques are essential for determining the overall and accurate size of NVs. SEM and TEM are the most commonly used electron microscopic techniques for characterization of NVs. Both instruments require a high vacuum for the imaging of samples. A sample is mounted on an aluminum stub on a double-adhesive carbon tape and coated with gold or carbon using a sputter unit. The sample is then scanned with a focused beam of electrons. Scattered electrons carry information about the surface topography of the sample (Jores et al. 2004). TEM operates on a different principle than SEM. A sample for TEM imaging is coated on carbon-coated grids and negatively stained with heavy metal salts like phosphotungstic acid, uranyl acetate, and ammonium molybdate. The surface topography of the sample is obtained with an electron beam that is transmitted through an ultrathin sample (Molpeceres et al. 2000). However, both techniques can lead to image artifacts due to high vacuum and other harsh treatments during sample preparation. Energy-dispersive x-ray (EDX) spectroscopy is coupled with SEM and TEM to provide information about the chemical composition of a sample.

To avoid image artifacts due to electron beam, the most preferred technique is AFM. AFM is based on physical scanning of a sample surface using the tip of an atomic scale. The surface topography of the sample is based on the force between the tip and sample surface. AFM can also be used for imaging nonconducting samples in their native state without any chemical treatment (Polakovic et al. 1999).

Light scattering techniques such as DLS are commonly used for the size determination of nanomaterials. DLS is based on scattering of light at different intensities by Brownian motion of particles in suspension. Analysis of intensity fluctuations gives the velocity of the Brownian motion, and hence the particle size using the Stokes–Einstein equation. DLS gives information about the hydrodynamic diameter of nanomaterials (DeAssis et al. 2008).

X-ray-based methods such as x-ray absorption (XAS), x-ray fluorescence (XRF), x-ray photoelectron spectroscopy (XPS), and x-ray diffraction (XRD) provide information about the surface structure, crystallographic structure, and elemental composition (López-Serrano et al. 2014).

Spectroscopic techniques, such as ultraviolet–visible (UV-Vis) spectrophotometer, are used for the initial screening of nanomaterial synthesis. Metallic NPs such as gold (GNPs) and silver (SNPs) give characteristic peaks in the UV-Vis region due to the excitation of surface plasmons, which is dependent on the size of the nanomaterials (Asharani et al. 2008). Fourier transform infrared (FTIR) spectroscopy is used to obtain information about the encapsulation or conjugation of chemical compounds on nanomaterials (Kumari et al. 2011). Energy from a light source is transferred to the molecule, and the molecule is excited to a higher energy state. FTIR is concerned only with vibrations and stretching.

SYNTHESIS OF METALLIC NANOPARTICLES USING *S. CUMINI*

The scientific community is now giving great emphasis to biostabilizer-stabilized NPs (Ravindran et al. 2003; Albrecht et al. 2006; Dahl et al. 2007; Iravani 2011). This approach is environmentally friendly and sustainable and hence has long-term benefits (Klaus et al. 1999; Nair and Pradeep 2002; Ravindran et al. 2003; Konishi et al. 2007; Mohanpuria et al. 2008). Plant-mediated NP synthesis is preferred, as it is cost-effective, environmentally friendly, and safe for therapeutic use (Klaus et al. 1999; Nair and Pradeep 2002; Willner et al. 2006; Konishi et al. 2007; Mohanpuria et al. 2008; Kumar and Yadav 2009, 2011a, 2011b, 2011c; Kumar et al. 2010). Plants contain many phytochemicals, such as alkaloids, tannins, flavonoids, and polyphenols. They have nutritional value and disease-preventing properties. So they may add value to synthesized NPs.

Owing to its wide spectrum of pharmacological properties, *S. cumini* has been exploited recently for green synthesis of NPs (Figure 10.3). Phytochemicals present in the leaf and seed extract of *S. cumini* also showed good potential for the synthesis of metallic NPs (Kumar et al. 2010, 2014) and as capping agent and stabilizers for metallic NPs (Kumar and Yadav 2009). *S. cumini* has also been exploited for the synthesis of different metallic NPs. Various research groups have reported different methodologies for the synthesis of SNPs. *S. cumini* extract was screened for the synthesis of SNPs by UV-Vis spectroscopy (Kumar et al. 2013). *S. cumini* extract was prepared by boiling leaves in 100 ml of deionized water and filtering the solution (Logeswari et al. 2015). SNPs have been synthesized by mixing appropriate volumes of plant extract with 1 mM silver nitrate solution (Logeswari et al. 2015). SNPs have been characterized by UV-Vis, SEM, AFM, and XRD. Their size was in the range of 26.5 nm (Logeswari et al. 2015). Leaf extracts prepared by various methods have shown potential for synthesis of different-sized SNPs. This was confirmed by another study in which a different procedure was followed for the preparation of leaf extracts (Behera and Nayak 2013). The size of SNPs obtained by the use of such a preparation was in the range of 30–100 nm. Prasad and Swamy (2013) reported the synthesis of SNPs by the use of bark extract of *S. cumini* through the following procedure: bark extract was prepared by boiling the bark of *S. cumini* at 60°C in water bath for 1 h. SNPs have been prepared by mixing bark extract and silver nitrate

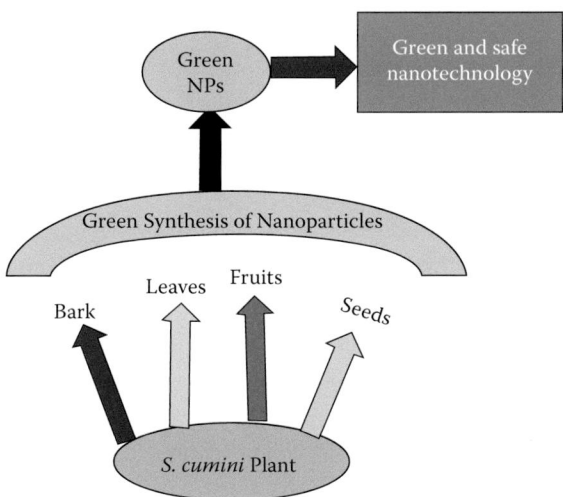

FIGURE 10.3   Use of *S. cumini* plant in nanotechnology for the synthesis of metallic NPs.

solution in a 1:9 ratio. SNPs were characterized by UV-Vis, SEM, and AFM. The particle size range of SNPs prepared using bark extract was 20–60 nm. Synthesis of SNPs using seed extract of *S. cumini* was made possible by boiling the fine powder of seeds with deionized water, and filtering the solution. SNPs were prepared by mixing seed extract and silver nitrate solution in a 1:9 molar ratio (Banerjee and Narendhirakannan 2011). SNPs were characterized by SEM, EDX, XRD, and FTIR. SNPs synthesized using seed extract are in the size range of 3.5 nm (Banerjee and Narendhirakannan 2011). SNPs synthesized by commercially available plant powder of *S. cumini* were also characterized by XRD, AFM, UV-Vis, and FTIR. The size of SNPs synthesized by this method was 53 nm (Logeswari and Silambarasan 2013). Prasad and Krishna (2014) demonstrated a facile green chemistry–based method for the synthesis of free-standing polymethyl methacrylate thin films by a green chemistry route and embedding them with GNPs and SNPs using leaves of three different plant species (*S. cumini*, *Bauhinia purpurea*, and Cymbopogon [lemon grass]) as substrates for the fabrication. The resultant films are shown to exhibit interesting plasmonic and catalytic properties. Mittal et al. (2014) described the biosynthesis of SNPs using *S. cumini* fruit extract and identified flavonoids as the biomolecules responsible for the synthesis of SNPs and demonstrated its anticancer activities by showing a 50% decrease in the viability of Dalton lymphoma (DL) cell lines with application of SNPs (100 μg/ml).

In our own lab, we have recently been able to synthesize SNPs using *S. cumini* leaf (Figure 10.4) and seed extract (Kumar et al. 2010). Extracellular synthesis of SNPs was carried out using leaf and seed extract of *S. cumini* with the following steps. Fractions of leaf and seed extracts of different polarities has been explored for the synthesis of SNPs. SNPs have been synthesized by mixing 10% (v/v) of extracts or fractions with 1.0 mmol/L aqueous silver nitrate solution for 24 h at room temperature. The change in color of the whole solution was an indication of synthesis of SNP. The solution was centrifuged at 12,000 rpm for 20 min to separate the SNPs.

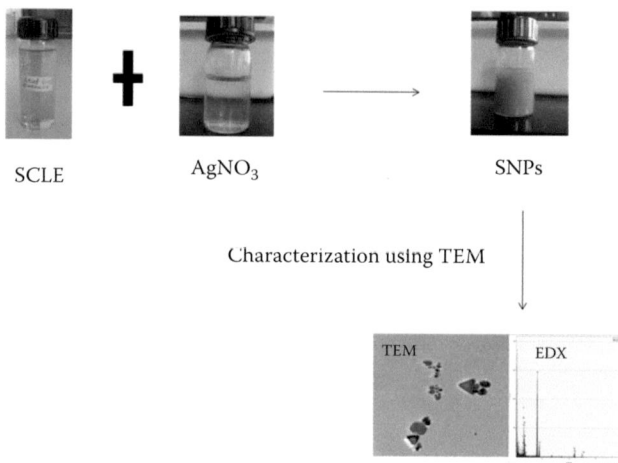

FIGURE 10.4   TEM and EDX of SNPs synthesized using leaf extract of *S. cumini*.

Synthesized SNPs have been characterized by SEM and AFM. FTIR and high-performance liquid chromatography analysis were attempted to find the molecules of plant extract involved in the synthesis of SNPs. Only leaf water fraction and seed water fraction of *S. cumini* have shown potential for the synthesis of SNPs. Colors of SNPs synthesized by leaf water fraction and seed water fraction were also different. Absorbance spectroscopy results revealed that seed extract has more potential for the synthesis of SNPs than leaf extract. The sizes of the SNPs synthesized by leaf water fraction and seed water fraction were 29 ± 1 nm and 82 ± 19 nm, respectively. Use of seed extract resulted in the formation of bigger NPs than those for leaf extract. The size of SNPs was dependent on the polyphenol content present in the leaf and seed extract. Leaf and seed extracts of *S. cumini* have shown potential for the synthesis of SNPs. Water fractions of leaf and seed extract showed a tendency for the formation of SNPs (Table 10.2). Interestingly, *S. cumini* leaf water fraction and seed water fraction showed potential for the synthesis of different-sized SNPs. These two tissues of this plant were also explored for the synthesis of GNPs.

Both leaf and seed extracts of *S. cumini* were found to catalyze the synthesis of GNPs (Kumar and Yadav 2012). Smaller-sized GNPs were formed with leaf and seed extracts of *S. cumini*. GNPs were synthesized by mixing 1.0 mM aqueous aurochloric acid solution with 10% (v/v) of leaf or seed extract and their polar fractions for 24 h at room temperature. The appearance of a purple color with an increase in incubation time was an indication of GNP synthesis. The reaction suspension was centrifuged at 10,000 rpm for 10 min to purify the GNPs. Synthesized GNPs were characterized by UV-Vis, SEM, AFM, and FTIR. Leaf and seed extracts, as well as their polar fractions, showed potential for the formation of GNPs (Kumar and Yadav 2012). Synthetic and morphological characterization revealed the synthesis of smaller-sized GNPs by leaf extract (23 ± 1 nm) than by seed extract (32 ± 3 nm). *S. cumini*

TABLE 10.2
Size of Metallic Nanoparticles Synthesized by Using Extracts of Different Parts of *S. cumini*

Name of Plant Part	Size (nm)	References
Silver Nanoparticles		
Bark extract	20–60	Prasad and Swamy 2013
Leaf extract	30	Kumar et al. 2010
Leaf extract	30–100	Behera and Nayak 2013
Leaf extract	26.5	Logeswari et al. 2015
Seed extract	92	Kumar et al. 2010
Seed extract	3.5	Banerjee and Narendhirakannan 2011
Commercial plant powder	53	Logeswari and Silambarasan 2013
Gold Nanoparticles		
Leaf extract	24	Kumar and Yadav 2012
Seed extract	35	Kumar and Yadav 2012

showed a different reducing capacity with silver and aurochloric salts (Figure 10.5). The size and amount of GNPs synthesized were found to be dependent on the amount of polyphenolic contents in the leaf and seed extract of *S. cumini*.

In another study, the effect of various physiochemical parameters like leaf extract ratio, metal ion concentration, and incubation temperature on the size of GNPs was examined (Kumar and Yadav 2013). An increase in the leaf extract ratio led to an increase in intensity of the reaction mixture at 550 nm (Kumar and Yadav 2013). The GNP synthesis rate was greater at a higher leaf extract ratio (Kumar and Yadav 2013). GNP synthesis also increased with an increase in incubation temperature. An increase in metal ion concentration up to 3 mM led to an increase in the synthesis of GNPs. But an increase in the concentration of metal ions beyond 3 mM led to aggregation of GNPs (Kumar and Yadav 2013).

Recently, Nath et al. (2015) synthesized iron NPs (FeNPs) from $FeSO_4$ by a rapid green method using aqueous leaf extract of *S. cumini*, and the green-synthesized FeNP supplementation in place of $FeSO_4$ showed a twofold increase in hydrogen production from glucose by a mesophilic bacterium, *Enterobacter cloacae*.

Biological activities of *S. cumini* extract–stabilized NPs have also been evaluated. Antibacterial activity (Figure 10.6) of SNPs synthesized by *S. cumini* leaf extract (SCLE) was evaluated against pathogenic bacteria, such as *Staphylococcus aureus*, *Pseudomonas aeruginosa*, *Escherichia coli*, and *Klebsiella pneumoniae*, using a well diffusion method (Logeswari et al. 2015). The SNPs synthesized by SCLEs were found to possess the highest antimicrobial activity against *S. aureus* (26 mm) and *E. coli* (26 mm) (Logeswari et al. 2015). In another study, the antibacterial property of the SNPs was determined using pathogenic bacteria, such as *E. coli*, *S. aureus*, *P. aeruginosa*, *Azotobacter chroococcum*, and *Bacillus licheniformis*, by the well diffusion method (Prasad and Swamy 2013). The zone of inhibition was highest for *B. licheniformis* even at lower concentrations of SNPs. The maximum zone of inhibition (19 mm) was obtained with *P. aeruginosa*, and *B. licheniformis* at higher concentrations of SNPs (Prasad and Swamy 2013). The antimicrobial activity

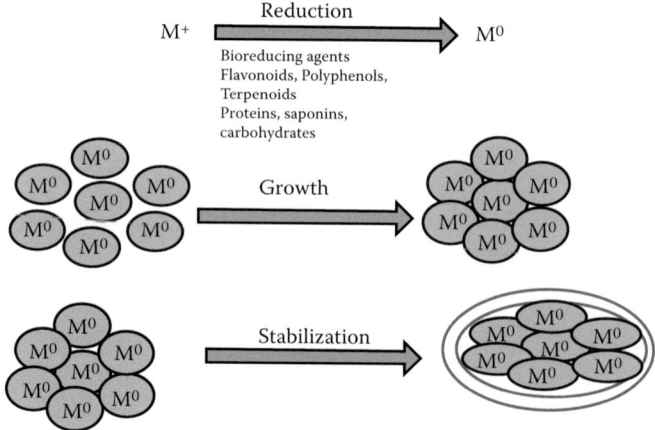

FIGURE 10.5 Mechanism of green synthesis of metallic NPs by *S. cumini* extract.

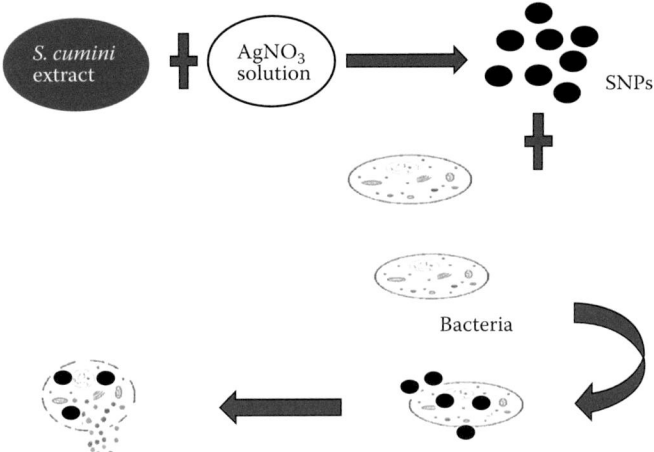

FIGURE 10.6   Pictorial representation of antibacterial activity of SNPs synthesized by *S. cumini* extract.

of SNPs synthesized by commercial plant powders of *S. cumini* was investigated against various pathogenic organisms, such as *S. aureus*, *P. aeruginosa*, *E. coli*, and *K. pneumoniae*, using the well diffusion method (Banerjee and Narendhirakannan 2011). These SNPs showed the highest activity against *S. aureus* (15 mm) and *P. aeruginosa* (15 mm). In another report, the antibacterial activity of *S. cumini*–synthesized SNPs investigated against pathogenic bacteria *E. coli*, *S. aureus*, and *S. pyrogenes* was reported. The maximum zone of inhibition was observed with *E. coli* (14 mm) (Banerjee and Narendhirakannan 2011).

The antioxidant activity of *S. cumini*–synthesized SNPs has also been evaluated by 1,1-diphenyl-2-picrylhydrazyl (DPPH). DPPH, a purple-colored stable free radical, is converted to a yellow-colored nonradical form after reaction with antioxidant molecules. The DPPH assay provides quick information about the antioxidant activity of nanomaterials. SNPs showed better antioxidant activity than seed extract alone (Logeswari and Silambarasan 2013).

SYNTHESIS OF POLYMERIC NANOPARTICLES USING *S. CUMINI*

Biodegradable polymers are used for the preparation of NPs. Poly-D,L-lactide (PLA) is a biodegradable polymer that breaks into monomeric units of lactic acid as a natural intermediate in carbohydrate metabolism. Among biodegradable NPs, PLA is extensively investigated for encapsulation of many therapeutic molecules due to its high hydrophobicity, biodegradability, biocompatibility, low toxicity, strong mechanical strength, and slow release (Kumari et al. 2010). PLA NPs are prepared by many methods, like solvent evaporation, salting out, solvent displacement, and solvent diffusion. Preparation of PLA NPs by solvent evaporation requires surfactants or stabilizers (Zambaux et al. 1998). Nowadays, a lot of attention is paid to the behavior of NPs in the environment and their toxicity.

There have been no conclusive studies regarding any type of NPs, but generally, it has been found that use of a toxic surfactant most of the time confers toxicity to synthesized NPs, especially when other chemicals used in the synthesis procedure are nontoxic or least toxic. PLA NPs are synthesized using PLA polymer, dichloromethane (DCM), and surfactants. The most commonly used surfactants for PLA NP synthesis are polyvinyl alcohol, polyethylene glycol, and polyvinyl pyrrolidone. These surfactants may confer toxicity (Zambaux et al. 1998; Soppimath et al. 2001), but PLA polymer does not possess toxicity (Kumari et al. 2010). DCM has some toxicity, but being an organic solvent with a very low boiling point, it is easily evaporated from the reaction during synthesis (Burek 1984; Hearne et al. 1990). However, the surfactant has the major role of stabilizing PLA NPs by binding on the NP surface. It has been reported that even several washings cannot remove surfactants from an NP surface. So they may confer toxicity to synthesized NPs (Gipps et al. 1987; Kumar et al. 2012).

Phytochemicals present in the leaf extracts can also act as stabilizers or surfactants for polymeric NPs. We have recently used SCLEs as a surfactant and stabilizer for the synthesis of safe and stable PLA NPs. This plant has already been tested for metallic NP synthesis and has shown very good stabilizing action (Kumar et al. 2010). However, the synthesis procedure of metallic and polymeric NPs varies to a great extent (Kumari et al. 2010; Kumar and Yadav 2011).

Interestingly, *S. cumini* extract was found to be a very useful stabilizer and surfactant in the synthesis of PLA NPs (Kumari et al. 2012). SCLE-stabilized PLA NPs were prepared by the solvent evaporation method by using leaf extracts as stabilizers or emulsifiers. PLA was dissolved in DCM and sonicated for 30 s at room temperature. One milliliter of SCLE solution was added and again sonicated to form emulsion. The emulsion was diluted by adding double-distilled water. The organic solvent was evaporated using a rotavapor. The NPs were purified by centrifugation and washed two times with distilled water by centrifugation. This method is based on emulsification of the organic phase (DCM + PLA + quercetin) and aqueous phase (SCLE solution). The emulsion droplets were stabilized by the SCLE molecules. These droplets were perfectly stable during the solvent evaporation stage. NPs were formed as a result from single-volume shrinkage of the initial emulsion droplet. SCLEs contain many valuable compounds having properties of stabilizers or emulsifiers (Kumar and Yadav 2009, 2013; Kumar et al. 2010). These compounds can act as very good stabilizers for PLA NPs. The exact mechanism of this process is unclear and very difficult to explain at this stage. The shape and size of PLA NPs were determined by SEM and TEM. DLS was also used to measure the zeta potential of SCLE-synthesized PLA NPs. TEM characterization of SCLE-synthesized PLA NPs revealed that all the NPs are stable, well dispersed, smooth, and spherical in shape. The size of SCLE-synthesized SNPs was 116 ± 27 nm. Zeta potential values of SCLE-synthesized PLA NPs were 53.6 ± 0.8 mV (Kumari et al. 2012).

PLA NPs of various sizes and shapes can be prepared by varying the concentration of SCLEs. Recently, synthesis of PLA NPs using *Curcuma longa* extract has been reported (Kumar et al. 2014). *S. cumini* extract can be explored for the synthesis of other polymeric NPs of variable shapes and sizes. Such biodegradable NPs prepared using *S. cumini* may prove to be promising candidates for the successful development of drug molecule or any biomolecule delivery vehicles and other important therapeutic applications.

THERAPEUTIC APPLICATIONS OF *S. CUMINI* EXTRACT–SYNTHESIZED NANOPARTICLES

There is great interest in exploring NPs for various therapeutic applications. Importantly, the size of NPs is similar to that of biological entities like DNA, protein, viruses, and bacteria. Biologically synthesized NPs find applications in different areas of nanotechnology. SNPs have been widely used as antibacterial, antifungal, anti-inflammatory, antiviral, and antifungal agents. So far, chemically synthesized SNPs have been used for these applications. But it is speculated that plant extract–synthesized SNPs have more advantages than chemically synthesized SNPs. SNPs synthesized using bark extract of *S. cumini* can be explored for these applications in a safer way. SNPs synthesized using leaf, seed, and bark extract have shown potent antibacterial activities (Prasad and Swamy 2013). Further studies are needed to evaluate other therapeutic potentials of *S. cumini*–synthesized SNPs.

GNPs have been widely used in drug delivery, gene delivery, and photodynamic therapy. So far, the potential of chemically synthesized GNPs has been explored for these applications. GNPs synthesized using leaf and seed extract of *S. cumini* can also be used in the area of nanomedicine. They can be used as carriers for delivery of drugs and small molecules. GNPs synthesized using *S. cumini* extract can also be used in photodynamic therapy (Lkhagvadulam et al. 2013). Photodynamic therapy is based on the destruction of cancer cells by laser-generated singlet oxygen that is toxic. GNPs are a highly efficient photodynamic therapy platform due to their chemical inertness and minimum toxicity. GNPs show size-dependent cellular uptake in tumor cells and photodynamic activity (Lkhagvadulam et al. 2013). Different-sized GNPs synthesized using *S. cumini* extract will find applications in photodynamic therapy.

GNPs synthesized using *S. cumini* extract also find applications in the area of biosensing (De et al. 2008). GNPs exhibit unique optical and physical properties. GNPs show an intense absorption peak from 500 to 550 nm arising from surface plasmon resonance (SPR). SPR occurs from collective oscillation of the conductive electrons owing to the resonant excitation by the incident photons. The SPR band is sensitive to the surrounding environment. This phenomenon leads to the widely applicable calorimetric sensing (De et al. 2008). GNPs can also be used as a multivalent receptor to enhance low-affinity interactions such as carbohydrate–protein interactions (De et al. 2008). Polymeric NPs synthesized using *S. cumini* extract will also find applications in the area of drug and small-molecule delivery (Kumari et al. 2012). They can also be explored for the delivery of *S. cumini*–based nutraceuticals and cosmeceuticals.

FUTURE PERSPECTIVES

Nanotechnology is an emerging branch of science that involves the intersection of physics, chemistry, and biology. It has made tremendous contributions in all spheres of human life. The biological synthesis of NPs has paved the way for better protocols in the medical field. Biological synthesis produces NPs of controlled morphology, size, and shape in less time. Nanotechnology has the potential to revolutionize the use of *S. cumini*, a rich source of phytochemicals having proven pharmacological activities. But use of these phytochemicals as a drug is limited due to problems of poor solubility, poor efficacy, and less bioavailability. These problems can be overcome with the help of nanotechnology. Nanotechnology provides a wide range of NVs that can be smartly synthesized and used for improving the efficacy and therapeutic index of phytochemicals present in the *S. cumini*. The extract of *S. cumini* has also shown good potential for the synthesis of SNPs and GNPs. Such SNPs and GNPs are biocompatible and safe for therapeutic purposes. The proven antibacterial activity of SNPs makes them suitable candidates for the development of antibacterial nanoformulations. *S. cumini*–synthesized NPs can be explored for various biomedical applications that so far have been explored for chemically synthesized NPs. Synthetic GNPs have so far been explored for drug delivery, photodynamic therapy, and imaging systems for diagnostics. *S. cumini*–synthesized GNPs can also be tested for such applications. Biodegradable NPs can find potential applications in the areas of drug delivery, gene delivery, and other small-molecule delivery. Chemically synthesized biodegradable NPs have been explored for these applications. Also, *S. cumini* extract–stabilized PLA NPs can be explored for use in food, agriculture, and pharma industries.

CONCLUSIONS

S. cumini possesses a wide spectrum of pharmacological properties. Different parts of *S. cumini* with proven pharmaceutical properties have been used for the synthesis of SNPs and GNPs. SNPs and GNPs prepared by *S. cumini* extracts have shown improved biological activities. Phytochemicals present in the extracts of the fruit, seed, and leaves of *S. cumini* play a vital role in the synthesis of different-sized SNPs and GNPs. Leaf extracts of *S. cumini* also have shown potential for the synthesis of biodegradable PLA NPs. Extracts of different parts of this plant can further be explored for the synthesis of other metallic and biodegradable NPs. Also, the therapeutic potential of pharmaceutically important phytochemicals of *S. cumini* can be improved with the help of smartly designed NVs.

ACKNOWLEDGMENTS

The authors are thankful to the director, CSIR-Institute of Himalayan Bioresource Technology, for continuous support and encouragement. Financial support from CSIR, Government of India, is duly acknowledged.

REFERENCES

Ahmad, N., S. Sharma, V. N. Singh, S. F. Shamsi, A. Fatma, and B. R. Mehta. 2011. Biosynthesis of silver nanoparticles from *Desmodium triflorum*: A novel approach towards weed utilization. *Biotechnol. Res. Int.* 2011: 454090.

Albrecht, M. A., C. W. Evans, and C. L. Raston. 2006. Green chemistry and the health implications of nanoparticles. *Green Chem.* 8: 417–432.

Amin, M., F. Anwar, M. R. Janjua, M. A. Iqbal, and U. Rashid. 2012. Green synthesis of silver nanoparticles through reduction with *Solanum xanthocarpum* L. berry extract: Characterization, antimicrobial and urease inhibitory activities against *Helicobacter pylori*. *Int. J. Mol. Sci.* 13: 9923–9941.

Ankamwar, B. 2010. Biosynthesis of gold nanoparticles (Greengold) using leaf extract of *Terminalia catappa*. *E-J. Chem.* 7: 1334–1339.

Ankamwar, B., C. Damle, A. Ahmad, and M. Sastry. 2005. Biosynthesis of gold and silver nanoparticles using *Emblica officinalis* fruit extract, their phase transfer and transmetallation in an organic solution. *J. Nanosci. Nanotechnol.* 5: 1665–1671.

Aromal, S. A., V. K. Vidhu, and D. Philip. 2012. Green synthesis of well-dispersed gold nanoparticles using *Macrotyloma uniflorum*. *Spectrochim. Acta A* 85: 99–104.

Asharani, P. V., Y. L. Wu, Z. Gongand, and S. Valiyaveettil. 2008. Toxicity of silver nanoparticles in zebrafish models. *Nanotechnology* 19: 1–8.

Ayyanar, M., and P. Subhash Babu. 2012. *Syzygium cumini* (L.) Skeels: A review of its phytochemcal constituents and traditional uses. *Asia Pac. J. Trop. Biomed.* 2: 240–246.

Banerjee, J., and R. T. Narendhirakannan. 2011. Biosynthesis of silver nanoparticles from *Syzygium cumini* (L.) Skeels seed extract and evaluation of their in vitro antioxidant activities. *Dig. J. Nanomater. Biostruct.* 6: 961–968.

Behera, S., and P. L. Nayak. 2013. *In vitro* antibacterial activity of green synthesized silver nanoparticles using jamun extract against multiple drug resistant bacteria. *World J. Nanosci. Technol.* 2: 62–65.

Bhandary, M. J., K. R. Chandrashekar, and K. M. Kaveriappa. 1995. Medical ethnobotany of the siddis of Uttara Kannada district, Karnataka, India. *J. Ethnopharmacol.* 47: 149–158.

Bhargava, K. K., R. Dayal, and T. R. Seshadri. 1974. Chemical components of *Eugenia jambolana* stem bark. *Curr. Sci.* 43: 645–646.

Bhatia, I. S., and K. L. Bajaj. 1975. Chemical constituents of the seeds and bark of *Syzygium cumini*. *Plant Med.* 28: 347–352.

Bhatia, I. S., S. K. Sharma, and K. L. Bajaj. 1974. Esterase and galloyl carboxylase from *Eugenia jambolana* leaves. *Indian J. Exp. Biol.* 12: 550–552.

Bhuiyan, M. A., M. Y. Mia, and M. A. Rashid. 1996. Antibacterial principles of the seeds of *Eugenia jambolana*. *Bangladesh J. Bot.* 25: 239–241.

Bindhu, M. R., and M. Umadevi. 2012. Synthesis of monodispersed silver nanoparticles using *Hibiscus cannabinus* leaf extract and its antimicrobial activity. *Spectrochim. Acta A* 101: 184–190.

Burek, J. D. 1984. Methylene chloride: A two-year inhalation toxicity and oncogenicity study in rats and hamsters. *Fundam. Appl. Toxicol.* 4: 30–47.

Chandran, S. P., M. Chaudhary, R. Pasricha, A. Ahmad, and M. Sastry. 2006. Synthesis of gold nanotriangles and silver nanoparticles using *Aloe vera* plant extract. *Biotechnol. Prog.* 22: 577–583.

Chang, S. S., C. W. Shih, C. D. Chen, W. C. Lai, and C. R. C. Wang. 1999. The shape transition of gold nanorods. *Langmuir* 15: 701–709.

Chauhan, A. S., S. Sridevi, K. B. Chalasani et al. 2003. Dendrimer-mediated transdermal delivery: Enhanced bioavailability of indomethacin. *J. Control. Rel.* 90: 335–343.

Cheng, J., Y. J. Gu, S. H. Cheng, and W. T. Wong. 2013. Surface functionalized gold nanoparticles for drug delivery. *J. Biomed. Nanotechnol.* 9: 1362–1369.

Costin, G. E., M. Trif, N. Nichita, R. A. Dwek, and S. M. Petrescu. 2002. pH-sensitive liposomes are efficient carriers for endoplasmic reticulum-targeted drugs in mouse melanoma cells. *Biochem. Biophys. Res. Commun.* 293: 918–923.

Dahl, J. A., B. L. S. Maddux, and J. E. Hutchison. 2007. Toward greener nanosynthesis. *Chem. Rev.* 107: 2228–2269.

Damasceno, D. C., P. H. O. Lima, M. S. Galhiane, G. T. Volpato, and M. V. C. Rudge. 2002. Evaluation of hypoglycemic effect of sapogenin extracted from *Eugenia jambolana* seeds. *Rev. Bras. Pl. Med.* 4: 46–54.

David, M. B., P. Martin, and A. S. William. 2005. Research strategies for safety evaluation of nanomaterials. Part III. Nanoscale technologies for assessing risk and improving public health. *Toxicol. Sci.* 88: 298–306.

De, M., P. S. Ghosh, and V. M. Rotello. 2008. Applications of nanoparticles in biology. *Adv. Mater.* 20: 4225–4241.

DeAssis, D. N., V. C. Mosqueira, J. M. Vilela, M. S. Andrade, and V. N. Cardoso. 2008. Release profiles and morphological characterization by atomic force microscopy and photon correlation spectroscopy of 99m Technetium–fluconazole nanocapsules. *Int. J. Pharm.* 349: 152 –160.

Deshpande, R. 2010. Rapid biosynthesis of irregular shaped gold nanoparticles from macerated aqueous extracellular dried clove buds (*Syzygium aromaticum*) solution. *Colloids Surf. B* 79: 235–240.

Dinesh, S., S. Karthikeyan, and P. Arumugam. 2012. Biosynthesis of silver nanoparticles from *Glycyrrhiza glabra* root extract. *Arch. Appl. Sci. Res.* 4: 178–187.

Fazaludeen, M. F., C. Manickam, I. M. A. Ashankyty, M. Q. Ahmed, and Q. Z. Beg. 2012. Synthesis and characterizations of gold nanoparticles by *Justicia gendarussa* Burm. f. leaf extract. *J. Microbiol. Biotechnol. Res.* 2: 23–34.

Freeman, R. G., K. C. Grabar, K. J. Allison et al. 1995. Self assembled metal colloid monolayers: An approach to SERS substrates. *Science* 267: 1629–1631.

Gardea-Torresdey, J. L., E. Gomez, J. R. Peralta-Videa, J. G. Parsons, H. Troiani, and M. Jose-Yacaman. 2003. Alfalfa sprouts: A natural source for the synthesis of silver nanoparticles. *Langmuir* 19: 1357–1361.

Geethalakshmi, R., and D. V. L. Sarada. 2010. Synthesis of plant mediated silver nanoparticles using *Trianthema decandra* extract and evaluation of their anti-microbial activities. *Int. J. Eng. Sci. Technol.* 2: 970–975.

Gelperina, S., K. Kisich, M. D. Iseman, and L. Heifets. 2005. The potential advantages of nanoparticle drug delivery systems in chemotherapy of tuberculosis. *Am. J. Resp. Crit. Care Med.* 172: 1487–1490.

Gericke, M., and A. Pinches. 2006. Biological synthesis of metal nanoparticles. *Hydrometallurgy* 83: 132–140.

Ghosh, S., S. Patil, M. Ahire et al. 2012. Synthesis of silver nanoparticles using *Dioscorea bulbifera* tuber extract and evaluation of its synergistic potential in combination with antimicrobial agents. *Int. J. Nanomed.* 7: 483–496.

Gipps, E. M., P. Groscurth, J. Kreuter, and P. P. Speiser. 1987. The effects of polyalkylcyanoacrylate nanoparticles on human normal and malignant mesenchymal cells in vitro. *Int. J. Pharm.* 40: 23–31.

Glyphis, J. P., and G. M. Puttick. 1998. Phenolics in some southern African Mediterranean shrub land plants. *Phytochemistry* 27: 743–751.

Gnanajobitha, G., G. Annadurai, and C. Kannan. 2012. Green synthesis of silver nanoparticles using *Elettaria cardamomom* and assessment of its antimicrobial activity. *Int. J. Pharm. Sci. Res.* 3: 323–330.

Gopalakrishnan, K. 2012. Antibacterial activity of copper oxide nanoparticles on *E. coli* synthesized from *Tridax procumbens* leaf extract and surface coating with polyaniline. *Dig. J. Nanomater. Biostruct.* 7: 833–839.

Gupta, G. S., and D. P. Sharma. 1974. Triterpenoid and other constituents of *Eugenia jambolana* leaves. *Phytochemistry* 13: 2013–2014.

Haverkamp, R. G., and A. T. Marshall. 2009. The mechanism of metal nanoparticle formation in plants: Limits on accumulation. *J. Nanopart. Res.* 11: 1453–1463.

Hearne, F. T., J. W. Pifer, and F. Grose. 1990. Absence of adverse mortality effects in workers exposed to methylene chloride: An update. *J. Occup. Med.* 32: 234–240.

Hudlikar, M., S. Joglekar, M. Dhaygude, and K. Kodam. 2012. Latex-mediated synthesis of ZnS nanoparticles: Green synthesis approach. *J. Nano. Res.* 14: 1–6.

Iravani, S. 2011. Green synthesis of metal nanoparticles using plants. *Green Chem.* 13: 2638–2650.

Jain, M. C., and T. R. Seshadri. 1975. Anthocyanins of *Eugenia jambolana* fruits. *Indian J. Chem.* 3: 20–23.

Jain, S. K. 1991. *Dictionary of Indian Folk Medicine and Ethnobotany*. New Delhi: Deep Publications.

Jia L., Q. Zhang, Q. Li, and H. Song. 2009. The biosynthesis of palladium nanoparticles by antioxidants in *Gardenia jasminoides* Ellis: Long lifetime nanocatalysts for p-nitrotoluene hydrogenation. *Nanotechnology* 20: 385601.

Jin, R. C., Y. W. Cao, C. A. Mirkin, K. L. Kelly, G. C. Schatz, and J. G. Zheng. 2001. Photoinduced conversion of silver nanospheres to nanoprisms. *Science* 294: 1901–1903.

Joglekar S., K. Kodam, M. Dhaygude, and M. Hudlikar. 2011. Novel route for rapid biosynthesis of lead nanoparticles using aqueous extract of *Jatropha curcas* L. latex. *Mater. Lett.* 5: 3170–3172.

Jores, K., W. Mehnert, M. Drechsler, H. Bunjes, C. Johann, and K. Mäder. 2004. Investigation on the stricter of solid lipid nanoparticles and oil-loaded solid nanoparticles by photon correlation spectroscopy, field flow fractionisation and transmission electron microscopy. *J. Control. Rel.* 17: 217–227.

Kirtikar, K. R., and B. D. Basu. 1918. *Indian Medicinal Plants*. 3rd ed. Allahabad: Sudhindra Nath Basu, M.B. Panini Office, Bhuwanéswari Asrama, Bahadurganj, 898–900.

Klaus, T., R. Joerger, E. Olsson, and C. G. Granqvist. 1999. Silver-based crystalline nanoparticles, microbially fabricated. *Proc. Natl. Acad. Sci. U.S.A.* 96: 13611–13614.

Konishi, Y., K. Ohno, N. Saitoh et al. 2007. Bioreductive deposition of platinum nanoparticles on the bacterium *Shewanella algae*. *J. Biotechnol.* 128: 648–653.

Kopanski, L., and G. Schnelle. 1988. Isolation of bergenin from barks of *Syzygium cumini*. *Plant Med.* 54: 572.

Kora, A. J., and J. Arunachala. 2012. Green fabrication of silver nanoparticles by gum tragacanth (*Astragalus gummifer*): A dual functional reductant and stabilizer. *J. Nanomater.* 2012: 869765.

Kumar, A., R. Ilavarasan, T. Jayachandran et al. 2009. Phytochemical investigation on a tropical plant, *Syzygium cumini* from Kattuppalayam, Erode District, Tamil Nadu, south India. *Pakistan J. Nutr.* 8: 83–85.

Kumar, D. A. 2012. Rapid and green synthesis of silver nanoparticles using the leaf extracts of *Parthenium hysterophorus*: A novel biological approach. *Int. Res. J. Pharm.* 3: 169–173.

Kumar, S., J. Saini, D. Kashyap et al. 2013. Green synthesis of plant mediated silver nanoparticles using *Mangifera indica* and *Syzygium cumini* leaf extract. *Int. J. Pharm. Sci. Res.* 4: 3189–3191.

Kumar, V., A. Kumari, P. Guleria, and S. K. Yadav. 2012. Evaluating the toxicity of selected types of nanochemicals. *Rev. Environ. Contamin. Toxicol.* 215: 39–121.

Kumar, V., A. Kumari, D. Kumar, and S. K. Yadav. 2014. Biosurfactant stabilised anticancer biomolecule–loaded poly (D, L–lactide) nanoparticles. *Colloids Surf. B* 117: 505–511.

Kumar, V., S. C. Yadav, and S. K. Yadav. 2010. *Syzygium cumini* leaf and seed extract mediated biosynthesis of silver nanoparticles and their characterization. *J. Chem. Technol. Biotechnol.* 85: 1301–1309.

Kumar, V., and S. K. Yadav. 2009. Plant-mediated synthesis of silver and gold nanoparticles and their applications. *J. Chem. Technol. Biotechnol.* 84: 151–157.

Kumar, V., and S. K. Yadav. 2011a. Synthesis of different sized silver nanoparticles by simply varying reaction conditions with leaf extracts of *Bauhinia variegata* L. *IET Nanobiotechnol.* 6: 1–8.

Kumar, V., and S. K. Yadav. 2011b. Synthesis of stable, polyshaped silver and gold nanoparticles using leaf extract of *Lonicera japonica* L. *Int. J. Green Nanotechnol.* 3: 281–291.

Kumar, V., and S. K. Yadav. 2011c. Synthesis of variable shaped gold nanoparticles in one solution using leaf extract of *Bauhinia variegata* L. *Dig. J. Nanomater. Biostruct.* 6: 1685–1693.

Kumar, V., and S. K. Yadav. 2012. Characterisation of gold nanoparticles synthesized by leaf and seed extract of *Syzygium cumini* L. *J. Exp. Nanosci.* 7: 440–451.

Kumar, V., and S. K. Yadav. 2013. Influence of physiochemical factors on size of gold nanoparticles synthesized using leaf extract of *Syzygium cumini*. *J. Chem. Sci. Technol.* 2: 21–24.

Kumari, A., V. Kumar, and S. K. Yadav. 2012. Plant extract synthesized PLA nanoparticles for controlled and sustained release of quercetin: A green approach. *PloS One* 7: e41230.

Kumari, A., and S. K. Yadav. 2011. Cellular interactions of therapeutically delivered nanoparticles. *Exp. Opin. Drug Deliv.* 8: 141–151.

Kumari, A., S. K. Yadav, Y. B. Pakade et al. 2011. Nanoencapsulation and characterization of *Albizia chinensis* isolated antioxidant quercitrin on PLA nanoparticles. *Colloids Surf. B* 82: 224–232.

Kumari, A., S. K. Yadav, and S. C. Yadav. 2010. Biodegradable polymeric nanoparticles based drug delivery systems. *Colloids Surf. B* 75: 1–18.

Lal, S. S., and P. L. Nayak. 2012. Green synthesis of gold nanoparticles using various extract of plants and spices. *Int. J. Sci. Innov. Discov.* 2: 325–350.

Leela, A., and M. Vivekanandan. 2008. Tapping the unexploited plant resources for the synthesis of silver nanoparticles. *Afr. J. Biotechnol.* 7: 3162–3165.

Lewis, Y. S., C. T. Dwarakanath, and D. S. Johar. 1956. Acids and sugars in *Eugenia jambolana*. *J. Sci. Ind. Res.* 15C: 280–281.

Li, L., J. Hu, and A. P. Alivisatos. 2001. Band gap variation of size- and shape-controlled colloidal CdSe quantum rods. *Nano Lett.* 1: 349–351.

Li, S., Y. Shen, A. Xie et al. 2007. Green synthesis of silver nanoparticles using *Capsicum annuum* L. extract. *Green Chem.* 9: 852–858.

Liggins, R. T., and H. M. Burt. 2002. Polyether–polyester diblock copolymers for the preparation of paclitaxel loaded polymeric micelle formulations. *Adv. Drug Deliv. Rev.* 54: 191–202.

Lima, L. A., A. C. Siani, F. A. Brito, A. L. F. Sampaio, M. D. G. M. O. Henriques, and C. A. D. S. Riehl. 2007. Correlation of anti-inflammatory activity with phenolic content in the leaves of *Syzygium cumini* (L.) Skeels (Myrtaceae). *Quim. Nova* 30: 860–864.

Link, S., M. B. Mohamed, and M. A. El-Sayed. 1999. Simulation of the optical absorption spectra of gold nanorods as a function of their aspect ratio and the effect of the medium dielectric constant. *J. Phys. Chem. B* 103: 3073–3077.

Liu, Z., K. Chen, C. Davis et al. 2008. Drug delivery with carbon nanotubes for in vivo cancer treatment. *Cancer Res.* 68: 6652–6660.

Lkhagvadulam, B., J. H. Kim, I. Yoon, and Y. K. Shim. 2013. Size dependent photodynamic activity of gold nano particles conjugate of water soluble purpurin-18-*N*-methyl-D glucamine. *Biomed. Res. Int.* 2013: 720579.

Logeswari, P., and S. Silambarasan. 2013. Ecofriendly synthesis of silver nanoparticles from commercially available plant powders and their antibacterial properties. *Scientia Iranica F* 20: 1049–1054.

Logeswari, P., S. Silambarasan, and J. Abraham. 2015. Synthesis of silver nanoparticles using plants extract and analysis of their antimicrobial property. *J. Saudi Chem. Soc.* 19: 311–317.

López-Serrano, A., R. M. Olivas, J. S. Landaluze, and C. Cámara. 2014. Nanoparticles: A global vision. Characterization, separation, and quantification methods. Potential environmental and health impact. *Anal. Methods* 6: 38–56.

Maheswari, R. U., A. L. Prabha, V. Nandagopalan, and V. Anburaja. 2012. Green synthesis of silver nanoparticles by using rhizome extract of *Dioscorea oppositifolia* L. and their anti-microbial activity against human pathogens. *J. Pharm. Biol. Sci.* 1: 38–42.

Mahmoud, I. I., M. S. Marzouk, F. A. Moharram, M. R. El Gindi, and A. M. Hassan. 2001. Acylated flavonol glycosides from *Eugenia jambolana* leaves. *Phytochemistry* 58: 1239–1244.

Mallikarjuna, K., G. R. Dillip, G. Narasimha, N. J. Sushma, and B. D. P. Raju. 2012. Phytofabrication and characterization of silver nanoparticles from *Piper betel* broth. *Res. J. Nanosci. Nanotechnol.* 2: 17–23.

Manna, L., E. C. Scher, and A. P. Alivisatos. 2000. Synthesis of soluble and processable rod, arrow, teardrop, and tetrapod shaped CdSe nanocrystals. *J. Am. Chem. Soc.* 122: 12700–12706.

Marin, A., H. Sun, G. A. Husseini, W. G. Pitt, D. A. Christensen, and N. Y. Rapoport. 2002. Drug delivery in pluronic micelles: Effect of high-frequency ultrasound on drug release from micelles and intracellular uptake. *J. Control. Rel.* 84: 39–47.

Mary, E. J., and L. Inbathamizh. 2012. Green synthesis and characterization of nano silver using leaf extract of *Morinda pubescens. Asian J. Pharm. Clin. Res.* 5: 159–162.

Misra, R. D. K. 2008. Quantum dots for tumor targeted drug delivery and cell imaging. *Nanomedicine* 3: 272–274.

Mittal, A. K., J. Bhaumik, S. Kumar, and U. C. Banerjee. 2014. Biosynthesis of silver nanoparticles: Elucidation of prospective mechanism and therapeutic potential. *J. Colloid Interface Sci.* 415: 39–47.

Mohanpuria, P., N. K. Rana, and S. K. Yadav. 2008. Biosynthesis of nanoparticles: Technological concepts and future applications. *J. Nanopart. Res.* 10: 507–517.

Molpeceres, J., M. R. Aberturas, and M. Guzman. 2000. Biodegradable nanoparticles as a delivery system for cyclosporine: Preparation and characterization. *J. Microencapsul.* 17: 599–614.

Mondal, S., N. Roy, R. A. Laskar et al. 2011. Biogenic synthesis of Ag, Au and bimetallic Au/Ag alloy nanoparticles using aqueous extract of mahogany (*Swietenia mahogani* Jacq.) leaves. *Colloids Surf. B* 82: 497–504.

Morton, J. 1987. Jambolan. In *Fruits of Warm Climates*, 375–378. Miami: J. F. Morton.

Mukherjee, P. K., K. Saha, T. Murugesan, S. C. Mandal, M. Pal, and B. P. Saha. 1998. Screening of anti-diarrhoeal profile of some plant extracts of a specific region of West Bengal, India. *J. Ethnopharmacol.* 60: 85–89.

Nabikhan, A., K. Kandasamy, A. Raj, and N. M. Alikunhi. 2010. Synthesis of antimicrobial silver nanoparticles by callus and leaf extracts from saltmarsh plant, *Sesuvium portulacastrum* L. *Colloids Surf. B* 79: 2488–2493.

Nadagouda, M. N., and R. S. Varma. 2008. Green synthesis of silver and palladium nanoparticles at room temperature using coffee and tea extract. *Green Chem.* 10: 859–862.

Nair, B., and T. Pradeep. 2002. Coalescense of nanoclusters and formation of submicron crystallites assisted by *Lactobacillus* strains. *Cryst. Growth Des.* 2: 293–298.

Nair, A. G. R., and S. S. Subramanian. 1962. Chemical examination of the flowers of *Eugenia jambolana. J. Sci. Ind. Res.* 21B: 457–458.

Nath, D., A. K. Manhar, K. Gupta, D. Saikia, S. K. Das, and M. Mandal. 2015. Phytosynthesized iron nanoparticles: Effects on fermentative hydrogen production by Enterobacter cloacae DH-89. *Bull. Mater. Sci.* 38 (6): 1533–1538.

Niu, L., L. Meng, and Q. Lu. 2013. Folate-conjugated PEG on single walled carbon nanotubes for targeting delivery of doxorubicin to cancer cells. *Macromol. Biosci.* 13: 735–744.

Njagi, C. E., H. Huang, L. Stafford, H. Genuino, H. M. Galindo, J. B. Collins, G. E. Hoag, and S. L. Suib. 2011. Biosynthesis of iron and silver nanoparticles at room temperature using aqueous *Sorghum* bran extracts. *Langmuir* 27: 264–271.

Palaniselvam, K., A. A. J. Velanganni, S. N. Govindan, and Karthi. 2012. Leaf assisted bio-reduction of silver ions using leaves of *Centella asiatica* L. and its bioactivity. *Eur. J. Lipid Sci. Technol.* 1: 46–49.

Parashar, U. K., P. S. Saxena, and A. Srivastava. 2009. Bioinspired synthesis of silver nanoparticles. *Dig. J. Nanomater. Biostruct.* 4: 159–166.

Parashar, V., R. Parashar, B. Sharma, and A. C. Pandey. 2009. Parathenium leaf extract medi-ated synthesis of silver nanoparticles: A novel approach towards weed utilization. *Dig. J. Nanomater. Biostruct.* 4: 45–50.

Parida, U. K., B. K. Bindhani, and P. Nayak. 2011. Green synthesis and characterization of gold nanoparticles using onion (*Allium cepa*) extract. *World J. Nanosci. Eng.* 1: 93–98.

Patil, D. C., S. V. Patil, H. P. Borase, B. K. Salunke, and R. B. Salunkhe. 2012. Larvicidal activity of silver nanoparticles synthesized using *Plumeria rubra* plant latex against *Aedes aegypti* and *Anopheles stephensi. Parasitol. Res.* 110: 1815–1822.

Pavani, K. V., T. Swati, V. Snehika, K. Sravya, and M. Sirisha. 2012. Phytofabrication of lead nanoparticles using grape skin extract. *Int. J. Eng. Sci. Technol.* 4: 3376–3380.

Petla, R. K., S. Vivekanandhan, M. Misra, A. K. Mohanty, and N. Satyanarayana. 2012. Soybean (*Glycine max*) leaf extract based green synthesis of palladium nanoparticles. *J. Biomater. Nanobiotechnol.* 3: 14–19.

Philip, D., C. Unni, S. A. Aromal, and V. K. Vidhu. 2011. *Murraya koenigii* leaf-assisted rapid green synthesis of silver and gold nanoparticles. *Spectrochim. Acta A* 78: 899–904.

Polakovic, M., T. Görner, R. Gref, and E. Dellacherie. 1999. Lidocaine loaded biodegradable nanospheres. II. Modelling of drug release. *J. Control. Rel.* 60: 169–177.

Prasad, M. D., and M. G. Krishna. 2014. Facile green chemistry-based synthesis and prop-erties of free-standing Au– and Ag–PMMA films. *ACS Sustain. Chem. Eng.* 2 (6): 1453–1460.

Prasad, R., and V. S. Swamy. 2013. Antibacterial activity of silver nanoparticles synthesized by bark extract of *Syzygium cumini. J. Nanopart.* 2013: 431218.

Rajasekaran M., J. S. Bapna, S. Lakshmanan, A. G. N. Ramachandran, A. J. Veliath, and M. Panchanadam. 1988. Antifertility effect in male rats of oleanolic acid—A triterpene from *Eugenia jambolana* flowers. *J. Ethnopharmacol.* 24: 115–121.

Ramteke C., T. Chakrabarti, B. K. Sarangi, and R. A. Pandey. 2013. Synthesis of silver nanoparticles from the aqueous extract of leaves of *Ocimum sanctum* for enhanced antibacterial activity. *J. Chem.* 2013: 278925.

Rao, C. N. R., G. U. Kulkarni, P. J. Thomas, and P. Peter. 2002. Size-dependent chemistry: Properties of nanocrystals. *Chem. Eur. J.* 8: 28–35.

Rastogi, L., and J. Arunachalam. 2012. Microwave-assisted green synthesis of small gold nanoparticles using aqueous garlic (*Allium sativum*) extract: Their application as anti-biotic carriers. *Int. J. Green Nanotechnol.* 4: 163–173.

Rastogi, R. M., and B. N. Mehrotra. 1990. *Compendium of Indian Medicinal Plants.* Vol. 1. Lucknow: Central Drug Research Institute, 388–389.

Ravindran, P., J. Fu, and S. L. Wallen. 2003. Completely green synthesis and stabilisation of metal nanoparticles. *J. Am. Chem. Soc.* 125: 13940–13941.

Rouhollah, K., M. Pelin, Y. Serap, U. Gozde, and G. Ufuk. 2013. Doxorubicin loading, release, and stability of polyamidoamine dendrimer-coated magnetic nanoparticles. *J. Pharm. Sci.* 102: 1825–1835.

Ruan, Z. P., L. L. Zhang, and Y. M. Lin. 2008. Evaluation of the antioxidant activity of *Syzygium cumini* leaves. *Molecules* 13: 2545–2556.

Sable N., S. Gaikwad, S. Bonde, A. Gade, and M. Rai. 2012. Phytofabrication of silver nanoparticles by using aquatic plant *Hydrilla verticillata*. *Nus. Biosci.* 4: 45–49.

Sagrawat, H., A. S. Mann, and M. D. Kharya. 2006. Pharmacological potential of *Eugenia jambolana*: A review. *Pharmacogn. Mag.* 2: 96–105.

Sathyavathi, R., M. B. Krishna, S. V. Rao, R. Saritha, and D. N. Rao. 2010. Biosynthesis of silver nanoparticles using *Coriandrum sativum* leaf extract and their application in nonlinear optics. *Adv. Sci. Lett.* 3: 1–6.

Satyavani, K., T. Ramanathan, and S. Gurudeeban. 2011. Plant mediated synthesis of bio-medical silver nanoparticles by using leaf extract of *Citrullus colocynthis*. *J. Nanosci. Nanotechnol.* 1: 95–101.

Sengupta, P., and P. B. Das. 1965. Terpenoids and related compounds. Part IV. Triterpenoids from the stem-bark of *Eugenia jambolana* Lam. *Indian Chem. Soc.* 42: 255–258.

Shankar, S. S., A. Ahmad, and M. Sastry. 2003. Geranium leaf assisted biosynthesis of silver nanoparticles. *Biotechnol. Prog.* 19: 1627–1631.

Shankar, S. S., A. Rai, A. Ahmad, and M. Sastry. 2004a. Rapid synthesis of Au, Ag, and bimetallic Au core-Ag shell nanoparticles using Neem (*Azadirachta indica*) leaf broth. *J. Colloid Interface Sci.* 275: 496–502.

Shankar, S. S., A. Rai, B. Ankamwar, A. Singh, A. Ahmad, and M. Sastry. 2004b. Biological synthesis of triangular gold nanoprisms. *Nat. Mater.* 3: 482–488.

Sharma, J. N., and T. R. Sheshadri. 1955. Survey of anthocyanins from Indian sources. Part II. *J. Sci. Ind. Res.* 14: 211–214.

Sheny, D. S., J. Mathew, and D. Philip. 2011. Phytosynthesis of Au, Ag and Au-Ag bime-tallic nanoparticles using aqueous extract and dried leaf of *Anacardium occidentale*. *Spectrochim. Acta A* 79: 254–262.

Singh, C., V. Sharma, P. K. Naik, V. Khandelwal, and H. Singh. 2011. A green biogenic approach for synthesis of gold and silver nanoparticles using *Zingiber officinale*. *Dig. J. Nanomater. Biostruct.* 6: 535–542.

Sinico, C., M. Manconi, M. Peppi, F. Lai, D. Valenti, and A. M. Fadda. 2005. Liposomes as carriers for dermal delivery of tretinoin: In vitro evaluation of drug permeation and vesicle skin interaction. *J. Control. Rel.* 103: 123–136.

Sivakumar, P., C. Nethradevi, and S. Renganathan. 2012. Synthesis of silver nanoparticles using *Lantana camara* fruit extract and its effect on pathogens. *Asian J. Pharm. Clin. Res.* 5: 97–101.

Solomon, M., and G. G. M. D'Souza. 2011. Recent progress in the therapeutic applications of nanotechnology. *Curr. Opin. Pediatr.* 23: 215–220.

Song, J. Y., E. Y. Kwon, and B. S. Kim. 2010. Biological synthesis of platinum nanoparticles using *Diopyros kaki* leaf extract. *Bioprocess Biosyst. Eng.* 33: 159–164.

Soppimath, K. S., T. M. Aminabhavi, A. R. Kulkarni, and W. E. Rudzinski. 2001. Biodegradable polymeric nanoparticles as drug delivery devices. *J. Control. Rel.* 70: 1–20.

Soundarrajan, C., A. Sankari, P. Dhandapani et al. 2012. Rapid biological synthesis of platinum nanoparticles using *Ocimum sanctum* for water electrolysis applications. *Bioprocess Biosyst. Eng.* 35: 827–833.

Srivastava, H. C. 1953. Paper chromatography of fruit juices. *J. Sci. Ind. Res.* 12B: 363–365.

Sulochana, S., P. Krishnamoorthy, and K. Sivaranjani. 2012. Synthesis of silver nanoparticles using leaf extract of *Andrographis paniculata*. *J. Pharmacol. Toxicol.* 7: 251–258.

Sundaravadivelan, C., and M. Nalini. 2011. Biolarvicidal effect of phyto-synthesized silver nanoparticles using *Pedilanthus tithymaloides* (L.) Poit stem extract against the dengue vector *Aedes aegypti* L. (Diptera; Culicidae). *Asian Pac. J. Trop. Biomed.* 2012: 1–8.

Suri, S. S., H. Fenniri, and B. Singh. 2007. Nanotechnology-based drug delivery systems. *J. Occup. Med. Toxicol.* 2: 16.

Timbola, A. K., B. Szpoganicz, A. Branco, F. D. Monache, and M. G. Pizzolatti. 2002. A new flavonol from leaves of *Eugenia jambolana*. *Fitoterapia* 73: 174–176.

Ullman, A. 1996. Formation and structure of self-assembled monolayers. *Chem. Rev.* 96: 1533–1554.

Vaishnava, M. M., and K. R. Gupta. 1990. Isorhamnetin 3-O-rutinoside from *Syzygium cumini* Lam. *J. Indian Chem. Soc.* 67: 785–786.

Vaishnava, M. M., A. K. Tripathy, and K. R. Gupta. 1992. Flavonoid glycosides from roots of *Eugenia jambolana*. *Fitoterapia* 63: 259–260.

Vankar, P. S., and D. Bajpai. 2010. Preparation of gold nanoparticles from *Mirabilis jalapa* flowers. *Indian J. Biochem. Biophys.* 47: 157–160.

Venkateswarlu, G. 1952. On the nature of the colouring matter of the jambul fruit (*Eugenia jambolana*). *J. Indian Chem. Soc.* 29: 434–437.

Wang, R., J. Yang, Z. Zheng et al. 2001. Dendron-controlled nucleation and growth of gold nanoparticles. *Angew. Chem. Int. Ed.* 40: 549–552.

Warrier, P. K., V. P. K. Nambiar, and C. Ramankutty. 1996. *Indian Medicinal Plants: A Compendium of 500 Species*. Vol. 5. New Delhi: Orient Longman Ltd., 225–228.

Willner, I., R. Baron, and B. Willner. 2006. Growing metal nanoparticles by enzymes. *Adv. Mater.* 18: 1109–1120.

Xiong, X. B., Y. Huang, W. L. Lu et al. 2005. Enhanced intracellular delivery and improved antitumor efficacy of doxorubicin by sterically stabilized liposomes modified with a synthetic RGD mimetic. *J. Control. Rel.* 107: 262–275.

Yokoyama, M., S. Fukushima, R. Uehara et al. 1998. Characterization of physical entrapment and chemical conjugation of adriamycin in polymeric micelles and their design for in vivo delivery to a solid tumour. *J. Control. Rel.* 50: 79–92.

Zambaux, M. F., F. Bonneaux, R. Gref et al. 1998. Influence of experimental parameters on the characteristics of poly(lactic acid) nanoparticles prepared by a double emulsion method. *J. Control. Rel.* 50: 31–40.

Zhao, M., L. Sun, and R. M. Crooks. 1998. Preparation of Cu nanoclusters within dendrimer templates. *J. Am. Chem. Soc.* 120: 4877–4878.

Zheng, J. 2002. Influence of pH on dendrimer-protected nanoparticles. *J. Phys. Chem. B* 106: 1252–1255.

11 Genetic Resources of *Syzygium cumini* in India
Present Status and Management

S. K. Malik, Rekha Chaudhury,
Vartika Srivastava, and Sanjay Singh

CONTENTS

INTRODUCTION

The underutilized fruit crops of Indian origin, like bael (*Aegle marmelos*), chironji (*Buchnania lanzan*), jamun (*Syzygium cumini*), karaunda (*Carissa carandas*), ker (*Capparis decidua*), khirni (*Manilkara hexandra*), lasora (*Cordia dichotoma*), and mahua (*Madhuca longifolia*), are directly interwoven into the socioeconomic fabric of rural masses and especially of tribes dwelling in remote hot, arid, and fragile ecosystems. These potential crops of the future are awaiting their popularization and full utilization, as until now, they have remained of only local importance. Southeast Asia is represented by more than 500 species of fruits (Arora and Rao 1995), while the Hindustani region of diversity represents 344 species of fruits, having vast potential for new crops (Arora 1995).

The management of genetic resources of underutilized fruits like *S. cumini* has not received much attention due to their low commercial prospects. These underutilized fruits have been known mostly for local uses. Jamun is one of the most commonly available fruits of India, but it has not been given much importance for its diverse collection, characterization, and conservation in a systematic manner. Conservation of genetic resources of jamun is important to safeguard the existing diversity in this underutilized fruit tree for various uses. Organized production and value addition of jamun would help enhance the income of small and marginal farmers and also the on-farm conservation of jamun germplasm. The present status of genetic resource management of jamun in India is discussed in this chapter.

DISTRIBUTION

Jamun is native to tropical Asia, especially the Indian subcontinent: India, Bhutan, Nepal, and Sri Lanka; Malesia–Indonesia and Malaysia; east tropical Africa: Kenya, Tanzania, and Uganda; China (Fujian, Guangdong, Guangxi, and Yunnan); and Pacific Thailand, Philippines, and Madagascar. Jamun trees are naturalized in Africa (Mauritius), the Seychelles, Australia, the West Indies, the United States (California and Hawaii), and Israel (GRINOnline Database: https://npgsweb.ars-grin.gov/grin global/taxonomydetail.aspx?36128). Figure 11.1 shows the native areas and exotic ranges of distribution of jamun in the world.

S. cumini and other related species are widely distributed in north and south India. Maximum genetic and species diversity is represented in the two hot spots of plant diversity, namely, the Western Ghats and the northeast India (Table 11.1). The Indo-Gangetic plains also harbor rich genetic diversity of jamun, where trees with diverse types of fruits are available in wild and semiwild conditions. Jamun also occurs in the lower ranges of the northwestern Himalayas up to an elevation of 1300 m above

Native range

Exotic range

FIGURE 11.1 World distribution map of jamun. (From Orwa, C. et al., Agroforestry database: A tree reference and selection guide, Version 4.0, 2009, http://www.worldagroforestry .org/output/agroforestree-database.)

TABLE 11.1

Syzygium **Species of Economic Importance and Their Distribution in India**

Serial No.	Species	Distribution
1	*S. cumini*	Indo-Gangetic plains, Tamil Nadu, Maharashtra, Gujarat; widely distributed
2	*S. arnottianum*	Western Ghats; the Nilgiris, Palni, and Anamalai Hills
3	*S. bracteatum*	Western Ghats, eastern/northeastern India
4	*S. operculatum*	Grows wild in Nilgiri Hills of Tamil Nadu, Western Ghats
5	*S. aqueum*	Mainly in Assam, Sikkim, and Meghalaya; eastern/northeastern India
6	*S. fruticosum*	Grows as an avenue tree; widely distributed

mean sea level, and in the Kumaon hills up to 1600 m. It is widely grown in the larger part of India from north to south (Chaturvedi 1956; Singh et al. 1967). It is found as wild and semiwild in the states of Punjab, Haryana, Uttar Pradesh, Maharashtra, Gujarat, Madhya Pradesh, Bihar, Chhattisgarh, Jharkhand, Karnataka, Tamil Nadu, Andhra Pradesh, and Kerala. Jamun is widely cultivated in homestead gardens and backyards for its edible fruits and also as an avenue tree in all parts of India. Jamun is one of the most hardy fruit crops, and it can easily be grown in neglected and marshy areas where other fruit trees cannot be grown successfully.

Besides *S. cumini*, other species of economic importance present in India are *Syzygium arnottianum*, *Syzygium bracteatum*, *Syzygium operculatum*, *Syzygium aqueum*, and *Syzygium fruiticosum* (Table 11.1). These species are mostly distributed in the Western Ghats and northeast India. Some of the species are endemic to the Western Ghats of south India.

GENETIC VARIABILITY

Vast genetic and species diversity exists in the genus *Syzygium* in India. Being a cross-pollinated fruit crop, *S. cumini* is highly heterozygous and seedlings exhibit a wide range of variations, which eventually help in the selection of superior genotypes with desirable horticultural traits. Propagation through seeds also brings in much of the desired variability in jamun germplasm. A number of jamun seedling strains with considerable variation in fruit shape and size, pulp color, total soluble solids (TSS), acidity, and earliness are reported from Uttar Pradesh, Gujarat, and Maharashtra. This enormous diversity provides ample scope for selection of desirable genotypes and cultivars for diverse uses. Several surveys have been undertaken in India to collect and document the available genetic diversity of jamun. Surveys in the Pune and Ahmednagar districts of Maharashtra resulted in documentation of wide variation in fruit weight (3.5–16.5 g), pulp content (54.29%–85.71%), TSS (4.5%–17%), and acidity (0.16%–0.55%) and the selection of four promising types: nos. 4, 13, 14, and 15 (Keskar et al. 1989). Ashraf (1987) reported fruit shape in jamun varying from round to oblong, and the apex of fruits from flat to pointed, besides the variations in the physicochemical characteristic of fruits that could help in the selection of varieties suitable for fresh fruit market and processing. Small seed size and high pulp content with better chemical qualities are considered ideal characteristics for jamun. Singh et al. (1999) evaluated eight genotypes of jamun under Faizabad conditions and reported that the oblong types had more fruit weight and relatively less seed weight. Among the locally available genotypes of jamun in West Bengal, no. 1 (oval-shaped large fruit) and no. 2 (cylindrical-shaped medium-sized fruit) proved to be better in terms of yield and fruit quality attributes (Kundu et al. 2001). A survey was undertaken in Karnataka to investigate the nature and extent of variability present in jamun seedling progenies for morphological characteristics of trees, and a high degree of variability was observed with regard to plant girth, leaf area, petiole length, and ratio of leaf length to petiole length (Prabhuraj et al. 2002). In North Goa, Devi et al. (2002) reported morphological and physicochemical variations in jamun fruit, showing individual fruit weight (3.42–13.67 g), length (3.31–5.26 cm), girth (5.21–9.82 cm), pulp content (58.57%–84.55%), TSS (12.00%–26.80%), titratable acidity (0.59%–1.63%), total sugars (6.87%–25.31%), and sugar–acid ratio (15.39 to 27.92).

The documented genetic diversity in jamun has not been exploited for the genetic improvement program. The available genetic diversity is facing the threat of genetic erosion as a result of large-scale developmental activities related to urbanization and lack of arable land for agriculture. The genetic diversity of the wild-related species of jamun is of immense value for the search of resistance to physiological races of pathotypes of fungi, bacteria, viruses, and nematodes, besides winter hardiness, resistance to drought, salinity, and so forth.

In jamun area-specific locales, selections have been identified based on fruit traits. One of the most popular and widely grown natural selection types in north India is 'Ra Jamun' (Singh et al. 2007). A late-maturing selection with small, rounded fruits (1.5–2.0 cm length and 1.0–1.5 cm diameter) is also common in north India. The local farmers' jamun selections grown in Haryana and western Uttar Pradesh are 'Bedana' (large size, very juicy), 'Kaatha' or 'Kathua' (with small and acidic fruits), 'Jadhi' (maturing in June or Jadth), 'Ashada' (maturing in July or Ashad), and the late-ripening type 'Bhado' (maturing in August) (Malik et al. 2010). Similarly, several local-type selections are also found in the Konkan area and Pune and Ahmednagar districts of Maharashtra (Keskar et al. 1989) and in Gujarat and Rajasthan.

GERMPLASM COLLECTION

Jamun germplasm has been collected from throughout India, especially from the states with sizable diversity, namely, Maharashtra, Rajasthan, Gujarat, Uttar Pradesh, Haryana, West Bengal, and the Western Ghats region. Variability in tree phenology, flowering and fruiting period, and fruit characters in jamun has been documented. Collected germplasm has been characterized, and some of the genotypes are being maintained in the field genebanks at various Indian Council of Agricultural Research (ICAR) institutes and state agricultural universities (SAUs). In order to identify promising germplasm and support improvement and value addition programs, the National Network Project on Underutilized Fruits was undertaken by the ICAR from 2007 to 2010. The program involved systematic collection, characterization, and conservation of underutilized fruits, including jamun in India.

TABLE 11.2
Present Status of Jamun Collection in India

Area of Collection	No. of Accessions Collected	Reference
Pune and Ahmednagar districts of Maharashtra	–	Keskar et al. 1989
Eastern Uttar Pradesh	8	Singh et al. 1999
West Bengal	–	Kundu et al. 2001
Karnataka	–	Prabhuraj et al. 2002
North Goa	18	Devi et al. 2002
Uttar Pradesh, Haryana, Maharashtra, and Gujarat	54	Anonymous 2010
Gujarat	33	Singh and Singh 2005
Uttar Pradesh and Jharkhand	32	Patel et al. 2005
Bihar	–	Kumar et al. 2006
Haryana and western Uttar Pradesh	20	Malik et al. 2010
Gujarat, Rajasthan	22	Singh et al. 2007
Karnataka	–	Patil et al. 2009
Uttar Pradesh	–	Jai Prakash et al. 2010
Uttar Pradesh and Uttarakhand	25	Srivastava et al. 2010
Gujarat	16	Singh and Singh 2012

As part of the ICAR National Network Project on Underutilized Fruits, a limited number of jamun collections were made and promising germplasm accessions were identified, characterized, established, and conserved in the field genebanks at the ICAR-Central Institute for Subtropical Horticulture (CISH), Lucknow (Singh et al. 2007; Anonymous 2010), and the Central Horticultural Experiment Station (CHES), ICAR-Central Institute of Arid Horticulture (CIAH), Godhra, Gujarat (Anonymous 2010). Region-wise surveys and explorations were carried out in the last two to three decades in India for the collection and characterization of jamun germplasm (Table 11.2). However, these collections are not sufficient, and there is a need to develop a national program to collect the existing genetic diversity of jamun and other wild and semiwild species of *Syzygium* from various parts of India and to maintain them at a suitable place for future use in genetic improvement programs.

GERMPLASM CHARACTERIZATION AND EVALUATION

Characterization and evaluation are important to identify and add value to the germplasm for its further use in crop improvement, as well as to avoid its duplication in germplasm repositories and genebanks. Since most of the tropical fruit species, including jamun, are mostly heterozygous due to a high degree of outcrossing, systematic morphological characterization backed by molecular analysis is essential to assess the extent of genetic variability and sustainable utilization of existing germplasm. Systematic characterization of physicochemical characters of available germplasm will help estimate the amount of genetic variability within the species and facilitate identification of superior genotypes with desired characters.

In India, significant efforts have been made to characterize diversity in jamun that enabled selection of some area-specific genotypes for further utilization. In a physicochemical study of ripe fruits of grafted as well as seedling types of jamun, Bhardwaj and Yamdagni (2005) reported that the fruits of grafted jamun were longer than the seedling fruits, although the breadth of the fruits, seeds, and pulp contents did not show marked differences. The fruits of grafted jamun contained higher TSS, ascorbic acid, sugar, pectin, crude fiber, and carotenoids, but lower total acidity and chlorophyll contents. The tannin, protein, and anthocyanin contents were at par in both graft-type and seedling fruits. The grafted fruits had higher P, K, Mg, and Fe contents but showed no marked differences with respect to N, Zn, and Cu contents. In a similar study, Patel et al. (2005) observed good variability in jamun accessions from Uttar Pradesh and Jharkhand. They observed that the genotypes 'RNC-26' and 'RNC-11' had higher fruit and pulp weight, along with higher TSS. The highest pulp content (97.71%) was recorded in 'V-8,' followed by 'V-6' (95.84%) and 'V-7' (93.81%) collected from the Varanasi region. Thin seed with almost negligible seed weight (0.12 g) was observed in 'V-8,' and this accession may prove useful in breeding for seedless jamun. Among the locally available types of jamun in West Bengal, Kundu et al. (2001) reported fruits with oval shape and large size (JS-1) and cylindrical shape and medium size (JS-2), which showed high characteristics for yield, fruit size, and weight. On the other hand, the pear-shaped, medium-sized fruits (JS-2 and JS-3) showed high amounts of TSS, reducing sugar, and total sugar.

In Gokak Taluk of Belgaum district, Karnataka, Prabhuraj et al. (2002) reported high morphological variability in jamun trees and their seedling progenies. Prabhuraj et al. (2002) studied correlation and path coefficient analysis in jamun strains for yield and fruit characters and reported that the fruit yield was significant and positively correlated with seed volume and the pulp–seed ratio. They also reported positive and significant correlations of fruit weight with a number of parameters viz., pulp weight, fruit size, fruit length and breadth, pulp thickness, seed weight, pulp: seed ratio, and pulp percentage, whereas seed percent was significantly but negatively correlated with fruit weight. Path coefficient analysis of different characters contributing fruit yield revealed that pulp weight, pulp percent, and pulp–seed ratio had high positive direct effects (Prabhuraj et al. 2002).

In another correlation and path analysis conducted on jamun fruit characters, Inamdar et al. (2002) observed that the data for correlation coefficient had a high positive significant correlation with pulp weight, fruit volume, seed weight, seed volume, fruit length, seed length, pulp thickness, and seed breadth. Fruit weight showed a negative significant correlation with seed percent, while TSS and titrable acidity were nonsignificant and negatively correlated with fruit weight. The direct and indirect effects showed that pulp weight, seed weight, fruit breadth, fruit volume, pulp–seed ratio, seed volume, and titratable acidity had high positive direct effects on fruit weight. Seed size had a low positive direct effect on fruit weight. However, pulp percent, seed percent, pulp thickness, seed breadth, and fruit size exhibited negative direct effects on fruit weight. It was concluded that major emphasis in selection should be given for higher pulp weight, fruit volume, fruit size, pulp thickness, and pulp–seed ratio (Inamdar et al. 2002). Srivastava et al. (2012) observed a highly significant positive correlation of pulp weight, fruit volume, pulp–seed ratio, and titerable acidity with fruit weight of jamun. They also reported that seed percentage was highly significant but had a negative correlation with fruit weight, and concluded that major emphasis in selection should be given to higher pulp weight, fruit volume, fruit size, and pulp–seed ratio. Also, higher TSS and acidity, along with smaller seed size, should be considered for selecting superior genotypes.

A wide range of diversity in jamun was reported by Kundu et al. (2006) in different agroecological zones of West Bengal. Kumar et al. (2006) identified and released a new jamun variety, 'Rajendra Jamun-1,' a promising genotype selected among an exhaustive collection made around Sabour and Bhagalpur in Bihar, during five successive fruiting seasons from 2000 to 2005. In the same study, another promising selection, Type-4, showed its superiority in trunk girth (400 cm) and leaf dimension (20 × 7.5 cm). The lanceolate leaf in Type-4 with an acute apex and entire margin was dark green in color. This genotype was found profusely flowering comparatively early, during March and April, and fruits are harvested during May and June. The fruits of Type-4 were oval in shape with a blunt apex and smooth skin surface. The other traits in Type-4 were better fruit dimension (3.9 × 2.3 cm), fruit weight (12.86 g), and pulp (88.4%); maximum TSS (18.20°Bx), and lowest acidity (0.31%), which made it the sweetest among all types examined. The yield potential (450 kg/tree) in Type-4 was strikingly greater than in other types. Interestingly, this genotype was also found to be pest and disease resistant.

Ghosh et al. (2006) evaluated jamun germplasm of West Bengal for its quality parameters and the storage behavior of fruits and observed significant variations among the different types of jamun fruits for physical as well as chemical parameters. The ranges of variation in TSS, total sugar, reducing sugar, nonreducing sugar, acidity, and ascorbic acid contents in different types of jamun fruits were 10.60%–16.10%, 4.86%–11.10%, 3.92%–10.12%, 0.94%–1.61%, 0.86%–1.90%, and 1.28–7.63 mg/100 g, respectively. Storage was optimal up to the fourth day, with a considerable percentage of fruits retaining their edible condition. Patil et al. (2009) conducted an evaluation trial in Dharwad (Karnataka) for jamun collections from Krishnagiri, Devargudihal, and Badami, and reported Krishnagiri collections as having the highest fruit yield, followed by Devaragudihal collections. The income from trees was higher for the Krishnagiri collections than for the Badami and other local collections.

Jai Prakash et al. (2010) evaluated the jamun collection from eastern Uttar Pradesh maintained by the Department of Horticulture at the campus of Banaras Hindu University in Varanasi. They reported Selection-1 suitable for commercial and systematic orcharding due to its higher fruit weight (14.55 g), minimum seed weight (1.73 g), higher pulp content (90.05%), and higher TSS (21.23%) and total sugar (20.24%). Srivastava et al. (2010), in a similar study, reported huge variability among the genotypes of jamun collected from the Varanasi and Pantnagar regions. Although the genotypes collected from the Varanasi region were smaller in fruit size and physical parameters, they showed better quality than the fruits collected from the Pantnagar region. Singh et al. (2007) evaluated 22 promising accessions of jamun for fruit characters based on their physicochemical characters and cultivars; CISH J-17, CISH J-14, CISH J-19, and CISH J-20 showed maximum scores for physicochemical attributes. Recently, CISH also reported CISH J-37 and CISH J-42 as superior and promising jamun accessions (Anonymous 2010).

Preliminary studies on molecular characterization of jamun have also been attempted at ICAR-CISH, and eight accessions were analyzed using random amplification of polymorphic DNA (RAPD) techniques where 14 primers were used, revealing the distinctness of J-36 and J-22 from other accessions (Anonymous 2010). Shakya et al. (2010) analyzed eight jamun accessions being maintained at ICAR-CISH for genetic diversity using RAPD and simple sequence repeat (SSR) markers. The accessions belonging to different geographical regions were grouped in separate clusters indicating interpopulation differences. They reported that OPZ9 and OPA12 primers can detect very small differences between these selections. Khan et al. (2011) analyzed 16 genotypes of jamun collected from three agroecological regions using RAPD markers. The study revealed a high level of polymorphism (47.69%–74.87%) in the sample populations that was correlated with the population size.

Despite being a commercially important fruit crop with high potential for developing value-added nutritional products, jamun diversity in India has not been characterized and evaluated in a systematic manner. Most of the studies described above have been based on only a few accessions collected from only a particular region and evaluated for only a few characters. Therefore, there is an urgent need to establish a national jamun collection at a suitable place, and multilocation trials may be undertaken similar to the All India Coordinated Project for detailed systematic characterization and evaluation of the diverse germplasms. Developing good-quality planting material of suitable varieties of jamun with desirable horticultural attributes, such as short juvenile period, small tree size, high yield with desirable fruit characters (size, color, seed–pulp ratio, TSS, and shelf life), and simultaneous fruiting, is an urgent priority to help large-scale cultivation of jamun in different agroecological zones. Most of the produce of jamun fruits comes from trees grown in a traditional manner by the small and marginal farmers, which provides some livelihood support to them. There is an urgent need to promote commercial cultivation of this important fruit crop to benefit the farmers and diversify the Indian fruit basket, which would also benefit the consumers with highly nutritious and diverse jamun products with varied tastes and properties.

GERMPLASM CONSERVATION

Conservation of horticultural genetic resources (HGRs) and specifically the underutilized fruit species, some of which are still grown in natural wild and semiwild conditions, would require adoption of complementary conservation strategies involving suitable *in situ* and *ex situ* conservation methods. Germplasm conservation of a diverse genus like *Syzygium*, with several species and vast genetic variability spread over the arid to lower hills of the Himalayas to the Western Ghats and northeast India, is indeed a challenging task, which needs special efforts with dynamic strategies. Traditionally, germplasm conservation is practiced using *ex situ* and *in situ* conservation methods, where depending on the biological status, extent of diversity available, and propagation method, one or both strategies are recommended to be employed. It is worrisome that tropical tree species of the Indian subcontinent have not been taken up seriously for the conservation efforts. This lacuna in conserving native tree species diversity is partly due to the commonness of many of these trees species in and around open places and marginal and forest lands. Therefore, the partial loss of existing diversity and complete vanishing of some of the important genotypes from the existing genepool either go unnoticed or are not taken seriously in several of the most common tree species. However, the loss of genetic diversity of indigenous tropical fruits is of great concern. Jamun, being one of the very important underutilized fruits, needs urgent attention for germplasm conservation. *Ex situ* conservation of jamun in the laboratory has been difficult due to the desiccation sensitivity of seeds, which is generally common in tropical tree species. Various studies have confirmed the seed storage behavior of jamun as highly recalcitrant, as the seeds germinate well when fresh, but the viability is lost within two weeks of open storage at room temperature (Orwa et al. 2009). Therefore, such species require the use of more than one strategy to achieve comprehensive conservation of available germplasm. There is a need to adopt innovative approaches by complementing and integrating available *in situ* and *ex situ* methods to develop best practices in horticultural and forestry species to maintain existing diversity, especially in high-value and vulnerable populations (Pritchard et al. 2014). *Syzygium* with vast species and genetic diversity in India suitably falls under this category, and the various approaches suitable for jamun genetic resources conservation are discussed in detail.

In Situ Conservation

In situ conservation is important for underutilized fruit species still occurring in natural wild or semidomesticated conditions. Several species of *Syzygium*, including *S. cumini*, still occur as wild and semiwild in the forests of the Western Ghats and in northeastern India. Some of these species are endemic and endangered to these areas and need specific conservation appraches. *In situ* conservation of various *Syzygium* species is undertaken using the following approaches:

1. *In situ* conservation in natural habitats like protected areas and national reserves. Some of the *Syzygium* species are rare and endangered and require a specific habitat for their survival, which is only possible in their natural home available in the national parks, biosphere reserves, and wildlife sanctuaries. Some specific examples of *Syzygium* species protected *in situ* include *S. laetum* (Dev Jamun) in the Radhanagari Wildlife Sanctuary in Kolhapur; *S. lanceolatum (S. claviflorum)* (Shareef et al. 2010) and *S. sasidharanii* in the Agasthyamalai Biosphere Reserve of the southern Western Ghats (Sasidharan and Jomy 1999); *S. travancoricum* (Kulavetti), a rare species only found in some pockets in Kerala; and *S. agastymalayanum* in the Kalakkad-Mundanthurai Tiger Reserve in Tamil Nadu. Similarly, there are several genotypes of *S. cumini* and related species distributed in specific localities that need immediate conservation efforts. There is an urgent need to identify particular species and protected areas for their *in situ* conservation based on biological status, population size, diversity maps, and forest policy of respective state governments. These *in situ* conservation sites may

be notified for these very important indigenous *Syzygium* species so as to protect the genetic resources and their microclimatic habitats.

2. Being sacred to the Hindus, many jamun trees have been grown and protected in several sacred groves, especially in south India. One such example is *S. travancoricum*, a critically endangered species found only in the swampy wetland habitat of the sacred grove of Aickad in Kerala (Roby et al. 2013).

3. *In situ* on-farm conservation is more important for semidomesticated species having socioeconomic importance and providing substantial livelihood support for local people. Generally, natural selections, cultivars, and varieties of local importance for social, religious, and edible purposes are grown by the small and tribal farmers as backyard garden trees. In some underutilized fruits like jamun, the local selections or farmers' varieties have been developed or identified since time immemorial. These local selections are being grown as isolated plants or in small numbers in the homestead gardens, farmers' fields, backyards, or common Panchayat lands in villages and small towns. Such selections or genotypes need urgent attention for further characterization, evaluation, and on-farm conservation. Several local types of jamun described under the earlier sections also serve the purpose of on-farm in situ conservation. Similarly, several jamun seedling trees are grown under social forestry as multipurpose trees for shade, avenue, and ornamentals in villages and cities, especially in gardens and parks and as roadside plantations. This large variability of jamun existing in the rural and urban areas in India remains unattended for scientific studies and selection or protection for future use.

Ex Situ Conservation

Field Genebanks

Ex situ conservation of underutilized fruits is important to safeguard the genetic wealth and use germplasm for genetic improvement to develop desirable cultivars or varieties. Field genebanks have an important place in the conservation and maintenance of clonally propagated species, tree species with a long juvenile phase, and species that do not produce true-to-type seeds, or produce recalcitrant seeds whose laboratory conservation technology has not been developed or standardized so far. Presently, several field genebanks for diverse horticultural species are established throughout the world. In India, field genebanks are established for several major horticultural crops, namely, citrus, mango, mulberry, coconut, oil palm, and several other fruits, at state and ICAR horticultural institutions or SAUs. As far as underutilized fruits are concerned, field genebank conservation has been recently undertaken under the All India Coordinated Project on Arid Fruits at various ICAR institutes, that is, ICAR-CIAH, Bikaner; ICAR-CISH; ICAR-Central Arid Zone Research Institute (CAZRI), Jodhpur; the National Bureau of Plant Genetic Resources (NBPGR) regional station, Jodhpur; and the Indian Institute of Horticulture Research (IIHR), Bangalore, and at SAUs, such as the Chaudhary Charan Singh Haryana Agricultural University (CCSHAU), Hisar and Regional Station of CCSHAU, Bawal, Haryana;

Mahatma Phule Krishi Vidyapeeth (MPKV), Rahuri, Maharashtra; Sardarkrushinagar Dantiwada Agricultural University (GAU), Sardar Krushinagar, Gujarat; Shri Karan Narendra College of Agriculture, Jobner, Rajasthan; Maharana Pratap University of Agriculture and Technology (MPUAT), Udaipur, Rajasthan; Narendra Deva University of Agriculture and Technology (NDUAT), Faizabad, Uttar Pradesh; RAU, Bikaner, Rajasthan; and other state horticulture stations at Tamil Nadu, Andhra Pradesh, and other states.

Jamun germplasm is maintained at only a few places in the field genebanks at ICAR-CISH; CHES, ICAR-CIAH; and MPKV, where large numbers of accessions are preserved. IIHR is maintaining 12 accessions of jamun in the field genebank. There is a need to establish promising seedling selections and local farmers' selections in the field genebanks to conserve the elite germplasm as true to type and use them for crop improvement. Recently, under the National Network Project on Underutilized Fruits, CHES, established 66 genotypes of jamun collected from various parts of India. Based on evaluation of these accessions, CHES released six promising selections, GJ-2 (Goma Priyanka), GJ-8, GJ-19, GJ-23, GJ-40, and Katha Jamun. Under the same network project, CISH established 38 accessions of jamun collected from various parts of the country in the field genebank and evaluated these for various agronomic and fruit characters. ICAR-CISH reported two seedless accessions (CISH J-37 and CISH J-42) collected from Uttar Pradesh. These two selections were registered as promising germplasm with ICAR-NBPGR for bold fruits and high pulp quantity (IC0587715 and INGR11025) and for seedlessness and pulp content and high TSS (IC0587714 and INGR11024).

Genebank and Cryogenebank Conservation

Conservation of underutilized fruit species is being undertaken in the laboratory in genebanks and cryogenebanks at ICAR-NBPGR, New Delhi. Conservation of plant germplasm in the form of seeds is the most convenient and reliable method being practiced in genebanks. Germplasms of tropical underutilized fruit species, where the seeds are relatively larger and have a high moisture content at the time of shedding, pose problems in traditional conservation. Hence, there is a need to study their basic seed physiology, longevity, and seed storage behavior. Seed storage behavior in its simplest form is measured in terms of survival and longevity of seed under various storage conditions. Information on this is available for only about 3% of the higher plant species (Hong and Ellis 1996). Seed storage behavior in several cases is misinterpreted because of scanty data generated on the survival and longevity of seed and the lack of detailed information on physiological characteristics. Conservationists can recommend and adopt short-, medium-, and long-term seed storage only after correct identification of seed storage behavior. Seed storage behavior has broadly been divided into three categories. Initially, Roberts (1973) defined two categories: orthodox and recalcitrant. Later, another category of seed storage behavior was designated by Ellis et al. (1990) and termed intermediate, where the behavior is in between orthodox and recalcitrant. Orthodox seeds can be desiccated at the desired moisture contents and can be conserved in the conventional genebanks comprising cold storage modules maintained at –20°C. Nonorthodox (intermediate and recalcitrant) seeds are not amenable to conventional genebank regimes, being sensitive to desiccation and chilling injury, and thus require special conservation protocols. Many tree species, especially of tropical origin, for example, rubber, several *Citrus* species, *Garcinia* species, jackfruit, cocoa, *Madhuca* species, and *Syzygium* species, produce such seeds. Several methods based on cryopreservation have been developed for genebank conservation of such nonorthodox seeded species. Cryopreservation, storage of biological materials at ultralow temperatures (–196°C), is the only method available for the long-term conservation of nonorthodox seeds and several vegetative explants, such as shoot apices, meristems, dormant buds, and somatic embryos.

Conservation of germplasm in the form of seeds for underutilized fruit species, including *Syzygium* species, which are predominantly cross-pollinated, only ensures the genepool conservation due to the heterozygous nature of seeds. As most of these species are found in the natural wild or semiwild and propagated through seeds in nature, conservation of available genetic variability needs to be protected for the selection of desired genotypes. In most of these fruit species, farmers or local people are propagating progenies of these fruits using seeds, as no commercial cultivars are available, and even if a few have been identified, planting materials are not available. Once the promising genotypes or cultivars are identified in these species, conservation of their vegetative tissues to achieve true-to-type conservation can be attempted using *in vitro* methods. It should be emphasized here that conservation of vegetative tissues in these tropical woody species would be an enormous task, as most of the species are known for their recalcitrance as far as *in vitro* establishment and cryopreservation are concerned. It is, therefore, recommended to conserve the available genetic diversity of such economically important species in the best possible way to fulfill the objective of safeguarding these indigenous species from genetic erosion. For genetic improvement and genotype conservation, collected and characterized elite genotypes are presently being conserved in field genebanks at various horticultural organizations. It is therefore emphasized that a complementary conservation strategy (Rao 1998) involving the use of more than one relevant approach would be the best option for achieving safe conservation of these underutilized fruit species facing severe threat of extinction.

Jamun is commonly propagated through seeds (Singh et al. 2007). Seeds are highly recalcitrant in nature, and only freshly extracted seeds can be sown for raising seedlings. Seeds lose viability very rapidly, as by 30 days of storage at room temperature, 50% viability is lost. Freshly harvested seeds shed at a high moisture content of 50% showed rapid decline in viability at 35% moisture. Seeds germinate within two weeks and can be transplanted during monsoon season in the field. In jamun, up to 50% polyembryony is reported and true-to-type nucellar seedlings are produced (Singh et al. 2007). Although vegetative propagation using various budding methods has been found to be successful, patch budding has been reported to be the most successful in the month of March in semiarid areas (Singh and Singh 2006). Soft wood grafting has also been successful for multiplication in Karnataka and Gujarat in the months of June and August, respectively (Singh and Singh 2006). Jamun seeds have been categorized as having recalcitrant seed storage behavior based on various factors, like thousand-seed weight (Hong and Ellis 1996) and short storage life, with rapid viability decline within 17 days (Baxter et al. 2005). In India, accelerated aging of jamun seeds at a high temperature of 41°C and 100% relative humidity (RH) was studied by Vanangamudi et al. (2000). As expected, rapid decline in viability below 50% was apparent by the fourth day, necessitating sowing in the nursery immediately after collection.

SEED STORAGE IN NON-ORTHODOX SEED SPECIES

Non-orthodox seed species are metabolically active and are invariably desiccation and freezing sensitive. Such seeds need to be stored for the short term as long as preliminary experiments are being carried out or planting programs are planned. Seeds can be stored for longer periods unless they are subjected to cryopreservation, where the longest storage periods are achievable. There are several field collections of tropical fruit species that produce recalcitrant seeds, and jamun features among them. Two problems that disallow short- to long-term storage of these seeds are the high metabolic state leading to depletion of metabolic reserves and intense desiccation sensitivity, and the fast proliferation of microflora, especially the fungi. Surface decontamination of fruits and seeds is thus recommended. For jamun, the relatively large seed size makes it impossible to cryostore, and hence alternative explants, like embryos and embryonic axes, are excised and processed. Jamun seeds can contribute from 20% to 80% to the total fruit weight on a fresh seed weight basis (Sivasubramaniam and Selvarani 2012). Seed coats and cotyledons contribute approximately 6% and 94% to the total seed weight, respectively. The embryos are substantially small and the embryonic axes are much smaller.

SEED VIABILITY AND STORAGE

Jamun seed viability can be tested by germinability and also by triphenyl tetrazolium chloride (TTC) tests. Sivasubramaniam and Selvarani (2012) observed pink staining using TTC solution, indicative of viable regions only along the embryonic axis with no staining in cotyledons. This has been attributed to a physical barrier on the cotyledon surface that possibly prevented penetration of the tetrazolium solution. Polyembryony, where multiple embryos arise within the embryo sac by budding or by cleavage of the zygotic proembryo or from the synergids and antipodal cells, has been observed in jamun with a maximum of up to four embryos from each seed (Sivasubramaniam and Selvarani 2012). Each seed was shown to give rise to two or three seedlings. Seeds sown without their seed coats when compared with intact seeds exhibited faster and higher rates of seed germination without affecting the final germination count of 100%. The jamun seeds at fresh harvest, especially when freshly picked from trees, have been reported to have high viability exhibiting more than 95% germinability (Mittal et al. 1999; Abbas et al. 2003; Radhamani et al. 2005), and high moisture contents of 60% (Abbas et al. 2003). Seeds reportedly harbor mycoflora with a prevalence of several fungi. Attack by pathogens was reported as a serious cause for loss of viability in seeds with high moisture contents (Mittal et al. 1999; Baxter et al. 2005). Use of fungicides for improved longevity was successfully undertaken with 0.1%

thiram (Radhamani et al. 2005) and sodium hypochlorite and benomyl (Mittal et al. 1999). Seeds have short storability, as a decline in germinability is observed after a few weeks of storage at ambient temperatures (Figure 11.2). Reduction was apparent by 8–20 weeks of storage at 16°C (Mittal et al. 1999) and by 15 days at a laboratory temperature of 23°C at 75.6% RH (de Araújo et al. 2008). Seeds also exhibit a chilling sensitivity to a temperature of 5°C, and at 0°C to –5°C, seeds reportedly lose their viability completely after 20 days of storage. Extension of viability by these methods has been reported up to six months, especially by seed storage at temperatures between 15°C and 20°C (Anandalakshmi et al. 2005), and four months at 12.3°C and 75.6% RH (de Araújo et al. 2008). Storage temperatures were found more critical for maintaining the viability than the seed moisture content. A high germination of 96.5% could be achieved for up to 165 days for seeds stored at 20°C with 20% seed moisture, and much longer for those with 10% moisture content. Seeds are desiccation sensitive, as after desiccation to 30% moisture, viability fell to 10% (Abbas et al. 2003); this is also proved by loss of viability after slow drying for 17 days over silica gel to 14%–16% seed moisture content (Baxter et al. 2005).

It is essential to select suitable containers for effective seed storage so as to prevent their direct contact with the storage environment and to protect them from pests and diseases. Polybags. when compared with cloth bags, paper bags, and plastic containers, have been found to be the best containers for prolonging seed viability and vigor at 20°C for more than nine months. These conditions ensured the availability of viable seeds for the next planting season.

Cryopreservation studies of embryonic axes of jamun were conducted at the Tissue Culture and Cryopreservation Unit at NBPGR, New Delhi, which revealed that longevity of embryonic axes could be enhanced up to 90 days using high-sucrose media (Figure 11.3, Table 11.3). Embryonic axes maintained in the dark for two months showed 80%–100% survival after transfer to normal growth conditions.

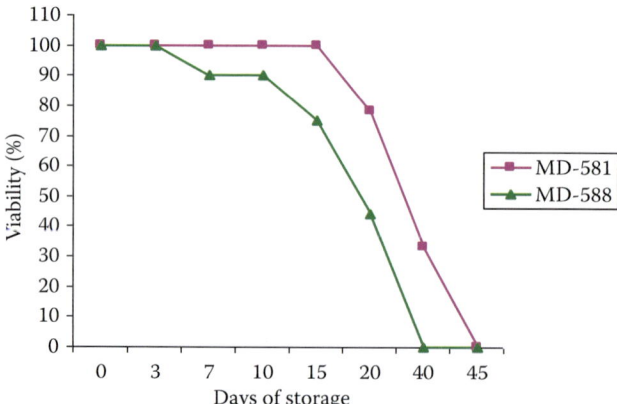

FIGURE 11.2 Seed longevity of two genotypes (MD-581 and MD-588) of Jamun at ambient temperature.

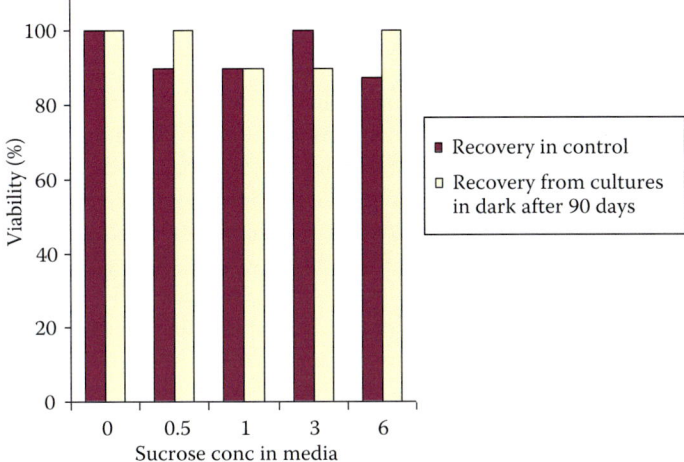

FIGURE 11.3 Longevity of embryonic axes of jamun in sucrose medium.

TABLE 11.3
Slow Growth Studies with Excised Embryonic Axes of Jamun

Culture Medium	Culture Conditions (Survival % of Embryonic Axes)	
	Normal Conditions (Cultured under Normal Light in Culture Room)	Slow-Growth Conditions (Cultured in the Dark in Culture Room)
0.5% sucrose	0	0
0.5% sucrose + After Cryopreservation	0	0
1.0% sucrose	80	90
1.0% sucrose + After Cryopreservation	0	0
3.0% sucrose	90	100
3.0% sucrose + After Cryopreservation	100	90
6.0% sucrose	100	90
6.0% sucrose + After Cryopreservation	0	0

UTILIZATION OF GERMPLASM AND CROP IMPROVEMENT IN *SYZYGIUM* SPECIES

Genetic resources of tropical underutilized fruits have not been much utilized for crop improvement purposes in India, as is the case with major fruits. Most of the attention has remained on farmers' selections, chance seedlings, and introduced germplasm rather than on breeding new types. The only report available is of Lancaster Percy and Bose (1965), who reported that an F1 hybrid of a cross between the Alba variety of water apple (*Syzygium javanicum*) and rose apple (*Syzygium jambos*) had prolific bearing, with larger fruits than those of the parents. The fruits contained the fragrance and sweetness of rose apple.

In such fruit species that are nutritionally very rich and mostly the fruits are directly harvested from the trees growing as wild or semiwild or in farmers' backyards, popularization and improvement in terms of higher productivity and quality of fruits in terms of percentage of edible portion, color, shape, taste, and flavor are important. However, crop improvement practices have totally been neglected in such crops due to less commercial value of fruits and also fruits being of only local importance. Due to the lack of advance propagation techniques and the absence of elite planting material, jamun is predominantly raised through seeds, leading to great variability, and therefore recognized cultivars are not available. Diverse local types are being grown in different places with great variation in size and shape of fruits, bearing habit, and quality of pulp; some of these have been given local names based on fruiting time, shape, size, and density of fruits available on the tree. There is a large genetic diversity present within and among the various populations of jamun throughout India. There are no standard varieties available in jamun; however, a common cultivar grown under north Indian conditions is "Ra Jamun." It produces large-sized (length 2.5–3.5 cm) and large-diameter (1.5–2.0 cm), oblong fruit, deep purple in color at the fully ripe stage, juicy, and sweet in taste. Fruit ripens in June and July and possesses small stone. Another late-maturing variety bears a small-sized (length 1.5–2.0 cm) and small-diameter (1–1.5 cm), slightly round fruit, deep purple or blackish in color at the fully ripe stage. The stone present in these cultivars is comparatively large in size. Fruits ripen in the month of August. The jamun is an indigenous minor fruit that offers a good scope for the selection of better varieties out of the innumerable seedling strains that exist in this country.

Considering the present status of jamun genetic resource conservation and evaluation in India, the following approaches are suggested for future crop improvement programs:

- Systematic germplasm management starting from the collection of available genetic variability, their thorough characterization and evaluation, followed by selection of elite genotypes
- Conservation of germplasm using complementary conservation methods *in situ* in protected areas and at farmers' fields and backyards, regional germplasm repositories, field genebanks, *in vitro* conservation, and cryo-genebanking, based on priority and need of the accessions
- Induction of genetic variability through combination breeding and mutation, followed by selection
- Selection of desired types from clonally propagated diverse populations
- Hybridization of selected genotypes with specific objectives

In view of the limitations of the conventional breeding techniques, biotechnological approaches may also be considered for jamun improvement. Several available strategies to fast-track the improvement work using plant biotechnological approaches of embryo rescue, somaclonal variation, haploidy, protoplast fusion, and recombinant DNA technology (genetic engineering) may be considered as per requirement. There is also a need to undertake detailed molecular characterization of germplasm using appropriate molecular markers to draw genetic maps of *S. cumini* and related species.

REFERENCES

Abbas, M., M. M. Khan, M. J. Iqbal, and B. Fatima. 2003. Studies on jamun (*Syzygium cumini* L. Skeels) seeds storage behaviour. *Pak. J. Agric. Sci.* 40: 164–169.

Anandalakshmi, R., V. Sivakumar, R. R. Warrier, R. Parimalam, S. N. Vijayachandran, and B. G. Singh. 2005. Seed storage studies in *Syzygium cumini*. *J. Trop. For. Sci.* 17: 566–573.

Anonymous. 2010. National Network Project on Underutilized Fruits: Final report. Lucknow: Central Institute for Subtropical Horticulture, 217.

Arora, R. K. 1995. Promoting conservation and use of tropical fruit species in Asia. In *Proceedings of Expert Consultation on Tropical Fruit Species of Asia*, ed. R. K. Arora and V. R. Rao. New Delhi: IPGRI Office for South Asia, 19–30.

Arora, R. K., and V. R. Rao, eds. 1995. *Proceedings of Expert Consultation on Tropical Fruit Species of Asia*, Kuala Lumpur, Malaysia, May 17–19, 1994, p. 116.

Ashraf, S. M. 1987. Studies on post-harvest technology of jamun fruits. PhD thesis, Department of Horticulture, NDUAT, Kumarganj, Faizabad.

Baxter, D., D. Erdey, P. Berjak, M. Sacande, D. Joker, M. E. Dullo, and K. A. Thomsen. 2005. Sensitivity of *Syzygium cumini* (L.) Skeels seeds to dessication. In *Comparative Storage Biology of Tropical Tree Seeds*, ed. M. Sacande, D. Joker, M. E. Dulloo, and K. A. Thompson. Rome: International Board for Plant Genetic Resources, 122–125.

Bhardwaj, R., and R. Yamdagni. 2005. Physico-chemical characteristics of jamun (*Syzygium cumini* Skeels). *Haryana J. Hortic. Sci.* 34: 54–55.

Chaturvedi, M. D. 1956. Jamun is another of our prized trees. *Indian Farming* 5: 17–19.

de Araújo, E. C., A. V. R. Mendonça, D. G. Barroso, and D. de A Ferreira. 2008. Desiccation and storage effect over physiological quality of seeds of *Syzygium jambolanum* Lam. *Rev. Cienc. Agron.* 39: 455–462.

Devi, S. P., M. Thangam, A. R. Desai, and P. G. Adsule. 2002. Studies on variability in physico-chemical characters of different jamun (*Syzygium cumini* (L.) Skeels) accessions from Goa. *Ind. J. Hortic.* 59: 153–156.

Ellis, R. H., T. D. Hong, and E. H. Roberts. 1990. An intermediate category of seed storage behaviour? I. Coffee. *J. Exp. Bot.* 41: 1167–1174.

Ghosh, D. K., S. Mitra, H. Pal, S. N. Ghosh, S. K. Mitra, B. C. Banik, M. A. Hasan, S. K. Sarkar, R. S. Dhua, J. Kabir, and J. K. Hore. 2006. Quality evaluation and storage behaviour of some local types of jamun (*Syzygium cumini* (L.) Skeels) fruits. In *Proceedings of the National Symposium on Production, Utilization and Export of Underutilized Fruits with Commercial Potentialities*, Kalyani, West Bengal, 266–270.

Hong, T. D., and R. H. Ellis. 1996. A protocol to determine seed storage behavior. In *IPGRI Technical Bulletin*, ed. J. M. M. Engels and J. Toll. No. 1. Rome: International Plant Genetic Resources Institute.

Inamdar, S., G. S. K. Swamy, P. B. Patil, and S. I. Athani. 2002. Correlation and path analysis studies in jamun (*Syzygium cumini* (L.) Skeels) for fruit characters. *J. Maharashtra Agric. Univ.* 27: 212–213.

Jai Prakash, A. N. Maurya, and S. P. Singh. 2010. Studies on variability in fruit characters of jamun. *Ind. J. Hortic.* 67: 63–66.

Keskar, B. G., A. R. Karale, B. C. Dhawale, and K. G. Chaudhary. 1989. Improvement of jamun (*Syzygium cumini* (L.) Skeels) by selection. *Maharastra J. Hortic.* 4: 117–120.

Khan, S., Vaishali, and V. Sharma. 2011. Inter and intra-population variability of *Syzygium cumini* using RAPD markers. *Prog. Agric.* 11: 17–22.

Kumar, R., J. Singh, H. P. Singh, S. N. Ghosh, S. K. Mitra, B. C. Banik et al. 2006. Release of new jamun variety—Rajendra Jamun 1. In *Proceedings of the National Symposium on Production, Utilization and Export of Underutilized Fruits with Commercial Potentialities*, Kalyani, 50–53.

Kundu, S., R. R. Chawdhury, S. N. Ghosh, A. Das, S. N. Ghosh, S. K. Mitra et al. 2006. Research on underutilized fruits in West Bengal. In *Proceedings of the National Symposium on Production, Utilization and Export of Underutilized Fruits with Commercial Potentialities*, Kalyani, 83–95.

Kundu, S., D. K. Ghosh, and S. C. Maiti. 2001. Evaluation of some local types of jamun (*Syzygium cumini* (L.) Skeels) of West Bengal. *Environ. Ecol.* 19: 872–874.

Lancaster Percy, S., and T. K. Bose. 1965. Studies on hybrids of *Syzygium cumini* Skeels. *Ind. J. Hortic.* 22: 87–88.

Malik, S. K., R. Chaudhury, O. P. Dhariwal, and D. C. Bhandari. 2010. *Genetic Resources of Tropical Underutilized Fruits in India*. New Delhi: NBPGR, 168.

Mittal, R. K., H. J. Hansen, and K. Thomsen. 1999. Effect of seed treatments and storage temperature on storability of *Syzygium cumini* seeds. In *Proceedings of IUFRO Seed Symposium 1998: "Recalcitrant Seeds,"* Kuala Lumpur, 53–63.

Orwa, C., A. Mutua, R. Kindt, R. Jamnadass, and S. Anthony. 2009. Agroforestry database: A tree reference and selection guide. Version 4.0. http://www.worldagroforestry.org /output/agroforestree-database.

Patel, V. B., S. N. Pandey, S. K. Singh, and B. Das. 2005. Variability in jamun (*Syzygium cumini* Skeels) accessions from Uttar Pradesh and Jharkhand. *Ind. J. Hortic.* 62: 244–247.

Patil, S. J., S. M. Mutnal, V. Maheswarappa, and G. Shahapurmath. 2009. Evaluation of jamun (*Syzygium cumini*) collections for productivity on black clay soils. *Karnataka J. Agric. Sci.* 22: 933–934.

Prabhuraj, H. S., G. S. K. Swamy, S. I. Athani, B. R. Patil, N. C. Hulamani, and P. B. Patil. 2002. Variability in morphological characteristics of jamun (*Syzygium cumini* Skeels) trees. *My For.* 38: 187–189.

Pritchard, H. W., J. F. Moat, B. S. Ferraz, T. R. Marks, J. L. C. Camargo, J. Nadarajan, and I. D. K. Ferraz. 2014. Innovative approaches to the preservation of forest trees. *For. Ecol. Manage.* 333: 88–98.

Radhamani, J., R. P. S. Dhaka, and A. K. Singh 2005. Post-collection care of jamun (*Syzygium cumini* L.) seeds for conservation. *New Bot.* 32: 237–242.

Rao, V. R. 1998. Complementary conservation strategy. In *Tropical Fruits in Asia: Diversity, Maintenance, Conservation and Use*, ed. R. K. Arora and V. Ramanatha Rao. Bangalore: Indian Institute of Horticultural Research, 142–151.

Roberts, E. H. 1973. Predicting the storage life of seeds. *Seed Sci. Technol.* 1: 499–514.

Roby T. J., J. Jose, and P. V. Nair. 2013. *Syzygium travancoricum* (Gamble)—A critically endangered and endemic tree from Kerala, India—Threats, conservation and prediction of potential areas; with special emphasis on Myristica swamps as a prime habitat. *Int. J. Sci. Environ. Technol.* 2: 1335–1352.

Sasidharan, N., and A. Jomy. 1999. A new species of *Syzygium* Gaertn. (Myrtaceae) from southern Western Ghats, India. *Rheedea* 9: 155–158.

Shakya, R., S. A. Siddiqui, N. Srivatava, and A. Bajpai. 2010. Molecular characterization of jamun (*Syzygium cumini* L. Skeels) genetic resources. *Int. J. Fruit Sci.* 10: 29–39.

Shareef, S. M., M. P. G. Kumary, E. S. S. Kumar, and T. Shaju. 2010. *Syzygium claviflorum* (Myrtaceae)—A new record for south India. *Rheedea* 20: 53–55.

Singh, A. K., A. Bajpai, A. Singh, A. Singhand, and B. M. C. Reddy. 2007. Evaluation of variability in jamun (*Syzygium cumini*) using morphological and physico-chemical characterization. *Ind. J. Agric. Sci.* 77: 845–848.

Singh, I. S., A. K. Srivastava, and V. Singh. 1999. Improvement of some under utilized fruits through selection. *J. Appl. Hortic.* 1: 34–37.

Singh, S., and A. K. Singh. 2005. Genetic variability in jamun (*Syzygium cumini* Skeels) in Gujarat. *Progressive Hortic.* 37: 44–48.

Singh, S., and A. K. Singh. 2006. Standardization of method and time of propagation in jamun (*S. cumini*) under semi-arid environment of western India. *Ind. J. Agric. Sci.* 76: 242–245.

Singh, S., and A. K. Singh. 2012. Studies on variability in jamun (*Syzygium cumini* Skeels) from Gujarat. *Asian J. Hortic.* 7: 186–189.

Singh, S. K., S. Krishnamurthy, and S. L. Katyal. 1967. *Fruit Culture in India*. New Delhi: ICAR, 255–259.

Sivasubramaniam, K., and K. Selvarani. 2012. Viability and vigor of jamun (*Syzygium cumini*) seeds. *Braz. J. Bot.* 35: 397–400.

Srivastava, V., P. Kumar, and P. N. Rai. 2012. Correlation study for physico-chemical characters of jamun (*Syzygium cumini* Skeels). *HortFlora Res. Spectrum* 1 (1): 83–85.

Srivastava, V., P. N. Rai, and P. Kumar. 2010. Studies on variability in physico-chemical characters of different accessions of jamun (*Syzygium cumini* Skeels). *Pantnagar J. Res.* 8: 139–142.

Vanangamudi, K., A. Venkatesh, B. Balaji, M. Vanangamudi, and R. S. V. Rai. 2000. Prediction of seed storability in neem (*Azadirachta indica*) and jamun (*Syzygium cumini*) through accelerated ageing test. *J. Trop. For. Sci.* 12: 270–275.

12 Horticultural Management of *Syzygium cumini*

S. K. Tewari, Devendra Singh, and R. C. Nainwal

CONTENTS

INTRODUCTION

Syzygium cumini, popularly known as jamun in India, is a highly nutritious fruit and an excellent tree of agroforestry and social forestry. In the wasteland development and dryland horticulture, jamun assumes great significance due to its multifarious uses and capacity to withstand adverse climatic conditions. The jamun, a medicinally important underutilized fruit, is adequately rich in antioxidants and phytochemicals, besides some essential nutritional components. The fruit is a good source of iron, sugars, minerals, protein, and carbohydrate. Fully ripe fruits are eaten fresh and can also be processed into many value-added products, like jelly, jam, squash, wine, and vinegar. The jamun fruit has subacid spicy flavor, and its squash is a very refreshing drink for quenching the thirst in the summer season. Seeds contain alkaloid jambosin and glycoside, jambolin, or antimellin, which reduce or stop the diastatic conversion of starch into sugars.

SOIL AND CLIMATE

The jamun is adapted to a wide range of soils. For high yield potential and good plant growth, deep loam and well-drained soils are needed. Such soils also retain sufficient moisture, which is beneficial for optimum growth and good fruiting. The jamun has been observed growing equally well on alluvial and lateritic soils. The west coast of India has lateritic soil, which is well drained and suitable for growing jamun. Likewise, medium black soils of peninsular India are also suitable for jamun growing. With little or no cultivation, it can flourish in poor soils. It tolerates sodic and saline soils and grows in the ravines and degraded lands and shallower water table conditions (Hebbara et al. 2002). Jamun trees may survive in alkali soils up to 10.5 pH (Singh et al. 1997). The soil having a pH of 7–8 and a water table below 2 m is most suitable for jamun.

Jamun tree is adapted to a wide range of ecological conditions, reflecting its wide geographical distribution in the subtropical and semiarid tropics. It can be grown in semiarid subtropical regions with an annual rainfall varying from 350 to 500 mm (Vashishtha 1991). The jamun requires dry weather at the time of flowering and fruit setting. In subtropical areas, early rain is considered to be beneficial for ripening of fruit and proper development of its size, color, and taste.

Gaur et al. (1998) carried out a study to evaluate the response of arbuscular mycorrhizal distribution by various tree species (*Terminalia arjuna*, *S. cumini*, and *Populus euphratica*, five years old), and the natural vegetation of *Typha elephantina* was found to be significantly different. Root length colonized by mycorrhizas was greatest in *P. euphratica* (56%), followed by *T. elephantina* (41%), *T. arjuna* (41%), and *S. cumini* (12.5%).

The tree occurs in the tropical and subtropical climates under a wide range of environmental conditions. Jamun can thrive on a variety of soils in low, wet areas and on higher, well-drained land (loam, marl, and sandy soils, and calcareous soils). It grows well in areas receiving heavy rainfall between 1500 and 10,000 mm per annum. It develops most luxuriantly in regions of heavy rainfall, as much as 400 in.

(1000 cm) annually. In India, it is usually found in areas receiving 900–5000 mm. The mean relative humidity in July varies from 70% to 100%, and in January from 40% to 90%. It can tolerate prolonged flooding. It also grows well on well-drained soils and, once established, can tolerate drought. The jamun tree grows well from sea level to 6000 ft (1800 m), but above 2000 ft (600 m), it does not fruit but can be grown for its timber. It prospers on riverbanks and has been known to withstand prolonged flooding. Yet, it is tolerant of drought after it has achieved some growth (Morton 1987).

REPRODUCTION

The new vegetative shoots in jamun emerge in two distinct flushes from February to May and from August to October (Mishra and Bajpai 1971). Flowering in jamun takes place in March and April. Flowers are bisexual. The inflorescence is terminal or lateral and develops mostly on current year growth, one-year-old shoots, and older branches. Before opening, the flower bud attains a size of 5.2 mm in length and 5 mm in diameter and requires 28–30 days from the appearance of the flower bud until the opening of flowers. The pollen grains are triangular in shape. The jamun is predominantly a cross-pollinated crop. The pollination is done by honeybees, houseflies, and wind. The maximum fruit set is obtained when pollination is done one day after anthesis, and thereafter, setting of fruit declines sharply (Mishra and Bajpai 1975).

The maximum anthesis (18.71%–43.08%) and dehiscence occur between 10 a.m. and 12 noon. The pollen fertility is higher in the beginning of the season. The maximum receptivity of stigma is observed one day after anthesis. The highest level of pollen germination in jamun was obtained in 20% sucrose solution (Singh 1978). Insects such as the honeybee and housefly are most active at noon. The housefly is active over a longer period than the honeybee, and its frequency of visits is also higher. Maximum insect activity during the day was between 11 a.m. and 3 p.m.

Bajpai et al. (2012) reported that the period of flowering ranged from the third week of February to the second week of March under subtropical conditions. The tree canopy had 5–13 panicles per shoot; the number of buds per panicle ranged from 19 to 73 in different directions of canopy. Panicle size had a clear relation with number of flowering buds. The total flowering phase culminating in a fruit set and later ripening lasted for 119–126 days, with a long phase of flower bud initiation, lasting for 45–50 days. Natural pollen transfer in the species was efficient, and fruit set following open pollination was quite high. In spite of the synchronous nature of anther dehiscence and stigmatic receptivity, selfing in a flower was found to coexist in nature with cross-pollination due to pollinator efficiency and pollen grain size. Selfing was promoted by the geitonogamous mode, and the species was suited to insect and wind pollination. It was found that reproductive phenology in jamun under Lucknow conditions was largely governed by seasonal climatic variables.

Under the semiarid ecosystem of western India, Singh et al. (2007a) reported that the peak period of panicle emergence was in February. The peak period of flowering and fruit set was recorded in the month of March in all genotypes.

VEGETATIVE PROPAGATION

Specialists in the fields of agriculture and horticulture take advantage of the regenerative ability of plants through such techniques as the rooting of cuttings; the grafting and budding of fruit trees; layering, or inducing the tips of branches to produce new plants; and the cutting apart of clusters of perennials, any plant that under natural conditions lives for several growing seasons, in contrast to an annual or a biennial. Perennial plants can be propagated in vegetative or generative ways. Fruit and nut trees are usually propagated by vegetative methods.

Rootstocks

Rootstock selection for vegetative propagation of jamun is important, as it controls the vigor and equilibrium between yield and quality. Dwarfing rootstocks induce dwarfness and facilitate easy management of the orchard. "Kala Jamun I" rootstocks are drought hardy.

Raising Rootstocks in Polythene Bags

Raising rootstocks in nursery beds and lifting budded plants with the earth ball in highly sandy soils are practically not feasible. Transportation of plants for a long distance may also cause high mortality, particularly under semiarid and arid environments. To reduce the time for raising rootstock and avoid damage during handling and transportation, polythene bags may be used on a commercial scale (Hiwale 2015). Polythene bags of sizes 25 × 10 cm, 25 × 15 cm, and 30 × 15 cm, as per the requirement, are used for raising the rootstocks in India. Small holes are made in the bottom and sides of bags for drainage and aeration, and they are filled with porous rooting medium or pot mixture for raising rootstocks. The seeds of jamun are sown in the polyethylene bags filled with soil and farmyard manure (FYM) (3:1) for raising the rootstocks. Seeds can also be sown in the raised nursery beds in open or polyhouses. Transplanting very young seedlings (two- to four-leaf stage) in the polythene bags gives 80% success. Sometimes coiling of the root becomes a problem; hence, root primer is also used for trimming of roots. Jamun seeds can also be sown in root trainers. Growth of the rootstocks can be improved with the application of 1 g/L urea solution. About 9- to 12-month-old seedlings of uniform size having a stem of pencil thickness are used as the rootstock for budding and grafting. Plants raised in the polythene bags can easily be transported to distant places and give higher transplanting success.

Rootstocks Raised in the Field

Sowing of seeds may be done directly in the field during monsoon at demarcated locations. This practice is followed under irrigated conditions, as proper irrigation is required to save the plants in the field during summer. The seedlings become ready for *in situ* budding during March and April, that is, eight to nine months after sowing. Since no transplanting is involved, in this case, the plants retain the deep taproots and thus become more hardy and vigorous after *in situ* grafting or budding.

Raising Rootstocks in Nursery Beds

Seed sowing is done at 30 × 30 cm spacing and 4–5 cm deep during June and July. The seedlings will become ready after six months for budding. Seedlings raised in nursery beds are handled carefully while lifting from the beds, packing, transporting, and transplanting in the field to obtain good survival. In this case, mortality is higher than for those raised in polythene bags because of root injury. The mortality of plant due to root damage may be reduced following the shifting of plants every two months, as the taproot does not penetrate too much in the soil. Seedlings raised in the seedbed are not preferred for planting under rain-fed conditions due to loss of their taproots while lifting from the nursery bed.

Patch Budding

Bud wood becomes available during the active growth period in spring or the rainy season. The buds stick, and well swollen and recently matured buds (but still not open) are collected. Immature and undeveloped buds from the upper part of the new shoots are not suitable. Similarly, overmature and inactive buds should not be used. The active growth period is indicated by easy and clear separation of the bark from the wood of scion sticks. After collection, the bud wood is often stored for some period or takes some time in transportation from one place to an other place for the budding purpose. During this period, considerable loss of survivability may take place. Bud woods retain good survival when kept under ventilated shade and wrapped in moist jute cloth. In this method, a healthy bud is selected from the axils of the leaf. The leaf blade is removed with the help of a sharp knife, leaving the petiole intact. The upper cut is given about 2.0 cm above the bud, which goes downward up to 1.0–1.5 cm below the bud without wood portion, and then a lower cut is given about 1.0 cm below the bud. The similar rectangle cut is made on the rootstock and the bud is placed at the juncture. The bud is pressed by hand to remove open space, if any, and tied tightly with a white polythene strip (200-gauge thickness and 2 cm wide). In case the cuts on the rootstock are wider, at least one side bark of scion and stock must be matched properly. The rootstock is de-topped just after budding, about 10 cm above the bud, to facilitate bud sprouting. After union, the top of the rootstock is cut a little above the bud union and polythene strips are carefully removed.

Studies revealed that the bud-take percent, days required for bud sprouting, bud sprouting percentage, linear scion growth, number of leaves, and final survival recorded maximum values under softwood grafting over the patch budding. With respect to time of propagation, March propagation showed better results with bud sprouting, scion diameter, and final survival of grafts and buddlings, while softwood grafting done during March recorded maximum vegetative growth with highest bud sprouting and final survival percentage of grafts and buddlings. With respect to age of rootstock, a seven-month-old seedling showed the maximum survival in patch budding, while a nine-month-old one showed the highest survival in softwood grafting (Bharad et al. 2006).

One-year-old, 10–12 mm thick rootstocks that budded in July and August showed better success (Singh et al. 1967). Patch budding was found to be successful when performed in June under Gujarat conditions (Chovatia and Singh 2000). Singh and

Singh (2006) conducted an experiment to standardize the suitable time to perform budding in jamun and reported that patch budding was successful in March and April under the semiarid environment of western India. In India, the patch, forkert, and shield methods of budding are generally employed.

In Situ Patch Budding

In jamun, the taproot system is very vigorous. The root system is therefore disturbed during the process of transplanting of grafts, which ultimately affects their growth and establishment adversely in field conditions. Therefore, *in situ* patch budding was tried at the Central Horticultural Experiment Station, Godhra. The plants propagated by *in situ* patch budding in March and April recorded good success (80.25% and 77.50%, respectively). In March and April, this may be practiced for multiplication of elite jamun genotypes, as well as establishment of orchards.

Use of Polycontainers for Budding

In jamun, the size of polythene standardized for optimum seedling vigor, root growth, and budding success at the Central Horticultural Experiment Station revealed that out of four types of polyethylene bags (25 × 15 cm, 25 × 10 cm, 30 × 8 cm, and 20 × 8 cm), the maximum number of buddable rootstocks (98.00%) and bud success (80.50%) were recorded in the 25 × 15 cm size bag, followed by 25 × 10 cm (94.50% and 75.10%, respectively). Looking into the transportation problem, the 25 × 10 cm bag is therefore recommended for use in the semiarid ecosystem of western India.

Softwood Grafting

About 15–20 cm long mature shoots (2–3 months old) are defoliated 12–15 days prior to the grafting operation. These shoots are detached from the mother plant with the help of a secateur or a sharp grafting knife for grafting by the cleft method. For this, a soft portion of seedling rootstock is cut at 20–35 cm height and the top portion is removed. With the help of a knife, a 5 cm long vertical downward incision is made in the center of the rootstock. A sharp cut of 5 cm is made on both sides on the base of the scion shoot to make a wedge shape. Thereafter, prepared scion is carefully inserted into the vertical slit of the rootstock and tightly secured with the help of 200-gauge-thick and 2 cm wide polythene strips. The polythene strips should carefully be removed after completion of the union. This method proved to be better for *in situ* grafting under rain-fed conditions. Madalageri et al. (1991) reported that softwood grafting was a successful method for multiplication of jamun under Karnataka conditions. Chovatia and Singh (2000) recorded 41.67% success in softwood grafting in June under Gujarat conditions. Softwood grafting was found successful in July and August in the semiarid environment of western India (Singh and Singh 2006; Singh et al. 2007b,c).

Mulla et al. (2011) carried out monthly softwood grafting of jamun from August to June under open and controlled conditions. The graft success under open conditions was (100%) in the months of November and May, whereas under controlled conditions, the maximum (100%) graft take was noticed in the months of October, November, and December. Gowda et al. (2009) reported that the softwood grafting during June recorded maximum grafting success on *S. cumini* rootstock (94%).

Studies on softwood grafting in jamun revealed that jamun can be very well grafted on its own and *Syzygium operculatum* rootstocks.

Thoke et al. (2011) reported that jamun seeds inoculated with microbial consortia or registered the highest germination (88.00%), graft take (49.68%), and graft survival (90.94%) compared with uninoculated control, wherein seed germination was 76.00%, graft success 41.42%, and graft survival 77.05%. In order to overcome the pest and disease problems around the basal part of the tree, clove (*Syzygium aromaticum*) was grafted onto seedlings of *S. cumini*. Grafting was performed at 30 cm using clove scions from semimature shoots (Mathew et al. 1999).

In Situ Softwood Grafting

Softwood *in situ* grafting also performs very well. The seeds are extracted from ripened fruits of jamun during May and June and sown in raised nursery beds at 2–3 cm depth and watered. Seedlings are shifted to 15 × 25 cm, 200-gauge-thick perforated polythene bags at the three- to five-leaf stage. The polythene bag is filled with mixtures of soil and FYM (3:1). After planting, seedlings are irrigated thoroughly and thereafter as per need. The seedlings are ready after one year for planting in the field after commencement of the monsoon. Seeds may also be sown directly in the field if enough irrigation water is available for summer irrigation.

Pits of maximum size should be prepared at the 10 × 10 m spacing during March and April and left open during summer for solarization, which kills most of the soil-dwelling insect pests and disease-causing agents. The pits are filled with a mixture of topsoil and 20–30 kg of well-rotten FYM. In order to ward off the seedlings from the attack of termites, the pit may be drenched with chlorpyrifos (3 ml/L water). The pit filling is raised up to 15–20 cm above the ground level, which settles down after rainfall or irrigation.

With the onset of the monsoon season in the middle of June, seedlings grown in polythene bags are planted in the pits with well-settled soil during the first week of July. The polythene bags are carefully removed and seedlings are planted in the center of the pit. After planting, the soil around the plants is firmly pressed to avoid formation of air pockets. Thereafter, the basin is prepared around the plant for irrigation or collection of rainwater. If there is no rain, the seedlings are thoroughly irrigated immediately after planting.

In situ softwood grafting through the cleft method is done in the months of July and August in the semiarid environment of western India. The growth below the graft union is removed regularly to encourage the sprouting and subsequent growth of the scion shoots. The bud sprouts within 14–21 days of grafting with proper bud-established vascular connection with the rootstock. The polythene strips are carefully removed after completion of the union. The plants are given support with the help of stakes to protect them from stormy winds. High temperature and relative humidity during July and August help in early sprouting and better graft success because of fast establishment of vascular connection with the rootstock. The quick and strong union formation and better nutrient uptake might cause higher plant growth and a greater number of leaves per plant in July. Based on this result, *in situ* softwood grafting may be practiced in July and August for the multiplication of jamun.

Veneer Grafting

Proper selection of scion is very important for success in this method. The scion stick should be three to four months old with some activated bud, either axillary or terminal. This is secured by defoliating the scion shoot about 10–12 days before the grafting operation. This method can be practiced on seedling stocks grown in polythene bags or nursery beds. About a 5 cm long slanting cut on one side of the stock stem is made, and the bark, along with wood, is removed, giving an oblique cut. Now, a slanting cut on one side of the scion is made that will just fit with the notch of the stock. The scion is then placed in position in such a way that the cambium layers of both the stock and scion come in close contact. It is then wrapped tightly with 1.5 cm wide tape of 200- to 300-gauge alkathene film, keeping the terminal ends free. When the scion begins to grow at the top after about 20–25 days, the upper part of the stock is removed, thus forcing the buds to grow more rapidly. The plastic wrap is removed after two to three months. Bose et al. (2001) recorded 31% success in veneer grafting when one-year-old seedlings were used as rootstock. The shoots are taken from spring flush and the operation is done in July.

Inarching

Jamun can be propagated by inarching, but it is not adopted commercially. In July, one-year-old rootstocks are inarched with the matching thickness scion. At the time of inarching, the seedling stock has the thickness of a pencil. The rootstocks are placed near six-month-old scion shoots of equal thickness, on the mother tree. At 25 cm above ground level, a 5–8 cm long slice of bark and wood is removed from the rootstock. This cut removes about one-third of the thickness of the stem and tapers gently towards the top and bottom. A corresponding cut is made on the scion shoot, so that the two cuts fit perfectly without leaving a chink. The two cuts are placed face-to-face and tied firmly with gunny string. July is the appropriate time for inarching in jamun. There should be no rain at the time of operation. Otherwise, water gets inside the union and causes rotting of tissues. The plants are regularly watered if there is no rain, until the union takes place in about six weeks. The scion shoot is given a knife cut through half its thickness below the union. After 10 days, if the scion shows no wilting, the cut is completed. After another two weeks, the top of the rootstock is cut off above the union. About 80% success is obtained with this method. The grafts are hardened in partial shade for another two months—before planting.

Cutting

There is no commercial practice of propagating jamun by cuttings. Bose et al. (2001) reported 45% rooting with 100 ppm indole-3-butyric acid (IBA) and 40% with 100 ppm indole-3-acetic acid (IAA) treatments. It was observed that tip cuttings of *Syzygium javanicum* rooted well under intermittent mist, and treatment with IBA at 5000 ppm produced 100% rooting.

Air Layering

Air layers are prepared during June, July, August, September, and October and treated with IBA at different concentrations (2500, 5000, and 10,000 ppm). The rooting behavior of air layers revealed that those prepared during the month of July recorded the maximum percentage of rooting (66.70), followed by June (60) and August (53.33). Air layers treated with 10,000 ppm IBA during September layering produced the maximum number of adventitious roots (2.24). However, root length (3.07 cm), root girth (0.94 mm), and number of tertiary roots (5.95) were highest in 10,000 ppm IBA-treated layers during the month of July (Gowda et al. 2009).

Transplanting of Budded and Grafted Plants

Budding is the insertion of the mature bud (scion) with a piece of bark underneath the bark of the stock plant. Buddings prepared in the nursery beds are lifted with large earth balls, 9–12 months after budding for transplanting in the field. A number of these plants may be lost due to damage while lifting, packaging, transporting to the field, and transplanting. These operations are also cumbersome and incur high costs due to the large size of the earth balls. Buddings prepared in polythene bags become ready for transplanting after 60 days of budding or grafting. The buddings in bags are removed from the nursery and kept in the shade for a week. These can then easily be transported and transplanted in the field with more than 90% survival. The roots of the plants do not coil, and therefore retain drought-hardy character and vigor, almost similar to the plants raised *in situ*. An experiment was conducted to see the effect of bud wood storage and grafting success during the month of July. The scion shoots wrapped in moist jute cloth may be used for grafting up to the third day from the date of detachment from the mother plant.

MICROPROPAGATION

There are several methods of culturing plant tissues, such as meristem culture, embryo culture, callus culture, protoplast culture, and cell culture. Yadav et al. (1990) induced multiple shoots from nodal and shoot tip segments of 10- to 15-day-old seedlings of *S. cumini* on a modified Murashige and Skoog (MS) medium supplemented with β-alanine (BA) singly and in combination with 1-naphthaleneacetic acid (NAA), IAA, or IBA. Excised shoots were placed for root induction on MS medium containing NAA or IBA and then transferred to MS basal medium to form complete plantlets. The regenerated plantlets were acclimatized and successfully transferred to the soil. Roy et al. (1996a,b) induced multiple shoots from nodal explants of 10-year-old elite trees and also from *in vitro* proliferated shoots of *S. cumini* on MS medium supplemented with 2.5 mg kinetin/L. Repeated subculture resulted in rapid shoot multiplication at an average of 10 shoots per subculture. Jain and Babbar (2000) obtained multiple shoots from the epicotyl segments bearing scaly leaves, excised from *in vitro*–grown seedlings of *S. cumini*, on MS medium supplemented with different concentrations of IBA. On average, 8.6 shoots per explant were produced in 60 days after inoculation, following transfer to fresh medium after 30 days. The shoots were excised and the residual explants were transferred to fresh medium, where they developed shoots again. Thus, a protocol was developed to raise plants of *S. cumini* at any time in the year. Somatic embryogenesis has also been found to be successful for multiplication of jamun plants.

For cloning of adult trees of *S. cumini*, explants were harvested from rejuvenated shoot sprouts produced by lopped trees. Multiple shoots regenerated by activation of axillary meristems of the explants on MS + 9.0 μM benzyladenine (BAP) + growth additives. Shoots were proliferated in culture by (1) repeated transfer of original explants and (2) subculture of shoots on amended medium with a combination of BAP (4.50 μM) + kinetin (2.25 μM). The cloned shoots were rooted (1) *in vitro* on half-strength MS + 0.1% activated charcoal supplemented with IBA (10 μM) or NAA (15.0 μM) and (2) *ex vitro* by pulse treatment with 2.50 mM IBA. The shoots root 100% under *ex vitro* conditions. The *ex vitro* rooting of cloned shoots was highly effective; it saved time and resources, and can be used for cloning of jamun (Rathore et al. 2004).

SEED PROPAGATION

Seed propagation is the most common method of propagating Jamun plants. Jamun seeds are recalcitrant in nature; hence, fresh seeds can be sown 4–5 cm deep in the nursery within 10–15 days. The seeds germinate 10–15 days after sowing (DAS). The seedlings become ready for transplanting in spring or the next monsoon. If the seeds are sown too deep, seedling emergence is delayed and there may be some rotting due to poor aeration. Seeds may also be sown in polythene bags, as this facilitates easy handling of rootstocks and grafted plants. There is occurrence of polyembryony in jamun to the extent of 20%–50%; hence, nuceller seedlings may be utilized to produce true-to-type plants. The seeds of jamun took 24–61 days for total germination under Bihar conditions (Singh and Thakur 1977). Studies revealed that seed extraction of jamun after heaping the fruits for a single day was better for getting good-quality seeds than extraction of seeds immediately after collection (Srimathi et al. 2003). Sasthri et al. (2001) recorded that large-sized seeds had a higher germination percentage (98%–99%) than small-sized seeds (79%–89%).

Hasan et al. (2006) studied the physiology of fruit growth and development of rose apple (*Syzygium jambos*) in seven-year-old plants. The fruit weight of rose apple at maturity was 21.98 g, while the fruit size was maximum (3.78 × 3.32 cm) at 65 days after fruit set. The seed size at harvesting was 2.46 × 2.42 cm. The increase of pulp thickness was maximum during the last phase of fruit growth. The specific gravity was maximum (1.09) at maturity. The total soluble solids were 8.2 at 65 days after fruit set. The total, reducing, and nonreducing sugars were 7.75%, 4.15%, and 3.6%, respectively, at maturity. The seed also showed an increase in sugar with the development of the fruit. The acidity of the whole fruit of rose apple varied between 0.70% and 0.86% during 10–30 days of fruit growth. The pulp acidity was estimated to be 0.53% at maturity, while the seeds contained 0.54% acidity. The ascorbic acid content of the pulp was 28.18 mg at harvest maturity, and that of seed was 19.48 mg.

The effects of arbuscular mycorrhizal fungi (AMF) (*Acaulospora laevis*, *Ambispora leptoticha*, *Gigaspora margarita*, *Glomus bagyaraji*, *G. fasciculatum*, *G. intraradices*, *G. leptotichum*, *G. monosporum*, *G. mosseae*, and *Sclerocystis dussii*) on the performance of S. cumini rootstock were studied (Gaur et al. 1998). Seeds were sown in soil inoculated with AMF. Seed germination was recorded daily from 65 DAS onward. Jamun responded positively to the soil inoculation of

AMF. *G. fasciculatum* and *G. intraradices* resulted in the greatest seed germination (89% each). Increases in plant height, stem diameter, and number of leaves were observed at 90 and 180 DAS in plants treated with *G. fasciculatum* (199.50 and 319.40 mm for stem height, 5.38 and 8.96 mm for stem diameter, and 19.75 and 27.67 for number of leaves, respectively). Rootstock vigor index was highest in rootstocks inoculated with *G. fasciculatum* (2842.45) and lowest in the control (1739.27). The number of spores at 180 DAS was highest in soil inoculated with *G. intraradices*, *G. fasciculatum*, *G. monosporum*, and *S. dussii* (889.75, 880.25, 855.75, and 849.25/50 g of soil, respectively). Root colonization at 180 DAS was greatest in plants inoculated with *G. fasciculatum* (94.0%), *S. dussii* (91.0%), *G. intraradices* (90.75%), and *G. monosporum* (90.0%). The results indicated that the most suitable AMF for jamun were *G. fasciculatum*, *S. dussii*, *G. intraradices*, and *G. monosporum* (Devachandra et al. 2008).

Katiyar et al. (2009) conducted the experiment to examine vesicular arbuscular mycorrhizal (VAM) association in about 20-year-old jamun (*S. cumini*), the tasar food plants, in the natural forest of Khapa, Bhandara (Maharashtra), India. Soil was laterite with acidic pH, ranging between 5.39 and 5.46. All the plants of jamun were found to be associated with mycorrhizal fungi. Root colonization percentage was higher in jamun (62.45%). The VAM fungal spore density was also more in jamun (175/20 g of dried soil). The root surface was found with aspersorium and running external hyphae, sometimes attached with spores. Different types of fungal spores belonging to the genera *Glomus*, *Acaulospora*, *Gigaspora*, and *Sclerocysts* were observed in the soil samples, of which the *Glomus* species was found to be dominant in both host plants (arjuna and jamun). The cortical region of the roots possessed arbuscules, vescules, or both in various intensities.

Seed viability, seed germination in response to various presowing treatments, and polyembryony were studied. Seeds stored at low temperature showed 100% viability for about four months. Acid (H_2SO_4) scarification for a 10 min duration resulted in better germination percentage in both diffused light and dark conditions (Stephens et al. 2012).

Various soaking, chemical, and growth regulator treatments were tested for increasing the germination and seedling growth and vigor of *S. cumini* collected as ripe fruits from a 15-year-old woodlot in Mettupalayam, Tamil Nadu, India. Germination was best enhanced by IBA treatment—soaking in 100 ppm IBA for 12 or 24 h increased germination to 65%–67% compared with a control value (dry seeds) of 46%. Seedling growth and vigor were also increased by the IBA treatment (12 h was the best period) and by 12 h soaking in 1% NaH_2PO_4 (Vanangamudi et al. 1999).

ESTABLISHMENT OF ORCHARD

The land is prepared by plowing, harrowing, and leveling. There should be a gentle slope to facilitate proper irrigation and prompt drainage to avoid the harmful effects of water stagnation during the rainy season. Jamun can be grown under various cropping systems, that is, as an orchard crop in a pure land or as an agroforestry species in mixed cropping systems. After marking the places for the plants, pits of 90 × 90 × 90 cm are usually dug out during the summer months. Digging pits (1–1.2 m depth) is very essential for somewhat heavy types of soil or soils with shallow hard pans. While digging, it is necessary to keep the topsoil and subsoil separate in two heaps near each pit in sun for about two weeks. This helps in exposing harmful soil organisms to weathering agencies, providing better aeration in the future rooting zone. Well-decomposed organic matter is mixed with soil and pits are filled. Planting is done during the rainy season when the soil in the pits has already settled. During planting, it should be ensured that the earth ball around the roots does not break and the graft union remains well above the ground level. The planting should preferably be done during cloudy days and preferably in the evening. The plants should be irrigated immediately after planting. In the initial two or three years, it is advisable to protect plants against low-temperature injury by covering them with some sort of cover, leaving the southeastern side open for the entrance of light. Jamun is planted at a distance of 10 × 10 m. It is normally planted along land borders, canal banks, field boundaries, village groves, or home gardens as scattered tree.

PLANTING SYSTEM

The main objective to follow a particular planting system is to accommodate the maximum number of trees per unit area without affecting the yield efficiency and fruit quality adversely. Some of the popular systems of planting are the square, rectangular, quincunx, hexagonal, contour, hedgerow, double hedgerow, paired, and cluster planting of these; the square system is the most popular. The hedgerow system of planting may be followed for high-density orchards of jamun. This aspect needs to be standardized under various agroclimatic conditions of the country.

Concept of High-Density Planting System

A high-density planting system ensures better utilization of the land, labor, and solar radiation and higher yields in the initial years of planting, because it accommodates more plants per unit area. High-density orcharding appears to be the most appropriate answer to the problem of low productivity and for early economic returns from the orchards. Considerations of the soil condition, planting geometry, and manipulation in the spacing are important means in obtaining higher production and productivity.

Rejuvenation of Old and Unproductive Orchards

Old, unproductive damaged trees or young plants can easily be rejuvenated by top working, employing scion wood or buds from the superior clone to raise productivity and improve fruit quality.

Top Working

The top of unproductive or inferior-quality trees of jamun can be improved by budding or grafting with the desired scion cultivar. The top of the old tree is headed back at 60–100 cm height from the ground. Several new shoots emerge from the trunk in 15–30 days. Of these, only four to five vigorous shoots are retained and budded with a suitable scion cultivar. The date of heading back is adjusted so that suitable new shoots become available for budding at the appropriate time.

AGROTECHNIQUES

Irrigation

Irrigation is not normally practiced in jamun cultivation, but it promotes better growth during establishment and the early stages of growth, especially during the dry seasons. At an early age, plants require 8–10 irrigations in a year, while bearing trees require 4–5 irrigations during the time of fruit development and ripening. In dry areas, the use of water-harvesting techniques during the rainy season ensures proper irrigation to improve subsequent growth and yield.

Use of mulch is very beneficial for successful cultivation of jamun. It reduces the loss of moisture from the soil, enhances the rate of penetration of rainwater or irrigation in the soil, and controls the growth of weed. Mulching can be done with black polythene or any suitable organic material. Mulching with grasses, paddy straw, and rice husk reduces the weed population and conserves the moisture in the soil. Organic mulching material improves soil pH. The microbial and earthworm population in the basin soil increases with the use of paddy straw and grasses as mulch. Leaf litter of jamun under the canopy is effective for retaining soil moisture during the summer.

Nutrient Management

An annual dose of about 20 kg of FYM during the prebearing period and 50–80 kg per tree at the bearing stage is considered beneficial (Bose et al. 2001). Lai and Lai (2000) reported the beneficial effect of jamun leaf litter in the soil. Sometimes in rich soils, the trees have a tendency to put on more vegetative growth, with the result that the fruiting is delayed. Under such conditions, the trees should not be manured and irrigation should also be given sparingly and withheld in September and October and again in February and March. This helps proper bud formation and blossoming and good fruit setting.

Integrated nutrient management refers to maintaining soil fertility and plant nutrient supply at an optimum level for sustaining the desired crop productivity through optimization of the benefits from all possible sources of plant nutrients in an integrated manner. Therefore, it is a holistic approach where we first know what exactly is required by the plant for an optimum level of production, in what different forms these nutrients should be applied in soil, at what different timings in the best possible method, and how best these forms should be integrated to obtain the highest productive efficiency in economically acceptable limits in an environmentally friendly manner. In soil application methods, fertilizer should be applied in the active root zone. However, with the advent of the drip irrigation system and availability of liquid fertilizers, the nutrient application technology through fertigation is considered the most efficient method. Increased fertilizer cost and the awareness of water pollution due to fertilizer runoff necessitated the use of biofertilizers for maintaining fertility more efficiently.

The diagnosis and recommendation integrated system (DRIS) represents a holistic approach to the mineral nutrition of a crop, and it has an impact on the integrated set of the norms, resulting in calibrations of the plant tissue, soil composition, environmental parameters, and farming practices as functions of the yield (Bhargava and

Singh 2001). The most important advantage of the DRIS approach is its ability to make a diagnosis at any stage of crop development and to list the nutrient elements in order of importance, which are responsible for limiting the yield. DRIS norms, however, need to be standardized in jamun.

Principle of Leaf Analysis

Leaf analysis is based on the plant behavior that is related to the concentrations of the essential minerals in leaf tissue. Leaf analysis, as a method for assessing the nutrient requirements of the crop, is based on the assumption that the certain limits, there is a positive relationship between the doses of the nutrient supplied, leaf content, and yield.

In general, 5 kg of FYM, 125 g of N, 50 g of P_2O_5, and 50 g of K_2O per plant per year may be applied to a one-year-old plant. This dose should be increased every year in the same proportion up to 10th year, after which the fixed dose should be applied each year. A full dose of FYM should be applied during July or August. A half dose of N, P, and K should be applied in July, and the remaining dose should be applied by the end of August or the first week of September under rain-fed conditions. Application of 6 kg of castor cake per plant per year, along with a standard dose of FYM and NPK after the fifth year of plantation, has been found to be very effective in improving the vegetative growth, yield, and fruit quality attributes. The earthworm and microbial population in the soil beneath the plant canopy has been found to increase with the use of FYM and cakes. The manure and mixture of fertilizer should be spread in the canopy of plants and incorporated into the soil.

Under rain-fed conditions, foliar feeding is very useful in supplementing nutrients, particularly nitrogen and micronutrients. The optimum concentration of urea for foliar application seems to be 1.00%, and one spray in April is effective in improving fruit retention, growth, and productivity. Supplementing micronutrients through foliar sprays of 0.2%–0.1% zinc sulfate has been observed to improve fruit quality attributes in terms of total soluble solids, total sugars, and vitamin C content and resulted in the development of a deep purple color in fruit.

Canopy Management

Canopy management of the crop deals with the development and maintenance of its structure in relation to the size and shape for maximum productivity and optimum fruit quality. The basic concept in canopy management of a perennial tree is to make the best use of the land and climatic factors for increased productivity in a three-dimensional approach. Tree vigor, light, temperature, and humidity play a vital role in the production and quality of the fruits. The crux of canopy management lies in how best to manipulate the tree vigor, sunlight, and temperature to improve the productivity and quality of the fruit and minimize the adverse effects of weather parameters. In other words, the basic principles in canopy management are maximum utilization of light, avoidance of the buildup of a microclimate congenial for diseases and pests, convenience in carrying out the cultural operations, and maximization of the productivity and quality.

Athani et al. (2009) examined the canopy structure in 10 jamun strains, of which 7 were found large and 3 medium with respect to height of the plant. The shape of the canopy was round in five strains, while five strains had an oval-shaped canopy. Five strains had a spreading habit and five had an erect branching habit. All 10 strains had a grayish-white bark color. The maximum tree girth was noticed in strain KLV-9 (94.4 cm). With respect to orientation, shape, and color of leaf, all 10 selected strains showed an intermediate, oblong-lanceolate, dark green leaf. The highest leaf length was noticed in strain DPD-25 (14.7 cm), while the highest leaf breadth was in strain DPD-20 (6.5 cm). The ratio of leaf length to breadth was maximum in strain GLH-85 (2.7). Strain KLV-9 had the highest leaf area (85.3 cm^2). The maximum petiole length was observed in strain GLH-58 (2.0 cm), while strain DPD-24 had a maximum ratio of leaf length to petiole length (8.7).

Training

Basically, training is a potential tool to manage the canopy architecture of the plant. Young plants should be allowed with three to five well-spaced branches to develop into the main scaffold structure of the tree. The framework of branches is allowed to develop above 60–100 cm from the ground level. Jamun plants may be trained through three training systems:

1. **Central leader:** The trees are trained on a main central leader with 6–10 strong lateral scaffold branches, spread in all directions. This system of training results in an unnecessarily large tree, which creates harvesting problems.
2. **Open center:** In this type, the central leader is removed about 1 m above ground level and four to six well-spaced tertiary branches are retained. Secondary scaffold branches are encouraged to develop the fruiting canes. Both primary and secondary scaffold branches produce fruit-bearing laterals. Training by this system keeps the center of the tree open, permitting entry of adequate sunlight throughout the tree, but due to development of a weak framework, plants do not resist high-velocity winds, resulting in the breaking of branches.
3. **Modified leader:** This is also known as delayed open center. In this system, the main central leader is allowed to grow for a few years, until 8–10 scaffolds develop around the central leader. The central leader is then cut to develop the side laterals, which in due course will grow as a modified leader. In this type of training, the tree develops well-spaced limbs with strong crotches. The top being open allows more sunlight to penetrate deep inside the tree canopy. This system is appropriate for training of jamun plants.

Pruning

Pruning is a tool to regulate tree size and shape to achieve a desired architecture of the canopy and also to reduce the foliage density by removing the unproductive branches to make the tree open. Regular pruning in jamun plant is not required. However, dry, weak, and diseased branches should regularly be removed. To maintain the dwarf framework of the jamun tree, topping of main stem (4–6 m) is needed. It will facilitate easy harvesting of the fruits. It is also observed that pruning of 50%

annual extension growth after harvesting is effective in reducing the plant canopy and improving the fruit quality attributes.

WEED MANAGEMENT

The productivity of jamun orchards can be increased by proper weed management. Weeds affect plant growth and yield adversely and very slowly in a subtle way. Most weeds, however, complete their life cycle in a shorter period, but compete with plants for light, water, and nutrients and thereby reduce the yield. In the orchards, hoeing, hand weeding, and plowing of the land two or three times a year are done to suppress weed growth. Intercropping and mulching also help in controlling weeds.

INTERCROPPING OR MULTISTORIED CROPPING AND MIXED CROPPING SYSTEMS

The practice of growing annuals or relatively short-duration crops in the interspaces during the formative years in jamun plantation is referred to as intercropping, and the growing of another perennial crop is called mixed cropping. The term *multistoried cropping* refers to a multispecies crop combination involving both annuals and perennials within jamun orchard. Intercropping is intended to maximize land and space use efficiency to generate supplemental income, particularly during the initial unproductive phase of the orchard, to protect the interspace from losses through weeds, erosion, impact of radiation, temperature, wind, and water, and enriching it by nitrogen-fixing legume crops. A compatible crop combination is necessary with regard to species, cultivars, planting method, and sequence. Peas, grams, lentils, black grams, cowpeas, cluster beans, cucurbitaceous crops, and capsicum may be grown as intercrops in the jamun orchard.

PLANT PROTECTION

Insect Pest Management

- **Leaf-eating caterpillar** (*Corea subtilis*): Caterpillar is a serious pest that defoliates the tender leaves of young growing plants. As a result, the tree loses vitality and the yield declines. To control this pest, spray dimethoate 30 EC 0.05% or use malathion 0.1%.
- **Bark-eating caterpillars** (*Inderbela tetraonis* and *Inderbela quadrinotata*): The larva feeds on the live bark tissues and shelters under the covering of silken webbing during night. Later, it makes a tunnel into the branch and stem and remains in the hole during daytime. As a result, the tree loses vitality and yield declines. It can be controlled by maintaining sanitation in the orchard and injecting petrol in the hole and plugging with mud. Foliar sprays with dimethoate (0.05%) or monocrotophos (0.05%), followed by endosulfan (0.07%) at a triweekly interval, controls the pest effectively.
- **Fruit borer** (*Meridarchis reprobata*): The caterpillars bore into the fruits and feed within, rendering the fruits unfit for consumption. It can be managed by spraying of dimethoate 0.05% 20 days before harvest.

- **Jamun leaf miner** (*Acrocercops syngramma* and *Acrocercops phaeospora*): This pest causes damage during April to September. The newly hatched caterpillar mines a narrow thread like silvery gallery on the leaf along the midrib upward or latterly. The larval mine is later transformed into a tubular blister-like swelling on the dorsal surface of the leaf. The pest can be controlled by clipping and burning of affected leaves, followed by spraying of 1.2 ml/L dimethoate 30 EC.
- **Jamun leaf roller** (*Polychorosis cellifera*): The larvae web the leaves by folding the tip downward on both margins parallel to the midrib and feed inside. In the case of a severe attack, one-fourth of the lamina is eaten up. The pest undergoes three to four generations between March–April and September–October in north India, and the second generation is most harmful (Lakra 1997). Regular clipping and burning of affected leaves can keep the population under control. In the case of a severe attack, spray chlorpyriphos 20 EC (2 ml/L) or endosulfan 35 EC (2 ml/L).
- **Leaf webbers** (*Argyroploce aprobola* and *Argyroploce mormopa*): The newly hatched larvae in large numbers web together the tender leaves at shoot tips and feed within (Lakra 1997). Regular clipping and burning of affected leaves can keep the population under control. In the case of a severe attack, spray chlorpyriphos 20 EC (2 ml/L) or endosulfan 35 EC (2 ml/L).
- **Fruit fly** (*Bactrocera correctus*): It damages the tree all over India. It can be controlled by maintaining sanitation in the orchard, by picking up the affected fruits and burying them deep in the soil, and by digging th area under the tree, so that the maggots in the affected fruits and the pupae hibernating in the soil may be destroyed.
- **White flies** (*Dialeurodes eugeniae* and *Singhiella bicolor*): The common white fly was found on jamun leaves all over India. In the case of young trees, it is desirable to clip off and burn all the infested leaves. In the case of severe infestation, spraying with dimethoate 0.03% is effective.
- **Termites:** The plant is damaged by termites. It can be controlled by spraying with chlorpyriphos 20 EC (2 ml/L), followed by soil drenching around the plant canopy.
- **Birds:** Birds also damage the jamun fruits. For keeping them away, beating of a drum or flinging stones has been found to be useful.

Disease Management

The fungus anthracnose (*Glomerella cingulata*) incites leaf spot and fruit rot. Affected leaves show small-scattered spots, light brown or reddish-brown in color. Affected fruits show small, water-soaked, circular and depressed lesions. Fruits rot and shrivel. This disease can be controlled by spraying with Dithane Z-78 at 0.2%.

Harvesting Practices and Yield

The fruit ripens in June and July. The ripe fruit at full size is deep purple or black in color. The jamun fruit should be picked immediately when it ripens. It cannot be retained on the tree at the ripe stage. The ripe fruits are picked singly by hand, and in all cases, care is taken to avoid all possible damage to fruits. For harvesting, the picker climbs on the tree with a cotton bag slung on the shoulder. When the bag is full, the picker comes down from the tree and empties the bag in a basket or hangs the bag down to the ground with the help of a rope, and another person standing below the tree catches the bag and empties it into baskets. The fruits of jamun are generally harvested daily and sent to the market on the same day. The average yield of fruits from a full-grown seedling jamun tree is about 80–100 kg, and from a grafted one is 60–70 kg per year. Athani et al. (2009) recorded the maximum yield in strain GLH-85 (260 kg/plant).

POSTHARVEST PRACTICES

Considerable variation exists in the quality of harvested fruits due to genetic, environmental, and agronomic factors. Grading is therefore required to get suitable returns for the produce. Proper grading, coupled with scientific packaging and storage, reduces the postharvest losses substantially, which enables the producer to fetch a competitive price. Jamun should be graded on the basis of fruit size, ripening stages, fruit uniformity, and cleanliness to fetch better prices in the market.

Packaging is a vital component of postharvest management to put the produce in convenient units and protect it from deterioration during handling and marketing. Jamun fruits are highly perishable and are normally packed in bamboo baskets for transport to local markets. Ideal packaging protects the fruits from physiological, pathological, and physical deterioration in the marketing channels and retains the freshness of fruits. Fruits prepacked in leaf cups covered with perforated polythene showed less loss in weight and shriveling and better appearance than conventional prepackaging in green leaf cups (Singh and Pathak 1988). Depending on the distance of transportation, jamun fruits are transported as head loads and in trucks and trains. Among different types of containers, shallow plastic crates are better during the transportation of fruits from the field to the market to prevent bruising losses.

An experiment conducted to study the effect of different packaging materials for the transportation and storage of jamun fruits—that is, bamboo basket without any lining, bamboo basket with jamun leaf lining, bamboo basket with polyethylene lining, bamboo basket with newspaper lining, corrugated fiberboard (CFB) box with polyethylene lining, CFB box with newspaper lining, wooden crates with polyethylene lining, and wooden crates with newspaper lining at ambient temperature conditions in a semiarid environment—revealed that the minimum spoilage loss was recorded in the fruits kept in the CFB box with newspaper liner, followed closely by CFB box with polythene liner (200 gauge, 2% ventilation). CFB box with newspaper liner also showed the lowest respiratory activity and exhibited three days of economic shelf life, while the control had two days of economic life under ambient conditions. CFB box provided an appropriate atmosphere and ventilation inside the box, and also proved strong to withstand heavy stack loads; further, it can be recycled as pulp or paper. CFB box with newspaper liner was found to be the most suitable and economically viable packing container during transportation of jamun fruits under ambient conditions and may be adopted for the benefits of both consumers and processors (Singh and Pathak 1988).

FUTURE RESEARCH NEEDS

There is considerable potential for the expansion of jamun cultivation in India. Therefore, some suggestions for future research priorities are given below:

1. The plant genetic resource (PGR) research needs to be undertaken on the classification of the genetic diversity through use of the morphological, biochemical, and molecular techniques. Efforts may be made in using the molecular techniques for understanding the genetic structure of the crop. Promising genotypes having tolerance to biotic and abiotic stress should be selected.
2. Model nurseries for the local supply of quality plant material should also be established.
3. Dwarfing rootstocks are considered to be best because they induce dwarfness and enable facilitation of easy management of the orchard. Such types of work may be initiated and standardized for different agroclimatic conditions.
4. Information should be made available on jamun-based cropping systems for different normal and problematic soils.
5. Agrotechniques like integrated nutrient management, diversified farming systems, high-density planting systems, weed management, canopy management, and irrigation management should be standardized under different ecosystems of the country.
6. Since jamun fruits are highly perishable in nature, varieties with prolonged shelf life should be developed.
7. Maturity, harvesting, grading, packaging, and a storage system should also be standardized.
8. There is a need to develop new products from jamun pulp and popularize them not only in the domestic market but also in the international market.
9. Research information on integrated pest and disease management should also be made available.

REFERENCES

Athani, S. I., Revanappa, and T. B. Allolli. 2009. Tree architectural characters and yield of *S. cumini* (jamun) strains. *J. Ecobiol.* 25: 293–296.

Bajpai, A., A. K. Singh, and H. Ravishankar. 2012. Reproductive phenology, flower biology and pollination in jamun (*S. cumini* L.). *Indian J. Hortic.* 69: 416–419.

Bharad, S. G., Lalan Rajput, V. S. Gonge, and S. R. Dalal. 2006. Studies on time and method of vegetative propagation in jamun. In *Proceedings of the National Symposium on Production, Utilization and Export of Underutilized Fruits with Commercial Potentialities*, Kalyani, Nadia, West Bengal, India, 96–99.

Bhargava, B. S., and R. Singh. 2001. Leaf nutrient norms in fruit crops. *Indian J. Hortic.* 58: 41–58.

Bose, T. K., S. K. Mitra, and D. Sanyal. 2001. *Fruits: Tropical and Subtropical.* Vol. 2. Calcutta: Naya Udyog.

Chovatia, R. S., and S. P. Singh. 2000. Effect of time on budding and grafting success in jamun (*S. cumini* Skeels). *Indian J. Hortic.* 57: 255–258.

Devachandra, N., C. P. Patil, P. B. Patil, G. S. K. Swamy, and M. P. Durgannavar. 2008. Screening of different arbuscular mycorrhizal fungi for raising jamun (*S. cumini*) rootstocks. *J. Mycorrhiza News* 20: 5–7.

Gaur, A., M. P. Sharma, A. Adholeya, and S. Chauhan. 1998. Variation in the spore density and percentage of root length of tree species colonised by arbuscular mycorrhizal fungi at a rehabilitated waterlogged site. *J. Trop. For. Sci.* 10: 542–551.

Gowda, V. N., V. Kumar, and P. V. N. Reddy. 2009. Studies on vegetative propagation in jamun (*S. cumini*). Presented at II International Symposium on Pomegranate and Minor Including Mediterranean Fruits (ISPMMF), Dharwad, India, June 23–27.

Hasan, M. A., A. Ghosh, R. R. Chowdhury, S. Sarkar, and P. K. Chattopadhyay. 2006. Studies on the physiology of fruit growth and development of rose apple (*S. jambos*). Presented at Proceedings of the National Symposium on Production, Utilization and Export of Underutilized Fruits with Commercial Potentialities, Bidhan Chandra Krishi Viswavidyalaya, Mohanpur, West Bengal, India, November 22–24.

Hebbara, M., M. V. Manjunatha, S. G. Patil, and D. R. Patil. 2002. Performance of fruit species in saline-water logged soils. *Karnataka J. Agric. Sci.* 15: 94–98.

Hiwale, S. 2015. *Sustainable Horticulture in Semiarid Dry Lands*. New Delhi: Springer.

Jain, J., and S. B. Babbar. 2000. Recurrent production of plants of black plum, *S. cumini* (L.) Skeels, a myrtaceous fruit tree, from in vitro cultured seedling explants. *Plant Cell* 19: 519–524.

Katiyar, R. S., N. R. Singhvi, R. V. Kushwaha, Lal Ramji, and N. Suryanarayana. 2009. VA-mycorrhizal association in arjuna and jamun trees in forest of Bhandara region, Maharashtra, India. *Int. J. Plant Sci.* 4: 229–232.

Lai, B., and B. Lai. 2000. Beneficial allelopathic effect of *S. cumini* litter fall biomass on arable crops. *Indian J. Agrofor.* 2: 53–57.

Lakra, R. K. 1997. Plant protection measurement of *S. cumini* (L.). *Haryana J. Horti. Sci.* 26: 35–48.

Madalageri, M. B., V. S. Patil, and U. G. Nalawadi. 1991. Propagation of jamun (*S. cumini* Skeels) by soft wood wedge grafting. *My For.* 27: 176–178.

Mathew, P. A., J. Rema, and B. Krishnamoorthy. 1999. Soft wood grafting in clove (*Syzygium aromaticum* (L.) Merr. & Perry) on related species. *J. Spices Arom. Crops* 8: 215.

Mishra, R. S., and P. N. Bajpai. 1971. Vegetative growth studies in the jamun. *Indian J. Hortic.* 28: 273–277.

Mishra, R. S., and P. N. Bajpai. 1975. Studies on the floral biology of jamun (*S. cumini* Skeels). *Indian J. Hortic.* 32: 15–24.

Morton, J. 1987. Jambolan. In *Fruits of Warm Climates*, 375–378. Miami: J. F. Morton.

Mulla, B. R., S. G. Angadi, J. C. Mathad, V. S. Patil, and U. V. Mummigatti. 2011. Studies on softwood grafting in jamun (*S. cumini* Skeels). *Karnataka J. Agric. Sci.* 24 (3): 366–368.

Rathore, V., N. S. Shekhawat, R. P. Singh, J. S. Rathore, and H. R. Dagla. 2004. Cloning of adult trees of jamun (*S. cumini*). *Indian J. Biotechnol.* 3: 241–245.

Roy, P. K., M. M. Rahman, and S. K. Roy. 1996a. Clonal propagation of *Syzygium cumini* L. through in vitro culture. *Bangladesh J. Bot.* 25: 159–164.

Roy, P. K., M. Rahman, and S. K. Roy. 1996b. Mass propagation of *Syzygium cumini* from selected elite trees. *Acta Hortic.* 429: 489–495.

Sasthri, G., P. Srimathi, and K. Malarkadi. 2001. Effect of seed size on seed quality in jamun (*S. cumini* Skeels). *Madras Agric. J.* 88: 524–526.

Singh, G., J. C. Dagar, and N. T. Singh. 1997. Growing fruits trees in highly alkaline soil—A case study. *Land Degrad. Dev.* 8: 257–268.

Singh, H. L. 1978. Studies on pollen germination in jamun (*S. cumini*). *Plant Sci.* 10: 151–152.

Singh, I. S., and R. K. Pathak. 1988. Packing of jamun, aonla and bael fruits: A souvenir on peeling of fruits and vegetables in India. Hyderabad: Agri-Horticultural Society Hyderabad, 108–111.

Singh, R. K., and S. Thakur. 1977. Seed germination and seedling growth of jamun (*S. cumini*) types. *Proc. Bihar Acad. Agric. Sci.* 25: 139–142.

Singh, S., H. K. Joshi, A. K. Singh, V. Lenin, B. G. Bagle, and D. G. Dhandar. 2007a. Reproductive biology of jamun (*S. cumini* Skeels) under semi-arid tropics of western India. *Hortic. J.* 20: 76–80.

Singh, S., and A. K. Singh. 2006. Standardization of method and time of propagation in jamun (*S. cumini*) under semi-arid environment of western India. *Indian J. Agric. Sci.* 76: 142–145.

Singh, S., A. K. Singh, and B. G. Bagle. 2007b. Propagating jamun successfully. *Indian Hortic.* 52: 31–33.

Singh, S., A. K. Singh, B. G. Bagle, and H. K. Joshi. 2007c. Scientific cultivation of jamun (*S. cumini* Skeels). *Agric. Update* 2: 37–40.

Singh, S. K., S. Krishnamurty, and S. L. Katyal. 1967. *Fruits Culture in India*. New Delhi: ICAR, 255–259.

Srimathi, P., T. Ramandane, K. Malarkodi, and K. Natrajan. 2003. Seed extraction in jamun (*S. cumini*). *Progressive Hortic.* 35: 221–223.

Stephens, A., C. D. Reddy, and P. J. Christopher. 2012. Effect of pre-sowing treatments on seed germination of *S. cumini* (L.) Skeels. *Adv. Plant Sci.* 25: 297–299.

Thoke, S., D. R. Patil, G. S. K. Swamy, and V. C. Kanamadi. 2011. Response of jamun (*S. cumini* Skeels) to *Glomus fasciculatum* and bioformulations for germination, graft take and graft survival. *Acta Hortic.* 890: 129–134.

Vanangamudi, K., A. Venkatesh, J. Jayaprakash, V. Mallika, R. Umarani, and R. S. V. Rai. 1999. Effect of pre-sowing treatments on germination and seedling growth in *S. cumini*. *J. Trop. For. Sci.* 11: 846–848.

Vashishtha, B. B. 1991. Fruit production and utilization in wastelands. *Indian J. Hortic.* 48: 247–251.

Yadav, U., M. Lal, and V. S. Jaiswal. 1990. *In vitro* micropropagation of the tropical fruit tree *Syzygium cumini* L. *Plant Cell Tiss. Org.* 21: 87–92.

13 Potential of *Syzygium cumini* for Biocontrol and Phytoremediation

S. K. Tewari, R. C. Nainwal, and Devendra Singh

CONTENTS

INTRODUCTION

Syzygium cumini is a multipurpose tree (MPT) known for its antibacterial, anti-inflammatory, and antioxidant activities. All its parts, that is, fruits, leaves, dried seeds, and bark, are used in Ayurvedic medicines for the treatment of several disorders. Owing to the presence of a wide array of phytochemicals, several biological activities have been ascribed to various plant parts of *S. cumini*. However, besides several medicinal applications, the seed, leaf, and bark have also been reported to have other activities, such as pesticidal activities against pathogens (bacteria and fungus) and insects, weed suppression due to allelopathic effects, and phytoremediation of stressful or degraded sites. Such activities, reported for various plant parts of *S. cumini*, are summarized in this chapter.

PESTICIDAL (BIOCONTROL) ACTIVITY

Various workers have identified the pesticidal activity of *S. cumini* against several pathogens and insects. Extract from bark and leaves of *S. cumini* decreased local lesion production by turnip mosaic virus in *Chenopodium amaranticolor* within four hours of application in preinoculation treatments (Pandey and Mohan 1986). The methanolic extract of *S. cumini* showed maximum inhibitory effect against *Xanthomonas campestris* (Uma et al. 2012). Sunder et al. (2005) examined the effect of a crude extract of fruit of *S. cumini* against bacterial blight pathogen (*Xanthomonas oryzae* pv. *oryzae*) of rice in field trials in Haryana, India, during kharif and found that the extract decreased both the severity and incidence of the disease significantly. It showed antifungal activity against *Trichophyton tonsurans, T. rubrum, T. simii, T. beigelii, Microsporum fulvum*, and *M. gypseum*. Gupta and Bhadauria (2012) studied the antifungal potential of aqueous extracts of leaves, bark, seeds, and fruits of *S. cumini* against two important fungal plant pathogens, *Alternaria alternata* and *Fusarium oxysporum*. Results revealed that among various plant parts, aqueous extract of fruit was most effective against the growth of *F. oxysporum* compared with other extracts, whereas the aqueous extract of bark showed potential to inhibit the growth of *A. alternata*. The extract of leaves was not found to be very effective against both test organisms. Arshad and Samad (2012) studied the antifungal activity of methanolic extracts of leaves of *S. cumini* against two strains of *A. alternata*, isolated from dying-back trees. The methanolic extract significantly reduced the fungal biomass. There were reductions in the range of 82%–88% in the biomass of *A. alternata* strains due to different concentrations of the leaf extracts of *S. cumini*. The study concluded that aqueous and n-butanol fractions of methanolic leaf extract of *S. cumini* could be used as biofungicides for the management of *A. alternata*.

The antibacterial properties of extracts of stem bark and leaves of *Syzygium* spp. (*S. cumini, S. andamanicum*, and *S. samarangense*—collected from the Andaman Islands) were investigated against 10 bacterial strains (Chattopadhyay et al. 1998), and *S. cumini* was found to be the most potent antibacterial plant. Shafi et al. (2002) extracted the essential oil from the leaves of *S. cumini* and examined its antibacterial activity against *Bacillus sphaericus, Bacillus subtilis, Staphylococcus aureus, Escherichia coli, Pseudomonas aeruginosa*, and *Salmonella typhimurium*. The essential oils showed considerable antibacterial activity, especially against *S. typhimurium*. De Oliveira et al. (2007) also evaluated the antimicrobial activity of *S. cumini* leaf extract and found it active against *Candida krusei* (inhibition zone of 14.7 ± 0.3 mm and minimum inhibitory concentration [MIC] = 70 μg/ml), and against multiresistant strains of *P. aeruginosa, Klebsiella pneumoniae*, and *S. aureus*. Riaz et al. (2010) conducted an experiment to investigate the potential of *S. cumini* leaves against the corm rot disease of gladiolus (*Gladiolus grandiflorus* sect. Blandus cv. Aarti) caused by *F. oxysporum* f. sp. *gladioli* (Massey) Snyd. & Hans. All the leaf incorporation and spreading treatments significantly reduced the disease incidence and number of infection lesions on corms.

Elansary et al. (2012) assessed antibacterial and antioxidant activities of leaf essential oil of *S. cumini* by using MICs and 2,2-diphenyl-1-picrylhydrazyl (DPPH) methods, respectively. The main oil constitutes were α-pinene (17.53%), α-terpineol (16.67%), and alloocimene (13.55%). The oil demonstrated strong inhibition activity against the tested bacterial strains and the total antioxidant activity (TAA) was 11.13%. Thus, the work revealed that the studied leaf oils are new promising potential sources of antioxidants and antibacterial compounds and good future practical applications for human and plant health. Farrukh and Ahmad (2003) also reported broad-spectrum antimicrobial activity (both antibacterial and antifungal) in crude extracts of seeds of *S. cumini*.

Lei et al. (2012) extracted active material from *Syzygium aromaticum* by ultrasound-assisted extraction (UAE). *F. oxysporum* f. sp. *cucumerinum* and *F. oxysporum* f. sp. *niveum* were inhibited by the plant extract, and the inhibition rates were measured by the growth rate method. The results showed that mycelial growth was significantly inhibited by the extract, and the antibacterial effect was increased significantly when the extract concentration was increased. Olia and Goswami (2000) optimized the *S. cumini* leaf powder alone and with a combination of crushed rice grains, neem oil cake, and cotton leaf powder to determine the most suitable medium for the fungal bioagent, *Aspergillus niger* culture *in vitro* for the growth and sporulation. Flasks containing the fungal colony and the substrates, singly (40 g each) or in combination (2:1:1), were incubated for 15 days. *A. niger* grew well on all media. Tessmann and Dianese (2002) extracted leaves of *Syzygium jambos* and sequentially fractionated them using a silica gel flash column. A bioassay based on the numbers of urediniospores of *Puccinia psidii* that germinated in 2% water agar detected an active stimulant of germination when the fraction eluted with 100% n-hexane was used. The active fraction induced up to an 88% increase in germination when added to a spore suspension in mineral oil. The active fraction was characterized as a hydrocarbon by ^{1}H nuclear magnetic resonance, ^{13}C nuclear magnetic resonance, and infrared analysis. Gas chromatography–mass spectrometry analysis indicated that the fraction was a long-chain 436 MW hydrocarbon corresponding to $C_{31}H_{64}$, namely, hentriacontane. This was the first-ever report of such a compound proved to be involved in stimulation of fungal spore germination. These results may contribute to a better understanding on the infection process of rusts.

Water extract of bark of *S. cumini* caused 100% mortality in golden snails at 200 ppm, 24 hours after treatment (Morallo Rejesus and Punzalan 1997). A petroleum ether and ethyl acetate, 1:1 fraction of *S. cumini*, inflicted considerable larval mortality on different instars of adult metamorphosis. At very low concentrations, the active fractions of this plant extract extended the duration of various instars and populations. The aqueous extracts of the leaves and stem barks of *S. cumini* markedly inhibit the growth of *Staphylococcus epidermidis*. Claver and Jesudasan (2007) reported the presence of phenolic compounds in leaves of *S. cumini* that exhibited a chemical defense against herbivorous insects. They analyzed the chemical components present in the honeydew of nymphs of the whiteflies, that is, *Rhachisphora ixorae* and *Sphericaleyrodes indica*, and the leaves of host plants of whiteflies, that is, *Ixora pavetta*, *S. cumini*, and *Ochlandra travancorica*. Among the secondary metabolites in infested and noninfested host leaves, steroids, alkaloids, and catachins were absent, while phenolic compounds were present in whitefly-infested leaves of *O. travancorica* and *S. cumini*, which exhibited a chemical defense against herbivorous insects. Tannins, saponins, anthroquinones, and coumarin were found in whitefly-infested leaves, and the above compounds deterred insects from causing feeding injury.

Chandrasekaran and Venkatesalu (2004) examined the water and methanolic extracts of *S. cumini* seeds for antibacterial and antifungal activity *in vitro* using the disc diffusion method, MIC, minimum bactericidal concentration, and minimum fungicidal concentration. They discussed the activity against gram-positive bacteria (*B. subtilis* and *S. aureus*), gram-negative bacteria (*S. typhimurium*, *P. aeruginosa*, *K. pneumoniae*, and *E. coli*), and fungal strains (*Candida albicans*, *Cryptococcus neoformans*, *Aspergillus flavus*, *Aspergillus fumigatus*, *A. niger*, *Rhizopus* sp., *T. rubrum*, *Trichophyton mentagrophytes*, and *M. gypseum*). The study revealed that the seed extract of *S. cumini* was very effective against *C. albicans*, *A. flavus*, *A. fumigatus*, *A. niger*, *B. subtilis*, and *S. aureus* compared with the other strain tested.

Bhutta et al. (2001) reported that seed diffusate of *S. cumini* used for the treatment of sunflower seeds reduced the populations of seed-borne fungi: *A. alternata*, *Drechslera tetramera* (*Cochliobolus spicifer*), *Emericellopsis terricola*, *Fusarium moniliforme* (*Gibberella fujikuroi*), *Fusarium semitectum* (*Fusarium pallidoroseum*), *Macrophomina phaseolina*, and *Phoma oleracea* (*Leptosphaeria maculans*). The results indicated that seed diffusates could substitute for costly chemicals for safe control of seed-borne diseases, at the same time protecting the environment from chemical pollution.

WEED CONTROL

Weeds are extensively present in agricultural systems, competing with crop plants for space, nutrients, water, and so forth, and reducing the crop yield. Weeds are usually controlled by using mechanical methods and synthetic herbicides as well. The mechanical methods are labor-intensive and time-consuming, while application of chemical herbicides not only creates perceived hazardous impacts on agricultural products, but also enhances environmental pollution (Batish et al. 2007). Additionally, the risk of weed resistance development and a high cost–benefit ratio are other disadvantages of synthetic herbicide and pesticide usage (Kordali et al. 2009). Therefore, in recent years new approaches, such as plant allelopathic effects, have been considered to suppress weeds in agricultural systems (Dayan et al. 2009). The allelopathic potential of *S. cumini* and its effectiveness for weed control have been studied under both laboratory and field conditions.

Aqueous extracts of *S. cumini* dry leaves can be used to control the weeds in arable lands. Arshad et al. (2006b) examined the effect of 2%, 4%, and 8% (w/v) aqueous extracts of dry leaves of *S. cumini* against germination and seedling growth of one of the most serious weeds of wheat, that is, *Phalaris minor* Retz. The aqueous extract of *S. cumini* at its highest concentration of 8% exhibited a significant negative impact against the germination of *P. minor*. Aqueous extract bioassays were also conducted to evaluate the allelopathic potential of *S. cumini*, for its use to control *Parthenium hysterophorus*, one of the world's worst weeds (Shafique et al. 2005; Arshad et al. 2006a). Aqueous extracts of 2%, 4%, 6%, 8%, and 10% (w/v), obtained from dry leaves, were bioassayed on *P. hysterophorus* seeds. Toxicity of the aqueous extracts was assessed by recording their effect on germination, radicle and plumule length, and seedling biomass of the test weed species. Aqueous extracts of 8% and

10% concentrations invariably and significantly ($p \le 0.05$) suppressed germination of *P. hysterophorus* seeds.

PHYTOREMEDIATION

Plants have various adaptive mechanisms to strive and survive in stressful environments, such as high salinity, extreme heat, drought, and freezing temperatures. Modern environmental biotechnology researches are now focusing on such adaptive traits in plants and modifying these traits for developing ecofriendly and sustainable technologies to combat environmental pollution, ecosystem degradation, climate change, and other problems. Phytoremediators are plants that are used for cleaning up soil in contaminated areas. They not only function as salt tolerant, but also can reduce some of the negative effects of soil salinity and sodicity by working as ion accumulators or excretors, and tend to promote soil permeability. Combined with accurate water management strategies, they can also remove heavy metals, arsenic, lead, aluminum (Al), and many other toxic elements from the soil. A study revealed that high population density of monoculture plantations may increase the C and N contents, up to six times in surface soil (0.15 m) in eight-year-old plantations (Garg 1998; Garg and Jain 1992). The soil properties are largely influenced by the dynamics of litter and fine roots in forest ecosystems, and both fluxes are equally important. Litter performs a major role in soil fertility, and fine roots contribute substantially in improving soil structure, pH, resource acquisition, and water permeability (Singh 1996). Phytoremediation, often referred to as bioremediation, botanical bioremediation, or green remediation, is the use of plants to make contaminants nontoxic. Phytoremediation includes rhizofiltration (absorption, concentration, and precipitation of heavy metals by plant roots), phytoextraction (extraction and accumulation of contaminants in harvestable plant tissue such as roots and shoots), and phytostabilization (absorption and precipitation of contaminants by plants) (Miller 1996). Phytoremediation is characterized by the use of vegetative species for *in situ* treatment of land areas polluted by a variety of hazardous substances (Sykes et al. 1999). The ideal type of phytoremediator is a species that creates a large biomass, grows quickly, has an extensive root system, and must be easily cultivated and harvested (Clemens et al. 2002).

The success of initial tree seedling establishment is related to the capture and use of primary resources such as light and nutrients. The selection of tree species with a greater potential to assimilate carbon and a capacity to efficiently utilize nutrients and light would facilitate the revegetation of degraded areas, primarily where irradiance is high and soil nutrient availability is low. Dos Santos et al. (2006) analyzed soil physical and chemical characteristics, survival, growth, photosynthesis, chlorophyll *a* fluorescence, leaf macro- and micronutrient content, and photosynthetic nutrient use efficiency in young tropical tree species planted in degraded areas in central Amazonia (Amazonas, Brazil). The predominant soil in the study site was an alic red-yellow podzolic (ultisol). The species studied were *Bellucia grossularioides*, *Bombacopsis macrocalyx*, *Cecropia ficifolia*, *C. sciadophylla*, *Chrysophyllum sanguinolentum*, *S. cumini*, *Inga edulis*, and *Iryanthera macrophylla*. Photosynthesis varied between 34 and 264 nmol/g/s for the eight species. Leaf macronutrient concentrations varied from 16 to 29, 0.4 to 1.0, 6 to 13, 7 to 22, and 1.6 to 3.4 g/kg for N, P, K, Ca, and Mg, respectively. Trees on these degraded soils are primarily limited by P or micronutrients. Despite the removal of the O horizon, N does not appear to limit photosynthetic activity. From this, it was concluded that *S. cumini* possesses ecophysiological mechanisms associated with carbon assimilation and nutrient use that determine success in early establishment and has the potential to recuperate degraded areas. Thus, *S. cumini* can be used with success in forest plantings to recuperate degraded areas as this species is found (1) efficient in the utilization of excess energy for photosynthesis, (2) efficient in the use of limited soil nutrients, and (3) with high survival and growth rates.

Microbial characterization of the tree rhizosphere provides important information relating to the screening of tree species for revegetation of degraded land. Sinha et al. (2009) collected some rhizosphere soil samples from a few predominant tree species growing in the coal mining ecosystem of Dhanbad, India, and analyzed these samples for soil organic carbon (SOC), mineralizable N, microbial biomass carbon (MBC), active microbial biomass carbon (AMBC), basal soil respiration (BSR), and soil enzyme activities (dehydrogenase, urease, catalase, phenol oxidase, and peroxidase). Among the tree species studied, *S. cumini* recorded the highest value for MBC/SOC (8.03%). Principal component analysis was employed to derive a rhizosphere soil microbial index (RSMI), and accordingly, dehydrogenase, BSR/MBC, MBC/SOC, electrical conductivity (EC), phenol oxidase, and AMBC were found to be the most critical properties. The observed values for the above properties were converted into unitless scores (0–1.00), and the scores were integrated into RSMI. The tree species could be arranged in decreasing order of RSMI as *Aegle marmelos* (0.718), *Azadirachta indica* (0.715), *Bauhinia* sp. (0.693), *Butea monosperma* (0.611), *S. cumini* (*Eugenia jambolana* [0.601]), *Moringa oleifera* (0.565), *Dalbergia sissoo* (0.498), *Tamarindus indica* (0.488), *Morus alba* (0.415), *Ficus religiosa* (0.291), *Eucalyptus* sp. (0.232), and *Tectona grandis* (0.181). It was concluded that tree species in coal mining areas had diverse effects on their respective rhizosphere microbial processes, which could directly or indirectly determine the survival and performance of the planted tree species in degraded coal mining areas. Tree species with higher RSMI values could be recommended for revegetation of degraded coal mining areas.

PHYTOREMEDIATION OF SALT-AFFECTED SOILS

Salt-affected soils are widespread on many continents of the world, constituting about 831 million ha (Martinez and Manzur 2005). There are two major types of salt-affected soils: saline and alkali soils. In general, soil sodicity is a more widespread form of land degradation, occupying >2 billion ha throughout the world (Grainger 1988). Reclamation and revegetation of degraded lands under productive ecosystems are currently global priority issues. Proven afforestation techniques have been employed to control land degradation, as, for example, enhancing the forest cover vis-à-vis biodiversity conservation and pollution abatement. Restoration and rehabilitation of degraded wastelands has attracted worldwide attention in view of the shrinking arable land, particularly in developing countries (Barrett-Lennard et al. 1986; Brown and Lugo 1994; Garg 1999). There is fragmentary information available on the restoration of the productivity and fertility of degraded sodic soils through tree plantation (Bhojvaid et al. 1996). However, the development of mixed forest tree species on sodic wastelands, apart from providing diverse needs and services to society, ameliorates the soil to various degrees. This appears as a function of species diversity, productivity, and decomposition process (Garg 1998). The phytoremediation of sodic soils is achieved by the ability of plant roots to increase the dissolution rate of calcite, thereby resulting in enhanced levels of Ca^{2+} in soil solution to effectively replace Na^+ from the cation exchange complex. Holmes (2001) conducted extensive laboratory and field investigations of the ecology of plants in extreme environments in an effort to select plants that are suitable for phytoremediation in saline sites and reported that the content of sodium in the soil was decreased by 65% two years after planting with salt-accumulating plants. While salt-tolerant plants species can effectively phytoremediate a saline–sodic system by interacting with salts in the soil–water environment and reducing them through absorption, the physical characteristics of rooting can also increase soil permeability and result in leaching of salts beyond the root zone. Root decomposition frees channels for water movement, thereby increasing hydraulic conductivity of the soil.

Dagar and Singh (2000) evaluated 11 fruit trees—*Achras zapota (Manilkara zapota), Carissa carandas, Citrus reticulata, Ficus carica, Mangifera indica, Prunus domestica, Prunus amygdalus (Prunus dulcis), Psidium guajava, Punica granatum, Sapindus mukorossi,* and *Syzygium cumini* for sodicity tolerance. *Achras zapota, C. carandas, S. cumini, P. guajava,* and *P. granatum* were more tolerant to sodicity than the rest studied, as they accumulate more potassium ions than the others. Dagar et al. (2000) also conducted long-term experiments to find suitable forest and fruit tree species for highly alkaline soil (pH > 10). Thirty forest tree species, 15 strains of *Prosopis,* and 10 fruit tree species were planted in a semiarid region of Haryana in India. Among the fruit tree species, *Ziziphus mauritiana, S. cumini, P. guajava, Emblica officinalis (Phyllanthus emblica),* and *C. carandas* were successful for the highly alkaline soils, showing good growth and fruit setting initiation.

Singh et al. (2004) and Singh and Garg (2007) investigated tree species diversity and dominance and the associated changes in soil characteristics in a man-made forest established on formerly barren sodic land at Banthra Research Station (BRS), an experimental field station of CSIR-NBRI, Lucknow, over three decades. The results revealed that the forest scored a moderate value for the tree species diversity index (*H*). The tree species *Derris indica, Dalbergia sissoo, Azadirachta indica, Cassia siamea (Senna siamea), Terminalia arjuna, S. cumini,* and *Tectona grandis* were found to be the major dominant species that may be considered suitable for planting on such degraded wastelands. There was a perceptible reduction in soil pH and ESP and an increase in organic C and $Ca^{2+} + Mg^{2+}$ cation contents over the three decades, indicating that the sodicity has declined in the surface soil. This experience can be tried on similar sites of arid and semiarid regions of the world for the bioreclamation of sodic lands. A decrease in soil pH, with respect to secondary succession on any rehabilitated site, is a common phenomenon due to the leaching and accumulation of weak organic acids as vegetation development proceeds (Crawley 1986).

Establishing salt-tolerant tree plantations, utilizing the saline groundwaters, may also provide for an economic use of abandoned arid lands, but issues related to long-term sustainability of such plantations are unknown. Thus, a field trial with 31 tree species was conducted over nine years (1991–2000) on a calcareous soil in a semiarid part (annual rainfall about 350 mm) of northwest India (Hisar, Haryana) (Tomar et al. 2003). Tree saplings were planted at the sill of furrows and irrigated with saline water (EC 8.5–10.0 dS/m) for an initial three years (four to six times per year), and thereafter plantations were irrigated once during the winter only. Measurements were made on tree growth, water use, and biomass production. *Syzygium cumini*, *Crescentia alata*, *Samanea saman*, and *T. arjuna* showed satisfactory early growth and survival when they were supplied with supplemental saline irrigation, but proved to be sensitive after the cessation of irrigation, thus emphasizing the need for long-term evaluation trials. *Cassia javanica* and *Cassia alata* were observed to be very sensitive to frost, whereas *Casuarina equisetifolia* could not survive drought due to the prevailing arid conditions at the site. Salt storage in the soil profile increased substantially during the irrigation period (5.6–10.4 dS/m), but the added salts were distributed in the soil profile as a consequence of the seasonal concentration of rainfall during monsoons and some episodic events of rainfall during the following years. The soil was enriched with organic carbon (>0.4% in upper 30 cm) under the promising tree species. Total ionic composition of leaves ranged between 2.8% and 6.8%. Thus, rehabilitation of arid soils with the above-recommended tree species using the available saline waters would not only render these abandoned soils to be productive, but also ensure conservation and improvement in the environment for long-range ecological security in these lands.

Hebbara et al. (2002) planted saplings of 12 fruit species (mango, *M. indica*; sapota, *M. zapota*, cultivars Kalipatti and Cricket Ball; wood apple, *Limonia acidissima*; tamarind, *T. indica*; pomegranate, *P. granatum*; custard apple, *Annona squamosa*; fig, *F. carica*; guava, *P. guajava*; ber, *Z. mauritiana*; aonla, *Phyllanthus emblica*; jamun, *S. cumini*; and pummelo, *Citrus maxima*) during 1990 in Gangavati, Karnataka, India, in three naturally occurring salinity blocks (9.3 ± 2.8 dS/m, 65–70 cm water table; 16.4 ± 4.1 dS/m, 75–90 cm water table; and 25.4 ± 4.9 dS/m, 85–114 cm water table). Performance of the fruit species was evaluated during 1990–2000 in terms of survival percentage, plant height, and diameter at stump height. Based on survival percentage and growth, among fruit species, *S. cumini* survived and grew better under relatively lower salinity and shallower water table conditions.

PHYTOREMEDIATION OF CONTAMINATED SOILS

Typical plant-based strategies for contaminated soils, such as those containing elevated levels of metals and metalloids, work through the cultivation of specific plant species capable of hyperaccumulating target ionic species in their shoots, thereby removing them from the soil. Many human diseases result from the buildup of toxic metal in soil, making remediation of these areas crucial in the protection of human health. Lead is one of the most difficult contaminants to remove from the soil, as well as one of the most dangerous. The presence of lead in the environment can have devastating effects on plant growth and can result in serious side effects—including seizures and mental retardation—if ingested by humans or animals (Lasat 2000). Phytoremediation of lead-contaminated soil involves two of the aforementioned strategies: phytostabilization and phytoextraction. It is believed that a plant's ability to phytoextract certain metals is a result of its dependence on the absorption of many metals—such as arsenic, lead, nickel, and copper—to maintain natural function (Lasat 2000). Some plants like *S. cumini*, termed "hyperaccumulators," extract and store extremely high concentrations (in excess of 100 times greater than nonaccumu-lator species) of metallic elements. Research has shown that these hyperaccumula-tors often do not exclude nonessential metals in the absorption process, thus resulting in plants that can extract high levels (1%–2% of their biomass) of pollutants from contaminated soil. It is believed that plants initially developed this ability to hyper-accumulate nonessential metallic compounds as a means of protecting themselves from herbivorous predators, which would experience serious toxic side effects from the ingestion of the hyperaccumulator's foliage (Pollard and Baker 1997). *S. cumini* hyperaccumulated nonessential compounds, such as lead and chromium. King et al. (2007) studied biosorption of lead ions from aqueous solution by *S. cumini* in a batch adsorption system as a function of pH, contact time, lead ion concentration, adsor-bent concentration, and adsorbent size. The biosorption capacities and rates of lead ions onto *S. cumini* were evaluated. The Langmuir, Freundlich, Redlich–Peterson, and Temkin adsorption models were applied to describe the isotherms and isotherm constants. Biosorption isothermal data could be well interpreted by the Langmuir model, followed by the Temkin model with a maximum adsorption capacity of 32.47 mg/g of lead ion on *S. cumini* leaf biomass. The difficulty with these models, in terms of performing phytoremediation, is that they grow very slowly and have a very low biomass (Huang et al. 2006). Therefore, the bioremediation process would be extremely slow because the rate of bioremediation is directly proportional to the growth rate, and the total amount of bioremediation is correlated with a plant's total biomass. No plant yet discovered meets the ideal criteria of an effective phytoreme-diator (fast growing, deep and extensive roots, high biomass, easy to harvest, and hyperaccumulators of a wide range of toxic metals). However, trees such as *S. cumini* are ideal in the remediation of heavy metals because they can withstand higher con-centrations of pollutants due to their large biomass, accumulate large amounts of contaminants in their systems because of their size, reach a huge area and great depths due to their extensive root systems, and stabilize an area and prevent erosion, and hence the spread of the contaminant, because of their perennial presence. They can also be easily harvested and removed from the area with minimal risk, effec-tively taking with them a large quantity of the pollutants that were once present in the soil (Sykes et al. 1999).

Laxminarayana and Reddy (2012) studied the removal of Cr (VI) ions from aqueous solution by using *S. cumini* seeds and showed that seeds were able to adsorb Cr (VI) ions from aqueous solutions. The maximum adsorption capacity was found to be 253.16 mg/g, with a 0.1 g of sorbent dosage at pH 2. The effect of pH, contact time, sorbent dosage, and temperature was studied. The adsorption of *S. cumini* seeds was best fitted by the Langmuir adsorption isotherm, and it follows pseudo-second-order kinetics. King et al. (2008) studied the biosorption of zinc ions from aqueous solution by *S. cumini* was also studied in a batch adsorption sys-tem as a function of pH, contact time, zinc ion concentration, adsorbent dosage, and adsorbent size. The Langmuir and Freundlich adsorption models were applied to describe the isotherms and isotherm constants. Biosorption isothermal data could be well interpreted by the Langmuir model, followed by the Freundlich model, with a maximum adsorption capacity of 35.84 mg/g of zinc ion on *S. cumini* leaf biomass. Shao et al. (2012) in their investigation found that *S. cumini* was effective for amending the aluminum (Al)-contaminated soils. This study also examined the relationship among soil microbial community composition, nematode community structure, and soil Al content in different vegetated Al-rich ecosystems. The results demonstrated that there were greater soil bacterial, fungal, and arbuscular mycorrhizal fungal biomasses in *S. cumini* plantation. The concentration of water-soluble Al was normally greater in vegetated than nonvegetated soil. The residual Al and total Al concentrations showed a significant decrease after planting *S. cumini* plantation onto the shale dump. These studies suggested that *S. cumini* vegetation was the primary driver on soil nematodes and microorganisms, and it also could regulate the sensitivity of the bioindicator role, mainly through the alteration of soil Al and physicochemical characteristics.

Dhillon et al. (2008) examined *S. cumini*, along with seven agroforestry tree species, for its ion accumulator property and grown in a clay loam soil treated with different levels of selenite (Se), that is, 0, 1.25, 2.5, and 5.0 mg/kg, supplied through sodium selenate. After one year of growth, a progressive decrease in dry matter of leaves, stem, and roots was observed with increasing levels of applied Se. The selenium content of leaves, stem, and roots of all the tree species increased significantly with increasing levels of Se applied, although a large variation within species was observed. In the stem portion of different trees, the highest concentration of Se was found in dek (*Melia azedarach*) (5.1 mg/kg) and the lowest in mulberry (*Morus alba*) (2.6 mg/kg). The efficiency of selenium removal (including leaves, stem, and roots) was the highest in Arjun, followed by eucalyptus (*Eucalyptus hybrid*)—Clone 10, mulberry, jamun (*S. cumini*), dek, shisham, and acacia. Effective removal of Se takes place through the stem portion of different trees where it constitutes 30%–50% of the total Se. Large variation in Se uptake by different tree species suggests that trees vary in their potential for phytoremediation of seleniferous soils. Dek, Arjun, and jamun removed the greatest amount of Se at 2.5 mg Se/kg soil. On the basis of average values, the removal efficiency of Se was in the order arjun > eucalyptus > mulberry > jambolin > dek > shisham > acacia. In the greenhouse investigation, the whole tree biomass was harvested to investigate the contribution of different components in Se removal. At different levels of applied Se, the contribution of leaves in total removal of Se was 11%–20% in jamun, arjun, and eucalyptus. Of the total Se accumulated by one-year-old trees, roots constituted 40%–50% in the case of arjun, acacia, jamun, shisham, and eucalyptus. Effective removal of Se could take place in the stems of trees constituting 41%–50% of the total Se accumulated in eucalyptus, jamun, and shisham.

PHYTOREMEDIATION OF WATERLOGGED SOILS

Gaur et al. (1998) studied variation in the spore density and percentage of root length of tree species colonized by arbuscular mycorrhizal fungi at a rehabilitated waterlogged site. The arbuscular mycorrhizal distribution in all the plots represented by various tree species (*T. arjuna*, *S. cumini*, and *Populus euphratica*, five years old) and the natural vegetation of *Typha elephantina* was found to be significantly different. The distribution profile showed the dominance of the two genera, *Glomus* and *Gigaspora*. The presence of *Glomus* showed a positive correlation with available soil P, while the presence of *Gigaspora* showed a positive correlation with soil organic matter content. Both mycorrhizal genera also predominated in the naturally grown *T. elephantina* plot. Root length colonized by mycorrhizas was greatest in *P. euphratica* (56%), followed by *T. elephantina* (41%), *T. arjuna* (41%), and *S. cumini* (12.5%).

Kumar (2004) conducted an experiment to determine the growth performance of five tree species, *Eucalyptus tereticornis*, *T. arjuna*, *D. sissoo*, *S. cumini*, and *P. pinnata*, under waterlogging conditions. The study site was subjected to high water table levels, soil saturation, and periodic flooding at various times of the year. The tree species differed in their ability to withstand waterlogging stress, as shown by the differences in their growth behavior at 17 months after transplanting. The maximum plant height was recorded in *E. tereticornis*, while *P. pinnata* had a minimum plant height. *T. arjuna* attained a significantly higher basal diameter among the other species. The growth performance of *P. pinnata* and *S. cumini* was poor compared with that of the others. *E. tereticornis* and *T. arjuna* had an edge over the other species in tolerating waterlogging stress.

PHYTOREMEDIATION OF INDUSTRIAL POLLUTED SOILS

Phytoremediation is an important method for restoring bare soil or slag; however, the physiological traits of plants used for revegetation are poorly known, even though such traits are important to successful remediation. A study was carried out on oil shale waste in Maoming City, Guangdong Province, to screen for tree species with high photosynthetic potential, appraise the ability of these plants to acclimatize to oil shale waste, and provide valuable information for ecological restoration of similar waste sites (Huang et al. 2006). Diurnal variation of photosynthesis was measured for 12 tree species in summer and winter, using portable photosynthetic equipment. Other parameters, such as transpiration, stomatal conductance, and relative humidity, were also measured simultaneously, and water use efficiency (WUE) was calculated as net photosynthesis divided by transpiration. There were large seasonal differences in all parameters, with values of net photosynthesis, transpiration, and stomatal conductance higher in summer than winter (60.9%, 77.7%, and 85.7%, respectively), but WUE higher in winter than in summer (26.8%–77.2%). Diurnal variation of net photosynthesis also exhibited seasonal differences. Among the trees, *S. cumini* exhibited a bimodal peak in winter and a unimodal peak in summer. Trees with a higher mean net photosynthetic rate in both winter and summer should be more useful for phytoremediation than species with a high net photosynthetic rate in only one season. Accordingly, the net photosynthetic rate, which synthesizes all parameters examined, should be considered the most important parameter to appraise the ability of plants to acclimatize.

Zhang et al. (2006) examined changes in photosynthetic parameters of *Cassia alata* and *S. cumini* grown on waste residue of oil shale from Maoming Petrochemical

Company as a polluted site, and at Maoming Forestry Institute as a control site. Results showed that absolute values of net photosynthetic rate, stomatal conductance, and transpiration rate in *C. alata* were higher than those in *S. cumini* at most hours during the daytime at both sites. Reductions in the net photosynthetic rate and WUE in *C. alata* were greater than those in *S. cumini*, whereas stomatal conductance recorded a contradicting trend, which might be related to soil water content at different experimental sites. It was shown that *S. cumini* was more resistant to oil shale pollution than *C. alata*. It is suggested that more comprehensive physiological indices in selecting pollution-resistant plant species must be established.

Singh and Singh (2006) conducted a series of experiments on the rehabilitation of mine spoil in a dry tropical region of India for determining the suitability of tree species for plantation, growth performance of selected indigenous species in monoculture, and impact of the plantations on the restoration of biological fertility of soil. All 17 indigenous species examined could grow in the mine spoil, and the growth of a majority of them could be improved by amending the mine spoil with NPK fertilizer. Direct seeding showed the greatest height of *Zizyphus jujuba* and *P. pinnata* on a flat surface, and of *A. indica* on a slope. In terms of diameter, *S. cumini* performed best on a flat surface and *T. arjuna* on a slope. Total biomass in plantations of selected native tree species on mine spoil at five years of age varied from 7.2 to 74.7 t/ha, being minimum for *Shorea robusta* and maximum for *Dendrocalamus strictus*. The total net production ranged from 3.5 (for *S. robusta*) to 32.0 (for *D. strictus*) t/ha/year, respectively. Microbial biomass in the redeveloping soil was lower than that in natural forest soil, but immobilization of soil C in microbial biomass was greater in the mine spoil than in the natural forest. The study indicated that the net primary production of the plantations was a function of the amount of foliage, soil C was a function of the amount of litter fall, and biomass C was a function of soil C. Plantation of trees like *S. cumini* significantly accelerated the soil redevelopment process on the mine spoil.

ROLE IN PLANT GROWTH REGULATION

Channal et al. (2002) studied the allelopathic effect of seven tree leaf extracts, *S. cumini*, *Acacia arabica* (*A. nilotica*), *T. grandis*, *E. tereticornis*, *T. indica*, *S. saman*, and *A. indica*, each at 5% and 10% concentrations on sunflower and soybean. The study indicated that germination of sunflower was increased by *T. grandis*, *T. indica*, and *S. saman* each at 5% and 10% concentrations, while it was suppressed by *E. tereticornis* and *A. arabica*. Soybean germination was increased by *A. arabica*, *T. grandis*, *S. saman*, and *A. indica* at both concentrations, while it was decreased by *T. indica*. Similarly, seedling length, vigor index, and seedling dry matter were also influenced by tree leaf extracts at different concentrations. The seedling length of sunflower was significantly increased by *S. cumini*, *A. indica*, *A. arabica*, and *S. saman*, while that of soybean was increased by all tree leaf extracts, although the effect was not that significant compared with sunflower. Almost all the leaf extracts enhanced the vigor index in sunflower, only *T. grandis*, *A. arabica*, and *A. indica* increased the vigor index in soybean and decreased the seedling dry matter of soybean.

Suman (1990) studied the allelopathic effect of tree litter biomass on germination percentage and growth characteristics of range legumes at the Division of Agronomy, Indian Agricultural Research Institute (IARI), New Delhi, in 1994 and 1995. The litter biomass of mango (*M. indica*) and jamun (*S. cumini*) contained tannins, and subabool (*Leucaena leucocephala*) litter contained mimosin. These secondary chemicals stimulated germination and elongation of radicle and plumule at lower concentrations, that is, 25 g litter biomass/L water. At higher concentrations (50 or 75 g litter/L), they greatly reduced germination and seedling growth. Sarkar (2012) investigated the response of seed priming with *S. cumini* leaf extract and reported that it showed a positive effect on yield and yield-attributing parameters of rice under both nonflooding and early flooding conditions. The investigation was carried out with two near-isogenic lines, Swarna and Swarna-Sub1. Seed priming was done with water and 2% *S. cumini* leaf extract, and it improved seedling establishment under flooding. Acceleration of growth occurred due to seed pretreatment, which resulted in longer seedling and greater accumulation of biomass. Seed priming greatly hastened the activities of total amylase and alcohol dehydrogenase. Weed biomass decreased significantly under flooding compared with nonflooding conditions.

Lal (2000) conducted field and laboratory experiments on a sandy loam soil to study the effect of *S. cumini* litter biomass with and without N fertilizer on the germination, growth, and productivity of some graminaceous and leguminous grain and forage plants (i.e., *Zea mays*, *Oryza sativa*, *Glycine max*, *Sesbania sesban*, *Desmanthus virgatus*, *Stylosanthes hamata*, *Atylosia scarabaeoides*, *Clitoria ternatea*, *Macroptilium* sp., *Sorghum sudanensis* [*S. sudanense*], *Vigna unguiculata*, and wheat). The results indicated that a combination of 75% N through urea fertilizer + 25% N through tree litter biomass proved more beneficial in terms of final yield, which was significantly higher over sole litter biomass, sole urea fertilizer, and control without N application. The higher concentrations of litter extracts, that is, 5%, 7.5%, and 10%, significantly inhibited the germination of wheat by 32%, 47%, and 112%, respectively, in comparison with pure water. The lower concentration (2.5%) of litter increased the germination by 9% in rice and soyabean; 7% in maize and *Sesbania*; 5% in *Desmanthus*, *Stylo*, *Atylosia*, and *Clitoria*; and 3% in siratro, sweet sudan grass, and green gram.

Channal et al. (2000) conducted a study to evaluate the allelopathic effect of leaf extracts from *S. cumini*, applied at 5% and 10% concentrations, on seed germination, vigor index, seedling length, and seedling dry matter of sorghum and rice. Irrespective of concentration, leaf extracts promoted germination in sorghum (15%–32% over the control); however, seedling length was considerably decreased in sorghum due to *S. cumini*, and the vigor index was markedly decreased in sorghum and rice due to *S. cumini*. Leaf extracts decreased the seedling dry matter in sorghum and rice irrespective of concentrations.

ROLE IN MITIGATING CLIMATE CHANGE

Dhand et al. (2003) conducted an investigation to estimate the carbon content of some important tree species (*S. robusta*, *T. grandis*, *Terminalia alata*, *D. strictus*,

Pinus roxburghii, Diospyros malabaricum [*D. malabarica*], *Pterospermum acerifolium*, and *S. cumini*) and to have an idea about the responses of these species to the changing climate. *P. roxburghii* stored the highest amount of carbon, followed by *P. acerifolium, S. cumini*, and *T. grandis*. The carbon storing efficiency of some species, like *P. roxburghii, S. robusta*, and *S. cumini*, decreased with increasing concentration of CO_2, indicating that these plants release additional carbon to the atmosphere rather than storing carbon.

POTENTIAL IN AGROFORESTRY

Gill et al. (2003) worked on the agroforestry aspect and grown wheat (*Triticum aestivum*) crop cv. WH 147, under 12 important MPTs (*Acacia nilotica, A. cupressiformis* [*A. nilotica* subsp. *cupressiformis*], *C. equisetifolia, Madhuca latifolia* [*Madhuca longifolia*], *Melia azedarach, L. leucocephala, D. sissoo, Albizia lebbeck, S. cumini, E. tereticornis, E. officinalis* [*P. emblica*], and *Hardwickia binata*) to know the effect of transmitted photosynthetic active radiation (PAR) through the tree canopies on the grain yield of wheat. The study was conducted during 1988–1989 at the National Research Centre for Agroforestry in Jhansi, Uttar Pradesh, India, on a sandy loam soil. The highest light transmission of 58.7% was observed on *M. longifolia*, followed by 58.1%, 55.3%, and 51.7%, respectively, in *S. cumini, H. binata*, and *P. emblica*, and the least light transmission of 34.2% was observed in *A. nilotica* under the agroforestry situation. The light transmission ratios after pruning were much higher than those from before pruning under both agroforestry and the control system. Light transmission ratios in general were higher under wider tree spacing in all the MPTs than under lower tree spacing in association with crops, as well as under control. There was no significant difference in wheat yield under the different MPT canopies in the first year; however, in the second year onward, the production of wheat grain was drastically reduced under the tree canopies, with a highest grain yield of 26.84 q/ha under *M. longifolia*. The status of PAR transmitted was also correlated with the grain yield of wheat.

PRODUCTION OF BIOPLASTICS

Recent studies have revealed that *S. cumini* seed extract can be used as carbon source of polyhydroxyalkanoates (PHAs), polyesters of hydroxyalkanoates synthesized by various bacteria as intracellular carbon and energy storage compounds and accumulated as granules in the cytoplasm of cells. PHA-producing bacteria from soil were isolated, characterized, and screened by the Nile blue staining method. Screened organisms were subjected to fermentation with glucose as a carbon source and a low-cost raw material like jambul seed (*S. cumini*). The strain SPY-1 showed a higher amount of PHA accumulation than the other strains and was comparable with that of the reference strain *Ralstonia eutropha* (Preethi et al. 2012). Jeyaseelan et al. (2013) also used *S. cumini* seed as a carbon source for the production of PHA from soil microbial isolates. The efficiency of selected isolates for PHA production utilizing the hydrolyzed substrate as a carbon source was compared with that of *R. eutropha* (reference strain) using the same production medium. Jamun seed accumulated

PHA 42.2% as a sole carbon source in comparison with the best isolate SPY-1 and *R. eutropha*, which were able to accumulate 26.76% and 28.97%, respectively, of their dry cell weight.

OTHER APPLICATIONS

Kuncha and Devi (2012) screened the *in vitro* antioxidant activity of a methanolic extract of *S. cumini* bark. Preliminary phytochemical investigation indicated the presence of carbohydrates, amino acids, tannins, saponins, phytosterols, terpenoids, phenols, and flavones. The antioxidant activity was determined by *in vitro* methods such as DPPH scavenging assay, hydrogen peroxide scavenging assay, and ferric reducing antioxidant power (FRAP) assay. The half-maximal inhibitory concentration (IC_{50}) values of the methanolic extract of *S. cumini* for DPPH and hydrogen peroxide scavenging activity were found to be 53.3% at a concentration of 600 mg and 42.03% at 1.2 mg/ml, respectively. The FRAP value was found to be 810 µg Fe^{2+}/g. The extract showed significant antioxidant activity in all antioxidant assays when compared with ascorbic acid. The results of this research work are promising, thus indicating the utilization of the bark of *S. cumini* as a significant source of natural antioxidants.

REFERENCES

Arshad, J., R. Bajwa, S. Shafique, S. Shafique, C. Preston, J. H. Watts, and N. D. Crossman. 2006a. Chemical, phytochemical and biological control of *Parthenium hysterophorus* L. in Pakistan. In *Managing Weeds in a Changing Climate, Proceedings of the 15th Australian Weeds Conference*, Adelaide, South Australia, September 24–28, 876–879.

Arshad, J., and S. Samad. 2012. Screening of allelopathic trees for their antifungal potential against *Alternaria alternata* strains isolated from dying-back *Eucalyptus* spp. *Nat. Prod. Res.* 26: 1697–1702.

Arshad, J., S. Shafique, and S. Shafique. 2006b. Herbicidal potential of aqueous leaf extract of allelopathic trees against *Phalaris minor. Pak. J. Weed Sci. Res.* 12: 339–346.

Barrett-Lennard, E. G., C. V. Malcom, W. Stern, W. R. Stern, and S. M. Wilkins. 1986. *Forage and Fuel Production from Salt Affected Wasteland*. Amsterdam: Elsevier.

Batish, D. R., K. Arora, H. P. Singh, and R. K. Kohli. 2007. Potential utilization of dried powder of *Tagetes minuta* as a natural herbicide for managing rice weeds. *Crop Prot.* 26: 566–571.

Bhojvaid, P. P., V. R. Timmer, and G. Singh. 1996. Reclaiming sodic soils for wheat production by *Prosopis juliflora* (Swartz) DC afforestation in India. *Agrofor. Syst.* 34: 139–150.

Bhutta, R., M. H. R. Bhatti, and I. Ahmad. 2001. Effect of seed diffusates on fungal population and germination of sunflower seeds. *Helia* 24: 77–81.

Brown, S., and A. E. Lugo. 1994. Rehabilitation of tropical lands: A key to sustaining development. *Restor. Ecol.* 2: 97–111.

Chandrasekaran, M., and V. Venkatesalu. 2004. Antibacterial and antifungal activity of *Syzygium jambolanum* seeds. *J. Ethnopharmacol.* 91: 105–108.

Channal, H. T., M. B. Kurdikeri, C. S. Hunshal, P. A. Sarangamath, S. A. Patil, and M. Shekhargouda. 2002. Allelopathic effect of some tree species on sunflower and soybean. *Karnataka J. Agric. Sci.* 15: 279–283.

Channal, H. T., M. B. Kurdikeri, and P. A. Sarangamath. 2000. Allelopathic effect of tree leaf extracts on germination of sorghum and rice. *Karnataka J. Agric. Sci.* 13: 338–342.

Chattopadhyay, D., B. K. Sinha, and L. K. Vaid. 1998. Antibacterial activity of *Syzygium* species. *Fitoterapia* 69: 365–367.

Claver, M. A., and R. W. A. Jesudasan. 2007. Impact of whitefly honeydew exudation and host-plant leaf surface chemicals on whitefly-ant interaction. *Hexapoda* 14: 73–77.

Clemens, S., M. G. Palmgren, and U. Kramer. 2002. A long way ahead: Understanding and engineering plant metal accumulation. *Trends Plant Sci.* 7: 309–314.

Crawley, M. J. 1986. *Plant Ecology.* London: Blackwell Scientist Publications.

Dagar, J. C., and G. Singh. 2000. Evaluation of fruit tree species for sodicity tolerance. *Ind. J. For.* 23: 390–396.

Dayan, F. E., C. L. Cantrell, and S. O. Duke. 2009. Natural products in crop protection. *Bioorgan. Med. Chem.* 17: 4022–4034.

De Oliveira, G. F., N. A. J. C. Furtado, A. A. da Silva Filho, C. H. G. Martins, J. K. Bastos, W. R. Cunha, and M. L. de Andrade de Silva. 2007. Antimicrobial activity of *Syzygium cumini* (Myrtaceae) leaves extract. *Braz. J. Microbiol.* 38: 381–384.

Dhand, V., A. K. Tripathi, R. K. Manhas, J. D. S. Negi, and P. S. Chauhan. 2003. Estimation of carbon content in some forest tree species. *Indian For.* 129: 918–922.

Dhillon, K. S., S. K. Dhillon, and H. S. Thind. 2008. Evaluation of different agroforestry tree species for their suitability in the phytoremediation of seleniferous soils. *Soil Use Manage.* 24: 208–216.

Dos Santos, U. M., Jr., J. F. de C. Gonçalves, and T. R. Feldpausch. 2006. Growth, leaf nutrient concentration and photosynthetic nutrient use efficiency in tropical tree species planted in degraded areas in central Amazonia. *For. Ecol. Manage.* 226: 299–309.

Elansary, H. O., M. Z. M. Salem, N. A. Ashmawy, and M. M. Yacout. 2012. Chemical composition, antibacterial and antioxidant activities of leaves essential oils from *Syzygium cumini* L., *Cupressus sempervirens* L. and *Lantana camara* L. from Egypt. *J. Agric. Sci.* 4: 144–152.

Farrukh, A., and I. Ahmad. 2003. Broad-spectrum antibacterial and antifungal properties of certain traditionally used Indian medicinal plants. *World J. Microbiol. Biotechnol.* 19: 653–657.

Garg, V. K. 1998. Interaction of tree crops with a sodic soil environment: Potential for rehabilitation of degraded environments. *Land Degrad. Dev.* 9: 81–93.

Garg, V. K. 1999. Leguminous trees for the rehabilitation of sodic wasteland in northern India. *Res. Ecol.* 7: 281–287.

Garg, V. K., and R. K. Jain. 1992. Influence of fuel wood trees on sodic soils. *Can. J. For. Res.* 22: 729–735.

Gaur, A., M. P. Sharma, A. Adholeya, and S. Chauhan. 1998. Variation in the spore density and percentage of root length of tree species colonised by arbuscular mycorrhizal fungi at a rehabilitated waterlogged site. *J. Trop. For. Sci.* 10: 542–551.

Gill, A. S., M. J. Baig, A. K. Bisaria, and R. Debroy. 2003. Effect of transmission of photosynthetic active radiation (PAR) in tree-wheat agroforestry system. *Ind. J. For.* 26: 286–290.

Grainger, A. 1998. Estimating areas of degraded tropical lands requiring replenishment of forest cover. *Int. Tree Crops J.* 5: 31–61.

Gupta, M., and R. Bhadauria. 2012. Evaluation of anti fungal potential of aqueous extract of *Syzygium cumini* Linn. against *Alternaria alternata* Nees. and *Fusarium oxysporum* Schle. *Int. J. Pharma Bio Sci.* 3 (B): 571–577.

Hebbara, M., M. V. Manjunatha, S. G. Patil, and D. R. Patil. 2002. Performance of fruit species in saline-waterlogged soils. *Karnataka J. Agric. Sci.* 15: 94–98.

Holmes, P. M. 2001. Mycorrhizal colonization of halophytes in central European salt marshes. Referenced by E. P. Glenn in *Scientific American*, April 2001, 112–114.

Huang, J., W. Tong, G. H. Kong, C. Z. Dong, and J. Z. Zhang. 2006. Seasonal changes of photosynthetic characteristics in 12 tree species introduced onto oil shale waste. *Chin. J. Plant Ecol.* 30: 666–674.

Jeyaseelan, A., S. Pandiyan, and P. Ravi. 2013. Production of polyhydroxyalkanoate (PHA) using hydrolyzed grass and *Syzygium cumini* seed as low cost substrates. *J. Microbiol. Biotechnol. Food Sci.* 2: 970–982.

King, P., N. Rakesh, S. B. Lahari, Y. P. Kumar, and V. S. R. K. Prasad. 2007. Removal of lead from aqueous solution using *Syzygium cumini* L.: Equilibrium and kinetic studies. *J. Hazard. Mater.* 142: 340–347.

King, P., N. Rakesh, S. B. Lahari, Y. P. Kumar, and V. S. R. K. Prasad. 2008. Biosorption of zinc onto *Syzygium cumini* L.: Equilibrium and kinetic studies. *Chemical Eng. J.* 144: 181–187.

Kordali, S., A. Cakir, T. A. Akcin, E. Mete, A. Akcin, T. Aydin, and H. Kilic. 2009. Antifungal and herbicidal properties of essential oils and n-hexane extracts of *Achillea gypsicola* Hub-Mor. and *Achille abiebersteinii* Afan. (Asteraceae). *Ind. Crop Prod.* 29: 562–570.

Kumar, R. 2004. Performance of some forest tree species under water logging conditions of Punjab. *Environ. Ecol.* 22: 835–837.

Kuncha, J., and V. S. Devi. 2012. In-vitro antioxidant activity of methanolic extract of *Syzygium cumini* L. bark. *Asian J. Biomed. Pharm. Sci.* 2: 45–49.

Lal, B. 2000. Beneficial allelopathic effect of *Syzygiuym cumini* litterfall biomass on arable crops. *Ind. J. Agrofor.* 2: 53–57.

Lasat, M. M. 2000. Phytoextraction of metals from contaminated soil: A review of plant/soil/metal interaction and assessment of pertinent agronomic issues. *J. Hazard. Subst. Res.* 2: 1–25.

Laxminarayana, K., and K. L. Reddy. 2012. Removal of Cr (VI) ions from aqueous solution by using *Syzygium cumini* seeds. *Asian J. Chem.* 24: 5771–5774.

Lei, W., J. Liu, D. Wang, J. Li Jiao and H. Fan. 2012. Inhibition effect of four plants compound fungicides against *Fusarium oxysporum*. *Vegetables* 16: 80–85.

Martinez, B. J., and C. L. Manzur. 2005. Overview of the salinity problem in the world and FAO strategies to address the problem. In *Proceedings of the International Society Forum*, Riverside, CA, 311–313.

Miller, R. 1996. Ground water remediation technologies. TO-96-03. Ground-Water Remediation Technologies Analysis Center. Available at http://www.gwrtac.org/html/tech-over.html#PYTOREM (verified March 31, 2001).

Morallo Rejesus, B., and G. Punzalan. 1997. Molluscicidal action of some Philippine plants on golden snails, *Pomacea* spp. *Philipp. Entomol.* 11: 65–79.

Olia, M., and B. K. Goswami. 2000. Dosage optimization and mass culture of the fungal bio-agent *Aspergillus niger* against root-knot nematode (*Meloidogyne incognita*) infecting tomato. *Ind. J. Nematol.* 30: 253–255.

Pandey, B. P., and J. Mohan. 1986. Inhibition of turnip mosaic virus by plant extracts. *Indian Phytopathol.* 39: 489–491.

Pollard, J. A., and A. J. M. Baker. 1997. Deterrence of herbivory by zinc hyperaccumulation in *Thlaspi caerulescens* (Brassicaceae). *New Phytol.* 135 (4): 655–658.

Preethi, R., P. Sasikala, and J. Aravind. 2012. Microbial production of polyhydroxyalkanoate (PHA) utilizing fruit waste as a substrate. *Res. Biotechnol.* 3: 61–69.

Riaz, T., S. N. Khan, and A. Javaid. 2010. Management of *Fusarium* corm rot of gladiolus (*Gladiolus grandiflorus* sect. Blandus cv. Aarti) by using leaves of allelopathic plants. *Afr. J. Biotechnol.* 9: 4681–4686.

Sarkar, R. K. 2012. Seed priming improves agronomic trait performance under flooding and non-flooding conditions in rice with QTL *SUB1*. *Rice Sci.* 19: 286–294.

Shafi, P. M., M. K. Rosamma, J. Kaiser, and P. S. Reddy. 2002. Antibacterial activity of *S. cumini* and *S. travancoricum* leaf essential oils. *Fitoterapia* 73: 414–416.

Shafique, S., R. Bajwa, A. Javaid, and S. Shafique. 2005. Biological control of *Parthenium* IV: Suppressive ability of aqueous leaf extracts of some allelopathic trees against germination and early seedling growth of *Parthenium hysterophorus* L. *Pak. J. Weed Sci. Res.* 11: 75–79.

Shao, Y., W. Zhang, Z. Liu, Y. Sun, D. Chen, J. Wu, L. Zhou, H. Xia, D. A. Neher, and S. Fu. 2012. Responses of soil microbial and nematode communities to aluminum toxicity in vegetated oil-shale-waste lands. *Ecotoxicology* 21: 2132–2142.

Singh, A. N., and J. S. Singh. 2006. Experiments on ecological restoration of coal mine spoil using native trees in a dry tropical environment, India: A synthesis. *New For.* 31: 25–39.

Singh, B. 1996. Influence of forest litter on reclamation of semi-arid sodic soils. *Arid Soil Res. Rehab.* 10: 201–211.

Singh, B., and V. K. Garg. 2007. Phytoremediation of a sodic forest ecosystem: Plant community response to restoration process. *Not. Bot. Hort. Agrobot. Cluj.* 35: 77–85.

Singh, S. P., V. K. Garg, and R. S. Katiyar. 2004. Tree species diversity and dominance in a man-made forest on sodic wasteland of north India. *J. Forest Res.* 9: 15–21.

Sinha, S., R. E. Masto, L. C. Ram, V. A. Selvi, N. K. Srivastava, R. C. Tripathi, and J. George. 2009. Rhizosphere soil microbial index of tree species in a coal mining ecosystem. *Soil Biol. Biochem.* 41: 1824–1832.

Suman, B. L. 1990. Allelopathic effect of biomass of tree species on range legumes. *Bhartiya Krishi Anusandhan Patrika* 14: 1–6.

Sunder, S., R. Singh, and D. S. Dodan. 2005. Management of bacterial blight of rice with botanical extracts and non-conventional chemicals. *Plant Dis. Res.* (Ludhiana) 20: 12–17.

Sykes, M., V. Yang, J. Blankenburg, and A. Said. 1999. Biotechnology: Working with nature to improve forest resources and products. In *International Vetiver Conference*, Chiang Rai, Thailand, February 4–8, 631–637.

Tessmann, D. J., and J. C. Dianese. 2002. Hentriacontane: A leaf hydrocarbon from *Syzygium jambos* with stimulatory effects on the germination of urediniospores of *Puccinia psidii*. *Fitopatol. Bras.* 27: 538–542.

Tomar, O. S., P. S. Minhas, V. K. Sharma, Y. P. Singh, and R. K. Gupta. 2003. Performance of 31 tree species and soil conditions in a plantation established with saline irrigation. *For. Ecol. Manage.* 177: 333–346.

Uma, T., S. Mannam, J. Lahoti, K. Devi, R. D. Kale, and D. J. Bagyaraj. 2012. Biocidal activity of seed extracts of fruits against soil borne bacterial and fungal plant pathogens. *J. Biopest* 5:103–105.

Zhang, J. Z., N. Liu, G. Kong, T. Wu, Z. Lin, and C. L. Peng. 2006. Responses of photosynthesis in two woody plants to oil shale residue pollution. *J. Trop. Subtrop. Bot.* 14: 100–106.

14 The Usage of Selected *Syzygium* Species

Roland Hardman

CONTENTS

INTRODUCTION

The genus *Syzygium* comprises about 1200 species of tropical evergreen trees, mainly originating and found in Southeast Asia, and which belong to the volatile oil-yielding family Myrtaceae (1,2). The trees vary in height: the small ones, when bearing numerous flowers, are used in floral displays; those medium in height are the most common and are often used in windbreaks or in orchards when bearing edible fruit; the tall trees are used as timber for construction purposes. All the trees may be used as sources of medicinal compounds. The volatile oils provide antibacterial and antifungal activities leading to treatment for disease and also its avoidance by preservation of the food. Other important activities include antidiabetic and anticancer properties. *Syzygium cumini* is the most popular and widely explored species for its multifarious uses, including a broad spectrum of medicinal properties. The preceding chapters have exclusively dealt with several aspects of *S. cumini*. However, there are a few lesser known *Syzygium* species in the tropical Indo-Malesian and Australasian region with reported medicinal and other uses. Examples of eight such *Syzygium* trees and other usages are given below.

USAGES OF SELECTED INDO-MALESIAN
AND AUSTRALASIAN *SYZYGIUM* TREES

1. ***Syzygium antispeticum*** (Blume) Merr. & L. M. Perry
   a. *Caryophyllus antisepticus* Blume
   b. *Eugenia grata* Wight
   c. *Syzygium gratum* (Wight) S. N. Mitra

   Distribution: Northeast India to Philippines
   Usage
       Antioomycetic activity (1)
       Antioxidant (2–5)
       Antibacterial (2)
       Cytotoxicity (3), potential roles in protection of vascular dysfunction (4)

REFERENCES

1. Panchai, K., C. Hanjavanit, N. Rujinanont, and K. Hatai. 2015. Effects of four Thai herbs *in vitro* on *Achlya* spp. isolated from Nile tilapia (*Oreochromis niloticus*) in Thailand and the toxicity to Nile tilapia fry. *AACL Bioflux* 8: 761–771.
2. Thummajitasakul, S., L. Tumchalee, S. Koolwong, P. Deetae, W. Kaewsri, and S. Lertsiri. 2014. Antioxidant and antibacterial potentials of some Thai native plant extracts. *Int. Food Res. J.* 21: 2393–2398.
3. Stewart, P., P. Boonsiri, S. Puthong, and P. Rojpibulstit. 2013. Antioxidant activity and ultrastructural changes in gastric cancer cell lines induced by northeastern Thai edible folk plant extracts. *BMC Complement. Altern. Med.* 13: 60.
4. Kukongviriyapan, U., S. Luangaram, K. Leekhaosoong, V. Kukongviriyapan, and S. Preeprame. 2007. Antioxidant and vascular protective activities of *Cratoxylum formosum*, *Syzygium gratum* and *Limnophila aromatica*. *Biol. Pharm. Bull.* 30: 661–666.
5. Maisuthisakul, P., M. Suttajit, and R. Pongsawatmanit. 2007. Assessment of phenolic content and free radical-scavenging capacity of some Thai indigenous plants. *Food Chem.* 100: 1409–1418.

2. ***Syzygium australe*** (J. C. Wendl. ex Link) B. Hyland
   a. *Eugenia australis* J. C. Wendl. ex Link

   Common names: Creek-cherry, creek lilly-pilly, creek satin-ash
   Distribution: Native to Australasia and also cultivated
   Usage
       Treatment and prevention of rheumatoid arthritis (1)
       Anticancer (2)
       Antimicrobial (3,4)
       Source of essential trace metal supplements (5)

REFERENCES

1. Cock, I. E., V. Winnett, J. Sirdaarta, and B. Matthews. 2015. The potential of selected Australian medicinal plants with anti-proteus activity for the treatment and prevention of rheumatoid arthritis. *Pharmacogn. Mag.* 11 (42) (Suppl.): 190–208.
2. Jamieson, N., J. Sirdaarta, and I. E. Cock. 2014. The anti-proliferative properties of Australian plants with high antioxidant capacities against cancer cell lines. *Pharmacogn. Commun.* 4: 71–82.
3. Sautron, C., and I. E. Cock. 2014. Antimicrobial activity and toxicity of *Syzygium australe* and *Syzygium leuhmannii* fruit extracts. *Pharmacogn. Commun.* 4: 53–60.
4. Cock, I. E. 2012. Antimicrobial activity of *Syzygium australe* and *Syzygium leuhmannii* leaf methanolic extracts. *Pharmacogn. Commun.* 2: 71–77.
5. Ansari, T. M., N. Ikram, M. Najam-ul-Haq, I. Fayyaz, Q. Fayyaz, I. Ghafoor, and N. Khalid. 2004. Essential trace metal (zinc, manganese, copper and iron) levels in plants of medicinal importance. *J. Biol. Sci.* 4: 95–99.

3. **Syzygium cordatum** Hochst. ex Krauss

Common names: Water berry, waterbessie
Distribution: Uganda to South Africa
Usage
    Potential natural food preservative (1)
    Antibacterial and antidiarrheal infections (2,3,5,9–bark, 10)
    Anti-*Klebsiella* activity (leaf and bark effective against *K. pneumonia*) (4)
    Antidiabetic (5)
    Antifungal—against anthracnose (6)
    Against coughs and fevers (7,13)
    Dyes for textile and craft materials (8)
    Essential oils (9,14)
    Treatment of sexually transmitted disease—*Candida albicans* (11)
    Positive renal effects—leaf (12)

REFERENCES

1. Cock, I. E., and S. F. van Vuuren. 2015. South African food and medicinal plant extracts as potential antimicrobial food agents. *J. Food Sci. Technol.* (Mysore) 52: 6879–6899.
2. Sidney, M. T., S. J. Siyabonga, and B. A. Kotze. 2015. The antibacterial and antidiarrheal activities of the crude methanolic *Syzygium cordatum* [S.Ncik, 48 (UZ)] fruit pulp and seed extracts. *J. Med. Plants Res.* 9: 884–891.
3. Sidney, M. T., S. J. Siyabonga, and B. A. Kotze. 2015. Evaluation of the antibacterial activity of *Syzygium cordatum* fruit-pulp and seed extracts against bacterial strains implicated in gastrointestinal tract infections. *Afr. J. Biotechnol.* 14: 1387–1392.
4. Cock, I. E., and S. F. van Vuuren. 2015. The potential of selected South African plants with anti-*Klebsiella* activity for the treatment and prevention of ankylosing spondylitis. *Inflammopharmacology* 23: 21–35.

5. Deliwe, M., and G. J. Amabeoku. 2013. Evaluation of the antidiarrhoeal and antidiabetic activities of the leaf aqueous extract of *Syzygium cordatum* Hoscht. ex C. Krauss (Mytraceae) in rodents. *Int. J. Pharm.* 9: 125–133.
6. Masangwa, J. I. G., T. A. S. Aveling, and Q. Kritzinger. 2013. Screening of plant extracts for antifungal activities against *Colletotrichum* species of common bean (*Phaseolus vulgaris* L.) and cowpea (*Vigna unguiculata* (L.) Walp). *J. Agric. Sci.* 151: 482–491.
7. Mulaudzi, R. B., A. R. Ndhlala, M. G. Kulkarni, and J. van. Staden. 2012. Pharmacological properties and protein binding capacity of phenolic extracts of some Venda medicinal plants used against cough and fever. *J. Ethnopharmacol.* 143: 185–193.
8. Wanyama, P. A. G., B. T. Kiremire, P. Ogwok, and J. S. Murumu. 2011. Indigenous plants in Uganda as potential sources of textile dyes. *Afr. J. Plant Sci.* 5: 28–39.
9. Chalannavar, R. K., H. Baijnath, and B. Odhav. 2011. Chemical constituents of the essential oil from *Syzygium cordatum* (Myrtaceae). *Afr. J. Biotechnol.* 10: 2741–2745.
10. Sibandze, G. F., R. L. van Zyl, and S. F. van. Vuuren. 2010. The anti-diarrhoeal properties of *Breonadia salicina*, *Syzygium cordatum* and *Ozoroa sphaerocarpa* when used in combination in Swazi traditional medicine. *J. Ethnopharmacol.* 132: 506–511.
11. van Vuuren, S. F., and D. Naidoo. 2010. An antimicrobial investigation of plants used traditionally in southern Africa to treat sexually transmitted infections. *J. Ethnopharmacol.* 130: 552–558.
12. Mapanga, R. F., M. A. Tufts, F. O. Shode, and C. T. Musabayane. 2009. Renal effects of plant-derived oleanolic acid in streptozotocin-induced diabetic rats. *Ren. Fail.* 31: 481–491.
13. Pallant, C. A., and V. Steenkamp. 2008. In-vitro bioactivity of Venda medicinal plants used in the treatment of respiratory conditions. *Hum. Exp. Toxicol.* 27: 859–866.
14. Chisowa, E. H., G. Sakala, D. R. Hall, and D. I. Farman. 1998. Composition of the essential oil of *Syzygium cordatum* Hochst. ex Krauss. *J. Essential Oil Res.* 10: 591–592.

4. *Syzygium guineense* (Willd.) DC.
   a. *Calyptranthes guineensis* Willd.

Common names: Woodland waterberry, waterpear
Distribution: Tropical and South Africa, southwest Arabian peninsula
Usage
   Malaria treatment (1)
   Antivenom (2)
   Anthelmintic activity (3)
   Antihypertensive and vasodepressor activities (4,11)
   Antisickling activity (5)
   A source of defined essential oil (6)
   Antibacterial (7)

Reforestation (8)
Source of food and economic value (9)
Molluscicides (10)

REFERENCES

1. Galabuzi, C., G. N. Nabanoga, P. Ssegawa, J. Obua, and G. Eilu, 2015. Double jeopardy: Bark harvest for malaria treatment and poor regeneration threaten tree population in a tropical forest of Uganda. *Afr. J. Ecol.* 53: 214–222.
2. James, O., E. U. Godwin, and O. D. Agah. 2013. Anti-venom studies on *Olax viridis* and *Syzygium guineense* extracts. *Am. J. Pharm. Toxicol.* 8: 1–8.
3. Bekele, M., T. Gessesse, Y. Kechero, and M. Abera. 2011. *In-vitro* anthelmintic activity of condensed tannins from *Rhus glutinosa*, *Syzygium guineensa* and *Albizia gummifera* against sheep *Haemonchus contortus*. *Global Vet.* 6: 476–484.
4. Ayele, Y., K. Urga, and E. Engidawork. 2010. Evaluation of *in vivo* antihypertensive and *in vitro* vasodepressor activities of the leaf extract of *Syzygium guineense* (Willd) D.C. *Phytother. Res.* 24: 1457–1462.
5. Mpiana, P. T., V. Mudogo, D. S. T. Tshibangu, E. K. Kitwa, A. B. Kanangila, J. B. S. Lumbu, K. N. Ngbolua, E. K. Atibu, and M. K. Kakule. 2008. Antisickling activity of anthocyanins from *Bombax pentadrum*, *Ficus capensis* and *Ziziphus mucronata*: Photodegradation effect. *J. Ethnopharmacol.* 120: 413–418.
6. Noudogbessi, J. P., P. Yédomonhan, D. C. K. Sohounhloué, J. C. Chalchat, and G. Figuérédo. 2008. Chemical composition of essential oil of *Syzygium guineense* (Willd.) DC. var. *guineense* (Myrtaceae) from Benin. *Rec. Nat. Prod.* 2: 33–38.
7. Djoukeng, J. D., E. Abou-Mansour, R. Tabacchi, A. L. Tapondjou, H. Bouda, and D. Lontsi. 2005. Antibacterial triterpenes from *Syzygium guineense* (Myrtaceae). *J. Ethnopharmacol.* 101: 283–286.
8. Tesfaye, G., D. Teketay, and M. Fetene. 2002. Regeneration of fourteen tree species in Harenna forest, southeastern Ethiopia. *Flora (Jena)* 197: 461–474.
9. Tchiegang-Megueni, C. P., M. Mapongmetsem, C. H. A. Zedong, and C. Kapseu. 2001. An ethnobotanical study of indigenous fruit trees in northern Cameroon. *For. Trees Livelihoods* 11: 149–158.
10. Oketch-Rabah, H. A., and S. F. Dossaji. 1998. Molluscicides of plant origin: Molluscicidal activity of some Kenyan medicinal plants. *S. Afr. J. Sci.* 94: 299–301.
11. Malele, R. S., M. J. Moshi, J. W. Mwangi, K. J. Achola, and R. W. Munenge. 1997. Pharmacological properties of extracts from the stem bark of *Syzygium guineense* on the ileum and heart of laboratory rodents. *Afr. J. Health Sci.* 4: 43–45.

5. *Syzygium myrtifolium* Walp.
   a. *Syzygium campanulatum* Korth

Common names: Red lip, kelat oil
Distribution: Bangladesh to western and central Malesia
Usage
   Anticancer (2–4)
   Strong antiproliferative activity against human colorectal carcinoma cells (HCT 116) (1,2)

REFERENCES

1. Memon, A. H., Z. Ismail, F. S. R. Al-Suede, A. F. A. Aisha, M. S. R. Hamil, M. A. A. Saeed, M. Laghari, and A. M. S. A. Majid. 2015. Isolation, characterization, crystal structure elucidation of two flavanones and simultaneous RP-HPLC determination of five major compounds from *Syzygium campanulatum* Korth. *Molecules* 20: 14212–14233.
2. Memon, A. H., Z. Ismail, A. F. A. Aisha, R. A. S. Fouad Saleih, R. H. M. Shahrul, S. Hashim, M. A. S. Ali, M. Laghari, and A. M. S. A. Malik. 2014. Isolation, characterization, crystal structure elucidation, and anticancer study of dimethyl cardamonin, isolated from *Syzygium campanulatum* Korth. *J. Evid. Based Complementary Altern. Med.* 2014: 470179.
3. Aisha, A. F. A., Z. Ismail, K. M. Abu-Salah, J. M. Siddiqui, G. Ghafar, and A. M. S. A. Majid. 2013. *Syzygium campanulatum* Korth methanolic extract inhibits angiogenesis and tumor growth in nude mice. *BMC Complement. Altern. Med.* 13: 168.
4. Aisha, A. F. A., K. M. Abu-Salah, Y. Darwis, and A. M. S. A. Majid. 2009. Screening of antiangiogenic activity of some tropical plants by rat aorta ring assay. *Int. J. Pharmacol.* 5: 370–376.

6. *Syzygium paniculatum* Gaertn.

Common name: Magenta lilly-pilly
Distribution: East Australia
Usage
  Local food source (1)
  Anticancer benefits, particularly for pancreatic cancer (1)
  Cultivation aspect—shows some resistance to rust (2)

REFERENCES

1. Vuong, Q. V., S. Hirun, T. L. K. Chuen, C. D. Goldsmith, M. C. Bowyer, A. C. Chalmers, P. A. Phillips, and C. J. Scarlett. 2014. Physicochemical composition, antioxidant and anti-proliferative capacity of a lilly pilly (*Syzygium paniculatum*) extract. *J. Herbal Med.* 4: 134–140.
2. Rayachhetry, M. B., T. K. Van, T. D. Center, and M. L. Elliott. 2001. Host range of *Puccinia psidii*, a potential biological control agent of *Melaleuca quinquenervia* in Florida. *Biol. Control* 22: 38–45.

7. *Syzygium polyanthum* (Wight) Walp.
  a. *Eugenia polyantha* Wight

Common Name: Indonesian bayleaf, Indian bayleaf
Distribution: Indo-China to Philippines
Usage
  Antidiabetic (1,3–6,14,16)
  Larvicidal activity against *Aedes aegypti* larvae (2)
  Antioxidant sources/activity (7,10,11,13,15,17,19,20)
  Sporicidal activity (*Bacillus cereus* [8] and *Bacillus subtilis* [9])

Antiproliferative effect (11)
Hypertension (12)
Acetylcholinesterase inhibition (13)
Alzheimer's disease treatment (13)
Antiobesity (16)
Antiradical properties (16–18)
Cytotoxicity (11,17)
Photocytotoxicity, potential new source for photodynamic therapy (18)
Source of gallic and caffeic acids (19)
Anti-tumor-promoting activity (Epstein–Barr virus) (21)
Antinematodal activity (22)
Antifungal activity in plants (23)

REFERENCES

1. Elya, B., R. Handayani, R. Sauriasari, Azizahwati, U. S. Hasyyati, T. Permana, and Y. I. Permatasari. 2015. Antidiabetic activity and phytochemical screening of extracts from Indonesian plants by inhibition of alpha amylase, alpha glucosidase and dipeptidyl peptidase IV. *Pak. J. Biol. Sci.* 18: 279–284.
2. Tinneke, L. S. V., and N. T. Puput. 2015. Larvicidal activity of *Syzygium polyanthum* W. leaf extract against *Aedes aegypti* L larvae. *Progress Health Sci.* 5: 102–106.
3. Yuliana and T. Widarsa. 2014. Indonesian bay leaf decoction could lower fasting blood glucose level on hyperglycemic rats and lower Kupffer cell count. *J. Vet.* 15: 541–547.
4. Widyawati, T., W. W. Purnawan, I. J. Atangwho, N. A. Yusoff, M. Ahmad, and M. Z. Asmawi. 2015. Anti-diabetic activity of *Syzygium polyanthum* (Wight) leaf extract, the most commonly used herb among diabetic patients in Medan, North Sumatera, Indonesia. *Int. J. Pharm. Sci. Res.* 6: 1698–1704.
5. Widyawati, T., N. A. Yusoff, M. Z. Asmawi, and M. Ahmad. 2015. Antihyperglycemic effect of methanol extract of *Syzygium polyanthum* (Wight.) leaf in Streptozotocin-induced diabetic rats. *Nutrients* 7: 7764–7780.
6. Widharna, R. M., Ferawati, W. D. Tamayanti, L. Hendriati, I. S. Hamid, and E. C. Widjajakusuma. 2015. Antidiabetic effect of the aqueous extract mixture of *Andrographis paniculata* and *Syzygium polyanthum* leaf. *Eur. J. Med. Plants* 6: 82–91.
7. Safriani, N., N. Arpi, and N. M. Erfiza. 2015. Potency of curry (*Murayya koeniigi*) and Salam (*Eugenia polyantha*) leaves as natural antioxidant sources. *Pak. J. Nutr.* 14: 131–135.
8. Lau, K. Y., and Y. Rukayadi. 2015. Screening of tropical medicinal plants for sporicidal activity. *Int. Food Res. J.* 22: 421–425.
9. Lau, K. Y., N. S. Zainin, F. Abas, and Y. Rukayadi. 2014. Antibacterial and sporicidal activity of *Eugenia polyantha* Wight against *Bacillus cereus* and *Bacillus subtilis*. *Int. J. Curr. Microbiol. Appl. Sci.* 3: 499–510.
10. Othman, A., N. J. Mukhtar, N. S. Ismail, and S. K. Chang. 2014. Phenolics, flavonoids content and antioxidant activities of 4 Malaysian herbal plants. *Int. Food Res. J.* 21: 759–766.

11. Sulistiyani, S. Falah, W. T. Wahyuni, T. Sugahara, S. Tachibana, and Syaefudin. 2014. Cellular mechanism of the cytotoxic effect of extracts from *Syzygium polyanthum* leaves. *Am. J. Drug Discov. Dev.* 4: 90–101.

12. Ismail, A., M. Mohamed, S. A. Sulaiman, and W. A. N. Wan Ahmad. 2013. Autonomic nervous system mediates the hypotensive effects of aqueous and residual methanolic extracts of *Syzygium polyanthum* (Wight) Walp. var. *polyanthum* leaves in anaesthetized rats. *J. Evid. Based Complementary Altern. Med.* 2013: 716532.

13. Darusman, L. K., W. T. Wahyuni, and F. Alwi. 2013. Acetylcholinesterase inhibition and antioxidant activity of *Syzygium cumini*, *S. aromaticum* and *S. polyanthum* from Indonesia. *J. Biol. Sci.* 13: 412–416.

14. Lelono, R. A. A., and S. Tachibana. 2013. Preliminary studies of Indonesian *Eugenia polyantha* leaf extracts as inhibitors of key enzymes for type 2 diabetes. *J. Med. Sci. (Pakistan)* 13: 103–110.

15. Lelono, R. A. A., and S. Tachibana. 2013. Bioassay-guided isolation and identification of antioxidative compounds from the bark of *Eugenia polyantha*. *Pakistan J. Biol. Sci.* 16: 812–818.

16. Saifudin, A., S. Kadota, and Y. Tezuka. 2013. Protein tyrosine phosphatase 1B inhibitory activity of Indonesian herbal medicines and constituents of *Cinnamomum burmannii* and *Zingiber aromaticum*. *J. Nat. Med.* 67: 264–270.

17. Perumal, S., R. Mahmud, S. P. Piaru, L. W. Cai, and S. Ramanathan. 2012. Potential antiradical activity and cytotoxicity assessment of *Ziziphus mauritiana* and *Syzygium polyanthum*. *Int. J. Pharm.* 8: 535–541.

18. Wei, H. L., K. Shaari, B. Lee-Hong, F. A. Kamarulzaman, and I. S. Ismail. 2012. Two new phloroglucinol derivatives and five photosensitizing pheophorbides from *Syzygium polyanthum* leaves (Salam). *Nat. Prod. Commun.* 7: 1033–1036.

19. Wei, H. L., and I. S. Ismail. 2012. Antioxidant activity, total phenolics and total flavonoids of *Syzygium polyanthum* (Wight) Walp. leaves. *Int. J. Med. Arom. Plants* 2: 219–228.

20. Lelono, R. A. A., S. Tachibana, and K. Itoh. 2009. *In vitro* antioxidative activities and polyphenol content of *Eugenia polyantha* Wight grown in Indonesia. *Pak. J. Biol. Sci.* 12: 1564–1570.

21. Ali, A. M., L. Y. Mooi, K. Y. Yih, A. W. Norhanom, K. M. Saleh, N. H. Lajis, A. M. Yazid, F. B. H. Ahmad, and U. Prasad. 2000. Anti-tumor promoting activity of some Malaysian traditional vegetable (Ulam) extracts by immunoblotting analysis of Raji cells. *Nat. Prod. Sci.* 6: 147–150.

22. Mackeen, M. M., A. M. Ali, M. A. Abdullah, R. M. Nasir, N. B. Mat, A. R. Razak, and K. Kawazu. 1997. Antinematodal activity of some Malaysian plant extracts against the pine wood nematode, *Bursaphelenchus xylophilus*. *Pesticide Sci.* 51: 165–170.

23. Mohamed, S., S. Saka, S. H. El-Sharkawy, A. M. Ali, and S. Muid. 1996. Antimycotic screening of 58 Malaysian plants against plant pathogens. *Pesticide Sci.* 47: 259–264.

8. *Syzygium polycephalum* (Miq.) Merr. & L. M. Perry
   a. *Eugenia polycephala* Miq.

Common name: Gowok
Distribution: West and central Malesia

Usage
Anticancer (1)
Lowers high blood pressure and high cholesterol level (2)
Exhibits antioxidant activity (2)
Treatment of dysentery (3)
Local food source, construction material (4)

REFERENCES

1. Ragasa, C. Y., O. B. Torres, C. C. Shen, M. K. E. G. Lachica, A. B. Sulit, D. B. D. L. Chua, A. D. M. Ancheta, C. J. B. Ismail, F. T. E. Bernaldez, and D. D. Raga. 2014. Triterpenes from the leaves of *Syzygium polycephalum*, *S. cumini*, and *S. samarangense*. *Chem. Nat. Compd.* 50: 942–944.
2. Florido, H. B., and F. F. Cortiguerra. 2003. *Res. Inform. Ser. Ecosyst.* 15: 6.
3. Roosita, K., C. M. Kusharto, M. Sekiyama, Y. Fachrurozi, and R. Ohtsuka. 2008. Medicinal plants used by the villagers of a Sundanese community in west Java, Indonesia. *J. Ethnopharmacol.* 115: 72–81.
4. Jansen, P. C. M., J. Jukema, L. P. A. Oyen, and T. G. van Lingen. *Syzygium polycephalum* (PROSEA) http://uses.plantnet- project.org/en/Syzygium_polycephalum _(PROSEA).

REFERENCES

1. Govaerts, R. et al. 2008. *World Checklist of Myrtaceae*. London: Royal Botanic Gardens, Kew.
2. Mitra, S. K., T. K. S. Irenaeus, M. R. Gurung, P. K. Pathak, C. A. F. Santos, S. K. Mitra, and J. L. Griffis Jr. 2012. Taxonomy and importance of Myrtaceae. *Acta Hortic.* 959: 23–34.

Index

Page numbers followed by f and t indicate figures and tables, respectively.